Broadband Strategies Handbook

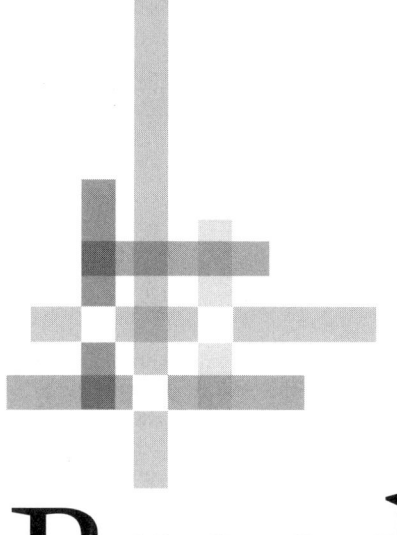

Broadband Strategies Handbook

Editors

Tim Kelly and Carlo Maria Rossotto

Coordinated by Telecommunications Management Group, Inc.

THE WORLD BANK
Washington, D.C.

Korean Trust Fund

*info*Dev
Innovate. Connect. Transform.

ISBN (paper): 978-0-8213-8945-4
ISBN (electronic): 978-0-8213-8946-1
DOI: 10.1596/978-0-8213-8945-4

Library of Congress Cataloging-in-Publication Data
Broadband strategies handbook / edited by Tim Kelly and Carlo Rossotto.
    p. cm.
  Includes bibliographical references and index.
  ISBN 978-0-8213-8945-4 — ISBN 978-0-8213-8946-1 (electronic)
  1. Telecommunication policy. 2. Broadband communication systems—Government policy. 3. Information technology—Government policy. I. Kelly, Tim (Tim John Charles) II. Rossotto, Carlo Maria, 1970-
  HE7645.B76 2012
  384—dc23

2011052001

Cover design: Naylor Design, Inc.

CONTENTS

Figures

Tables

FOREWORD

The world is shifting from narrowband to broadband. Services that were only available in the form of static, text-based websites 10 years ago are now offered in full-motion, high-definition video. Usage-based transmission prices that were once prohibitive are now bundled into an affordable monthly "all you can eat" charge. A decade after the dot.com bubble burst because network realities had not yet caught up with user aspirations, a whole new generation of Internet entrepreneurs is ready to take its ideas to the stock market.

Nevertheless, a gap remains between the developed and the developing world when it comes to broadband. For instance, not a single one of the top 10 economies by average broadband speed is in the Southern Hemisphere. The digital divide that was once measured in terms of differences in *access* to communications is now measured in terms of differences in *quality* of access. Slow speeds for download translate into lost economic opportunities. Yet the evidence seems to suggest that, where broadband is available in developing countries, it is a major contributor to economic growth. For instance, a 10 percent increase in the penetration rate of broadband in developing countries is associated with a 1.4 percent increase in gross domestic product (GDP) per capita, higher than the equivalent relationship for developed countries. The developing world has adopted mobile phones much more readily than tethered ones, so as mobile broadband becomes more readily available, a further boost to growth can be expected.

A decade ago, *info*Dev and the World Bank's Information and Communication Technology (ICT) Sector Unit joined forces with the International Telecommunication Union (ITU) to develop a handbook for regulators around the world on basic principles of telecommunication regulation. The *Telecommunication Regulation Handbook* subsequently became a bestseller and was updated and reissued in 2010. It formed the basis for the ICT Regulation Toolkit (http://www.ictregulationtoolkit.org), which now delivers around 1,000 downloads daily.

This new *Broadband Strategies Handbook* is intended as a next-generation tool for policy makers, regulators, and other relevant stakeholders as they address issues related to broadband development. It aims to help readers, particularly those in developing countries, by identifying issues and challenges in broadband development, analyzing potential solutions to consider, and providing practical examples from countries that have addressed broadband-related matters. It goes beyond the regulatory issues and looks more broadly at the challenges of promoting and universalizing broadband access. It will also form the basis for a toolkit—http://www.broadband-toolkit.org—that will complement the other toolkits and technical assistance guides available from the World Bank Group.

This new handbook has been made possible through the generous funding of the Korean Trust Fund for ICT for Development. We hope that it will meet the requirements of developing-country policy makers and regulators for sound advice on developing national strategies for broadband. But we also hope that it will provide incentives for users to share their own experiences, via the toolkit website, of what works well. Consider this handbook, then, as a living resource that will grow as the broadband market worldwide grows.

Valerie D'Costa
Program Manager
*info*Dev

Philippe Dongier
Sector Manager
ICT Sector Unit

ACKNOWLEDGMENTS

This report was produced by the Telecommunications Management Group, Inc. (TMG) and other consultants under the supervision of Tim Kelly (*info*Dev) and Carlo Maria Rossotto (ICT Unit) of the World Bank Group. The report has benefited from the inputs, ideas, and review of many World Bank Group colleagues and management as well as peer reviewers. The authors are grateful to Mohsen Khalil, former director of the World Bank's Global Information and Communication Technologies (GICT) Department, for his guidance and support throughout the preparation of this report. The authors also thank Valerie D'Costa, program manager, *info*Dev, and Philippe Dongier, sector manager, ICT Sector Unit, for their comments and support. For their review and comments, the authors and project team thank Francois Auclert, Kevin Donovan, Elena Kvochko, Wonki Min, Victor Mulas, James Neumann, Duncan Wambogo Omole, Christine Qiang, Siddhartha Raja, David Satola, Lara Srivastava, Mark Williams, and Masatake Yamamichi from the World Bank Group, as well as external reviewers Michael Best from Georgia Institute of Technology (United States), Yongsoo Kim from the Korea Communications Corporation (Republic of Korea), Mandla Msimang from Pygma Consulting (South Africa), and Paul de Sa from the Federal Communications Commission (United States).

As part of our work, we convened an advisory group comprising of Rodrigo Abdalla F. de Sousa, Ben Akoh, Jeff Eisenach, Torbjörn Fredriksson, Sverre Holt-Francati, Parvez Iftikhar, Lars Krogager, James Losey, Youlia Lozanova, Sascha Meinrath, Sam Patridge, Rohan Samarajiva, Nancy Sundberg, Sharil Tarmizi, and Marianne Treschow. We would like to thank these participants for their valuable input and ideas in shaping the initial table of contents and for attending our consultation meeting and "write-shop" in August 2009.

Case studies were prepared to support the work of chapter 7. We thank the authors of these case studies—Diane Anius (St. Kitts and Nevis), Samantha Constant (Morocco), Helani Galpaya (Sri Lanka), Michael Jensen (Brazil), Mandla Msimang (Kenya), Çağatay Telli (Turkey), and Tran Minh Tuan (Vietnam). In addition, we are grateful for the contributions made by Rob Frieden on financing strategies and technologies to support broadband; by Victor Mulas on absorptive capacity; and by Helani Galpaya and Rohan Samarajiva on measurement, monitoring, and evaluation. We also thank the Organisation for Economic Co-operation and Development (OECD) and the International Telecommunication Union (ITU) for their comments and ideas, especially on the measurement section of chapter 2. We gratefully acknowledge the support of Stephen McGroarty, Nora Ridolfi, and Dina Towbin from the World Bank Office of the Publisher.

This handbook, case studies, and other reports have been generously funded by the Korean Trust Fund (KTF) on Information and Communication Technology for Development (ICT4D). The KTF is a partnership between the government of Korea and the World Bank. Its purpose is to advance the ICT4D agenda, with the goal of contributing to growth and reducing poverty in developing countries. The report has also benefited from funding from the U.K. Department for International Development. The handbook is part of a longer-term project to create a broadband toolkit (see http://www.broadband-toolkit.org), an online resource for regulators and policy makers. Future updates of the handbook, as well as the full text of the case studies, practice notes, indicators, and training materials will be posted there.

The authors retain sole responsibility for any residual errors.

ABOUT THE AUTHORS

This report was prepared and input documents were coordinated by the Telecommunications Management Group, Inc. (TMG). TMG is a telecommunications and information technology consulting firm providing regulatory, policy, economic, technical, and financial advice. Established in 1992, TMG is composed of a team of regulatory experts, lawyers, economists, market analysts, business development and investment specialists, engineers, and spectrum management specialists.

TMG advises public and private sector clients on issues related to information and communication technology (ICT) and provides assistance to regulators and policy makers on regulatory and policy reform matters. TMG has advised more than 60 countries on regulatory reform issues in Africa, Asia, Europe, Latin America and the Caribbean, and the Middle East. In addition, TMG has worked on regulatory matters with international and regional organizations involved in ICT issues.

The TMG team that worked on this project includes Flavia Alves, Kari Ballot-Lena, Jeff Bernstein, Joel Garcia, Janet Hernandez, Daniel Leza, Sofie Maddens-Toscano, Jorge Moyano, William Wiegand, David Wye, and Amy Zirkle, as well as outside consultants Michael Minges, Calvin Monson, and Björn Wellenius.

ABBREVIATIONS

2G	second-generation mobile telecommunications system
3G	third-generation mobile telecommunications systems
3GPP	3G Partnership Project
4G	fourth-generation mobile telecommunications systems
ACE	Africa Coast to Europe
ADSL	asymmetric digital subscriber line
APEC	Asia-Pacific Economic Cooperation
API	application programming interface
App	application
ARCEP	Autorité de Régulation des Communications Électronique et des Postes (France) (French Electronic Telecommunications and Postal Sectors Regulator)
ASO	analog switch-off or switch-over of analog broadcast television
BAK	bill and keep
BEREC	Body of European Regulators of Electronic Communications
BPL	broadband over powerline
BRAND	Broadband for Rural and Northern Development (Canada)
BRIC	Brazil, the Russian Federation, India, and China
BSC	base station controller
BSNL	Bharat Sanchar Nigam Ltd.
BT	British Telecom
BTS	base transceiver station
CATV	cable television
CBI	capacity-based interconnection
CCI	centro de capacitación en informática (Dominican Republic) (local community computer training center)

CCK	Communications Commission of Kenya
CDMA	Code Division Multiple Access (family of mobile communication standards)
CERT	computer emergency response team
CII	critical information infrastructure
CLC	computer learning center
CMTS	cable modem termination system
CPE	customer premises equipment
CPEA	Cross-Border Privacy Enforcement Arrangement
CPNP	calling party network pays
CRT	cognitive radio technology
CSC	common service center (India)
DBKL	Kuala Lumpur City Hall
DMCA	Digital Millennium Copyright Act (United States)
DNS	domain name system
DOCSIS	Data Over Cable Service Interface Specification (cable modem standard)
DPI	Deep Packet Inspection
DSL	digital subscriber line
DSLAM	digital subscriber line access multiplexer
DTH	direct to home (satellite)
DTT	digital terrestrial television
DWDM	Dense Wave Division Multiplexing
EASSy	Eastern Africa Submarine Cable System
EC	European Commission
ECTEL	Eastern Caribbean Telecommunications Authority
EDGE	Enhanced Data Rates for GSM Evolution
EFM	ethernet in the first mile
EMS	element management system
EP2P	ethernet over point-to-point
EPON	ethernet passive optical network
ESCAP	Economic and Social Commission for Asia and the Pacific
ETTx	Ethernet to the home and business
EU	European Union
EV-DO	CDMA2000 Evolution Data Optimized (mobile communication standard)
FBO	facilities-based operator
FCC	Federal Communications Commission (United States)
FDD	Frequency Division Duplexing

FDMA	Frequency Division Multiple Access
FICORA	Finnish Communications Regulatory Authority
FIRST	Forum of Incident Response and Security Teams
FITEL	Fondo de Inversión de Telecomunicaciones (Peru) (universal access fund for telecommunications)
FRIENDS	Fast Reliable Instant Efficient Network for Disbursement of Services (India)
FTTx	Different types of access to fiber optic networks, including fiber to the node (FTTN), fiber to the cabinet or curb (FTTC), fiber to the premises (FTTP), which may be fiber to the home (FTTH), or fiber to the building or business (FTTB)
FUST	Fund for Universal Telecommunications (Brazil)
GB	gigabyte
Gbit/s	gigabits per second
GDP	gross domestic product
GHz	gigahertz
GICT	Global Information and Communication Technologies
GPOBA	Global Partnership on Output-Based Aid
GPON	gigabit passive optical network
GPRS	General Packet Radio Service
GPT	general-purpose technology
GSA	Global Mobile Suppliers Association
GSM	global system for mobile communications (mobile communication standard)
GSMA	GSM Association
HD	high definition
HFC	hybrid fiber coaxial (cable)
HKBN	Hong Kong Broadband Network
HSDPA	High-Speed Download Packet Access
HSI	H.323 Signaling Interface
HSPA	High-Speed Packet Access
HSUPA	High-Speed Upload Packet Access
ICT	information and communication technology
ICT4D	ICT for Development
IDA	Info-communications Development Authority of Singapore
IDC	International Data Corporation
IEEE	Institute of Electrical and Electronics Engineers
iFrame	inline frame

IMT-2000	International Mobile Telecommunications-2000 (family of mobile communication standards)
IMT-Advanced	International Mobile Telecommunications-Advanced (family of mobile communication standards)
IP	Internet Protocol
iPoDWDM:IP	IP (Internet Protocol) over DWDM (Dense Wave Division Multiplexing)
IPR	intellectual property right
IPTV	Internet Protocol television
IPv4	Internet Protocol version 4
IPv6	Internet Protocol version 6
ISP	Internet service provider
IT	information technology
ITU	International Telecommunication Union
IXP	Internet exchange point
kbit/s	kilobits per second
KCC	Korea Communications Corporation
KDN	Kenya Data Networks
kHz	kilohertz
KISA	Korean Information Security Agency
KIXP	Kenya Internet Exchange Point
KPN	Koninklijke PTT Nederland (Royal Dutch Telecom)
KTF	Korean Trust Fund
LAN	local area network
LDC	least developed country
LLDC	landlocked developing country
LLU	local loop unbundling
LTE	Long-Term Evolution (mobile communication standard)
MB	megabyte
Mbit/s	megabits per second
MCMC	Malaysian Communications and Multimedia Commission
MDG	Millennium Development Goal
M-health	mobile health
MHz	megahertz
MSE PE	Mobility Services Engine Provider Edge
MSPP	Multiservice Provisioning Platform
MPLS	Multiprotocol Label Switching
MVNO	mobile virtual network operator

MyICMS	Malaysian Information, Communications, and Multimedia Services
NBI	National Broadband Initiative (Malaysia, Indonesia)
NBN	national broadband network
NBP	National Broadband Plan (United States)
NBS	National Broadband Scheme (Ireland)
NCC	Nigerian Communications Commission
NGA	next-generation access
NGN	next-generation network
NPV	net present value
OBA	output-based aid
OECD	Organisation for Economic Co-operation and Development
OFDM	Orthogonal Frequency Division Multiplexing
OLPC	One Laptop per Child
OSP	online service provider
PC	personal computer
PDA	personal digital assistant
PII	personally identifiable information
PLMN	public land mobile network
PON	passive optical network
POP	point of presence
PPP	public-private partnership
PSTN	public switched telephone network
R&D	research and development
RAN	radio access network
RCDF	Rural Communications Development Fund (Uganda)
RCIP	Regional Communications Infrastructure Program
RFID	radio frequency identification
RTR	Rundfunk & Telekom Regulierungs (Austria)
SAR	Special Administrative Region
SAT3/SAFE	South Atlantic 3/South Africa Far East
SAT3/WASC	South Atlantic 3/Western Africa Submarine Cable
SBO	service-based operator
SC-FDMA	Single Carrier-Frequency Division Multiple Access
SD	standard definition
SDH	Synchronous Digital Hierarchy
SEACOM	Southern and East Africa Cable System
SE-ME-WE	South East Asia-Middle East-West Europe
SENA	Servicio Nacional de Aprendizaje (Colombia) (National Training Service)

SIDA	Swedish International Development Agency
SIDS	small island developing state
SLCERT	Sri Lanka computer emergency response team
SME	small and medium enterprise
SMS	short message service
SOFDMA	Scalable Orthogonal Frequency Division Multiple Access
SONET	Synchronous Optical Network
STB	set-top box
SUBTEL	Subsecretaría de Telecomunicaciones (Chile) (Telecommunications Regulator)
Tbit/s	terabits per second
TDD	Time Division Duplexing
TDF	Telecommunications Development Fund
TD-SCDMA	Time Division–Synchronous Code Division Multiple Access (mobile communication standard)
TEAMS	The East African Marine System
TESPOK	Telecommunications Service Providers Association (Kenya)
TLD	Top-level domain
TMG	Telecommunications Management Group, Inc.
TRA	Telecommunications Regulatory Authority
TRAI	Telecommunications Regulatory Authority of India
TV	television
UAF	Universal Access Fund (Jamaica)
UAS	universal access and service
UASF	universal access and service fund
U-CAN	Ubiquitous Canadian Access Network
ULL	unbundled local loop
UMTS	Universal Mobile Telecommunications System (see W-CDMA)
UNCTAD	United Nations Conference on Trade and Development
USF	universal service fund
USO	universal service obligation
USOF	universal service obligation fund
VANS	value added network service
VDSL	very high-speed DSL
VLE	virtual learning environment
VoB	voice over broadband
VoD	video on demand
VoIP	voice over Internet Protocol

VSAT	very small aperture terminals (satellite)
W-CDMA	Wideband Code Division Multiple Access (family of mobile communication standards)
WDM	Wavelength Division Multiplexing
Wi-Fi	Wireless Fidelity, a wireless local area network standard based on the IEEE 802.11 standards
WiMAX	Worldwide Interoperability for Microwave Access (fixed and mobile communications standard)
WSIS	World Summit on the Information Society
xDSL	includes different types of digital subscriber line, including ADSL and VDSL

EXECUTIVE SUMMARY

The *Broadband Strategies Handbook* is a guide for policy makers, regulators, and other relevant stakeholders as they address issues related to broadband development. It aims to help readers, particularly those in developing countries, by identifying issues and challenges in broadband development, analyzing potential solutions to consider, and providing practical examples from countries that have addressed broadband-related matters.

The handbook consists of seven chapters and two appendixes that look at how broadband is defined, why it is important, and how its development can be encouraged. Throughout the handbook, broadband is viewed as an *ecosystem* consisting of supply and demand components, both of which are equally important if the expansion of broadband networks and services is to be successful. In addressing the challenges and opportunities to which broadband gives rise, the handbook discusses the policies and strategies that government officials and others should consider when developing broadband plans, including what legal and regulatory issues to address, what broadband technologies to choose, how to facilitate universal broadband access, and how to generate demand for broadband services and applications.

Chapter 1, "Building Broadband," introduces the concepts of broadband by defining the term "broadband" more conventionally (that is, speed or functionality) as well as explaining how this handbook seeks to define the term as broadband comes to be seen as an enabling platform. This chapter examines why broadband, both as an information and communication technology (ICT) and as an enabling platform, is important. Chapter 1 focuses on how broadband can help to transform a country's economic development and improve employment growth, provided that effective policies are put in place that encourage the use of broadband as an essential input by all sectors of the economy. Chapter 1 also identifies the main trends fostering the deployment of broadband networks (supply side) and the adoption of broadband services and applications (demand side). Lastly, this chapter offers a framework—the broadband ecosystem—to assist policy makers and stakeholders in viewing broadband policies in a more holistic manner and as a means to ensure the greatest impact throughout the economy and society.

Chapter 2, "Policy Approaches to Promoting Broadband Development," identifies the issues that governments and the private sector will face when developing policies and programs to support broadband development. It discusses policies and strategies for promoting the build-out of broadband networks as well as ways to encourage the use of broadband services and applications, particularly in populations that may have limited knowledge of or interest in broadband. In that context, the impacts of broadband on other sectors (education, health, banking, environment, and cybersecurity) are discussed. The chapter also addresses the options for funding broadband development strategies and identifies the issues associated with measuring the effectiveness of policies designed to promote network build-out and user demand.

Chapter 3, "Law and Regulation in a Broadband World," discusses the key policies and regulatory trends that policy makers and regulators are considering to foster broadband. As the world moves to a converged ICT environment, countries are reforming their traditional legal and regulatory frameworks and developing new laws and regulations to address some of the supply and demand issues associated with broadband development. This chapter covers a wide range of policy issues, including liberalization of licensing frameworks, spectrum management policies to maximize wireless broadband, Internet Protocol (IP) interconnection regulation, policies to promote competition in the various segments of the broadband supply chain, vertical integration in a converged environment, network neutrality, cybersecurity and data protection, and regulation of online content.

Chapter 4, "Extending Universal Broadband Access and Use," discusses what roles governments should play in promoting universal broadband access when market mechanisms do not meet goals for broadband access and use on their own. The chapter seeks to define a broadband development strategy capable of addressing market failures, to provide an overview of what policy makers can do to address perceived shortfalls in the market, and to work toward achieving universal broadband service. It discusses the universal service objectives that a government strategy may pursue, the role of private-led competitive markets in achieving these objectives, the role of the government in narrowing or eliminating gaps between markets and the country's development needs, and how effective government strategies can be designed to meet such challenges. It finally examines the use of fiscal resources to support private supply of broadband, including the choice of instruments, use of subsidies, and use of different mechanisms to collect and disburse funds for subsidies.

Chapter 5, "Technologies to Support Deployment of Broadband Infrastructure," focuses on the supply side of the broadband ecosystem. It describes the various wireline and wireless technologies now being used to build out broadband infrastructure, including examples of broadband deployments throughout the world. The objective of the chapter is to provide policy makers with an overview of how broadband networks work and their components. It describes the broadband supply chain from a topological perspective, starting from international connectivity and progressing to regional, national, and, finally, local access deployment solutions. It describes the technologies being deployed in each of these segments, including fiber optics, satellite, microwave, mobile wireless, and traditional copper wire. Finally, chapter 5 addresses some of the implementation issues associated with these technologies, including open access, quality of service, and spectrum constraints.

Chapter 6, "Driving Demand for Broadband Networks and Services," recognizes that, although supply-side issues are important, simply building networks does not guarantee that they will be used or used most effectively. This chapter thus focuses on the issue of demand facilitation: what government and the private sector can do to spur the use and adoption of broadband networks and services by consumers. In particular, this chapter identifies various policies that may be implemented where demand is stifled because consumers are not aware of the benefits of broadband, broadband is not affordable, or broadband is not attractive or relevant to them. This chapter also highlights the importance of public-private cooperation to facilitate demand and increase broadband access to a wider number of users worldwide.

Chapter 7, "Global Footprints: Stories from and for the Developing World," addresses the main challenges that developing countries face in deploying broadband networks, including underdeveloped infrastructure, low income, significant differences between rural and urban areas, constrained inter- and intra-modal competition, and weaknesses in regulatory and legal frameworks. This chapter assesses the broadband bottlenecks and opportunities found in developing countries and discusses the importance of improving broadband infrastructure and leveraging existing infrastructure to create greater competition in the broadband market. Chapter 7 further highlights the status of broadband development in different developing regions around the world and summarizes broadband experiences in Brazil, Kenya, Morocco, Sri Lanka, St. Kitts and Nevis, Turkey, and Vietnam.

CHAPTER 1

Building Broadband

In just the past decade, the world of information and communication technology (ICT) has changed dramatically, evolving from a means by which information can quickly travel from point to point into an enabling platform for countless new and expanded personal, social, business, and political uses. In short, the Internet has become an integral part of people's lives. Consumers can use broadband networks to access the Internet at speeds up to or exceeding 100 megabits per second (Mbit/s) over wired connections in their homes and offices, and they can use their broadband-enabled mobile phones and other devices for a wide range of activities, including surfing the World Wide Web, engaging in two-way real-time video chats, purchasing goods and services online, streaming video or music, and conducting financial transactions.

But broadband is not just about improving the speed at which users can read online news, play video games, and engage in social networking, although these are useful drivers of demand and do provide benefits to users. It is also an enabling platform that allows developers and individual users to enhance existing services and to develop previously unimaginable tools that improve business and society. The benefits of broadband can expand beyond the ICT sector itself, reverberating throughout the economy and serving as an essential input for all other sectors, including education, health, transportation, energy, and finance. Its role as a transformative technology is similar to the impact that electricity has had on productivity, growth, and innovation over the last two centuries, with the potential to redefine how economies function. Broadband can also be a critical enabler of civic and political engagement and the exercise of fundamental rights such as freedom of expression and opinion. However, in order to achieve broadband's full potential, its reach must be expanded in both developing and developed economies. Governments must implement effective policies that spur construction of broadband networks as well as encourage the uptake of broadband services in all sectors of the economy.

The rollout of broadband requires significant investment from the private sector as well as support from the public sector. It also requires a long-term perspective because the benefits of broadband will not occur overnight. For developing countries with limited resources, it may be difficult to focus on broadband when many of their communities do not have schools for children, safe drinking water, or access to hospitals and health care. However, broadband offers countries an enabling platform and new tools to foster growth, extend public services, enhance businesses, and benefit their people. Making broadband a priority within a country's development agenda will be necessary to ensure that the digital divide between developed and developing countries does not extend further. In crafting a broadband strategy, however, countries should ensure that the use of public funds is supported by sound economic analysis and that the benefits of investing in broadband are weighed against the benefits of investing in other areas, such as energy, health, or education. Market-based solutions for the deployment and uptake of broadband are generally preferable to government investment in order to avoid straining public finances.

This first chapter of the handbook is designed to "set the stage" for the discussion in subsequent chapters of the various ways in which government policy makers and the private sector can promote greater deployment of broadband networks and services, particularly in developing countries. It first describes what broadband is and how it may be defined. Next, it explains why broadband is important by identifying how it contributes to the growth and development of a country's economy, noting, in particular, the findings of several studies pointing to broadband's impact on gross domestic product (GDP) and employment. Then, it considers the trends that characterize the development of broadband. Lastly, this chapter addresses the approaches that governments can use to support the development of broadband, by focusing on both the deployment of broadband networks (supply-side approaches) and the adoption of broadband services and applications (demand-side approaches). Overall, this chapter seeks to demonstrate that broadband can enable growth and productivity throughout the economy, provided that appropriate and specific policies are designed, developed, and effectively implemented.

What Is Broadband?

Despite its worldwide growth and promotion by policy makers, network operators, and content providers, broadband does not have a single, standardized definition. The term "broadband" may refer to multiple aspects of the network and services, including (a) the infrastructure or "pipes" used to deliver services to users, (b) high-speed access to the Internet, and (c) the services and applications available via broadband networks, such as Internet Protocol television (IPTV) and voice services that may be bundled in a "triple-play" package with broadband Internet access. Further, many countries have established definitions of broadband based on speed, typically in Mbit/s or kilobits per second (kbit/s), or on the types of services and applications that can be used over a broadband network (that is, functionality). Due to each country's unique needs and history, including economic, geographic, and regulatory factors, definitions of broadband vary widely.

Traditionally, however, broadband has often been defined in terms of data transmission speed (that is, the amount of data that can be transmitted across a network connection in a given period of time, typically one second, also known as the data transfer rate or throughput). Defining broadband in terms of speed has been an important element in understanding broadband, particularly since the data transfer rate determines whether users are able to access basic or more advanced types of content, services, and applications over the Internet.

However, attempts to define broadband in terms of speed present certain limitations. First, broadband speed definitions vary among countries and international organizations, generally ranging from download data transfer rates of at least 256 kbit/s on the low end, as in India, South Africa, the International Telecommunication Union (ITU), and the Organisation for Economic Co-operation and Development (OECD), to faster than 1.5 Mbit/s on the high end, as in Canada (see ITU 2009, 22). Second, definitions based on speed may not keep pace with technological advances or with the speeds, services, and applications required for the application to function properly. In other words, what is considered "broadband" today may be regarded as too slow in the future, as more advanced applications technologies are developed. Thus, any speed-based definition of broadband will need to be updated over time. Third, such definitions may not reflect the speeds realized by end users, so the speeds advertised by commercial broadband providers may be much higher than the speeds set by the government as broadband or vice versa. For example, while Colombia's broadband speed definition is 1 Mbit/s, its average broadband connection speed is already 1.8 Mbit/s.

Due to the limitations of definitions based on speed, some countries (Brazil) and international organizations (the OECD) have decided or proposed not to categorize broadband in terms of speed, but are instead looking at broadband in terms of functionality, focusing on what can and cannot be done with a certain type of connection.[1] However, establishing a definition of broadband based only on functionality may make the term overly subjective. A legal definition of broadband Internet access based on speed is easy to apply: if broadband is defined as at least 1.5 Mbit/s of download speed, then a 2 Mbit/s connection is broadband, while a 1 Mbit/s connection is not. When broadband is defined in terms of functionality, the distinction between what is and is not broadband becomes less straightforward. Is being able to watch a YouTube video equivalent to having a broadband connection? What if it takes minutes to buffer and starts and stops throughout?

In considering what broadband is and how it should be defined, this chapter and the handbook as a whole view broadband more holistically as a high-capacity ICT platform that improves the variety, utility, and value of services and applications offered by a wide range of providers, to the benefit of users, society, and multiple sectors of the economy. From a policy perspective, broadband should be viewed more broadly as an enabling ICT platform that can potentially influence the entire economy and thus may act as a general-purpose technology (GPT) that is used as a key input across sectors. To capture the full range of potential benefits, policy makers may find it useful to consider broadband as an ecosystem comprising both supply-side considerations (network platforms) and demand-side considerations (e-government initiatives, development of services and applications, promotion of broadband use). To encourage the diffusion of broadband-enabled innovations throughout the economy, policy makers should also consider the absorptive capacity of various sectors, including health, education, energy, and transportation. Unless all of these elements—supply, demand, and absorptive capacity—are coordinated, broadband's impact on the economy as a whole will be constrained.

Why Is Broadband Important?

With the appropriate policies in place, broadband is a transformative platform that affects the ICT sector as well as other sectors of the economy.

While some may disagree on the precise economic and social benefits that can be specifically attributed to broadband and may challenge the studies that have found a large impact, few can argue against the fact that broadband has dramatically changed our personal lives, our businesses, and our economies. Moreover, as an enabling ICT platform and potential GPT, broadband can facilitate growth and innovation in the ICT sector and throughout the economy, serving as a vital input for each sector that strengthens the economy as a whole. The multiplier effect of broadband can drive GDP, productivity, and employment growth; however, policies that support the supply and demand elements of the ecosystem as well as the absorptive capacity to learn and incorporate broadband capabilities into other sectors must all be in place in order to realize such benefits.

Impact of Broadband on Gross Domestic Product

Due to their potentially wide-ranging impacts and ability to provide easier access to information that increases efficiencies and productivity in the economy, it is unsurprising that increased use of broadband networks and services has been found to produce positive outcomes that reverberate throughout a country, particularly involving GDP. A frequently cited World Bank study found that low-income and middle-income countries experienced "about a 1.38 percentage point increase in GDP for each 10 percent increase in broadband penetration" between 2000 and 2006 (Qiang and Rossotto 2009, 45; see also Kim, Kelly, and Raja 2010). This study further found that the development impact of broadband is greater in emerging economies than in high-income countries, which "enjoyed a 1.21 percentage point increase in per capita GDP growth" for each 10 percent increase in broadband penetration. The study also demonstrated that broadband has a potentially larger growth effect than other ICTs, including wireline telephony, mobile telephony, and the Internet, as shown in figure 1.1. Broadband's predominance may be unexpected considering that, over the last decade, mobile telephony has been the fastest-growing ICT worldwide, with a global penetration rate in 2010 of 76.2 for every 100 persons.[2]

Other studies support the World Bank findings. Management consulting firm McKinsey and Company estimated that "a 10 percent increase in broadband household penetration delivers a boost to a country's GDP that ranges from 0.1 percent to 1.4 percent" (Buttkereit et al. 2009). Additionally, a study of OECD countries by consulting firm Booz & Company found, among high-income countries, a strong correlation between average annual GDP growth and broadband penetration, wherein "countries in the top tier of broadband penetration have also exhibited 2 percent higher

Figure 1.1   Effect of Various Information and Communication Technologies on GDP Growth in High- and Low-Income Economies, 2000–06

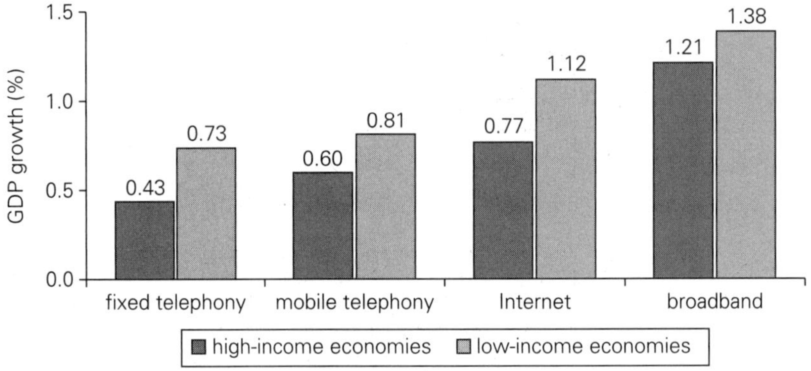

Source: Adapted from Qiang and Rossotto 2009, 45.

Note: Measures the percentage point increase in gross domestic product that is associated with a 10 percent increase in different information and communication technologies.

GDP growth than countries in the bottom tier of broadband penetration" (Friedrich et al. 2009, 4).

Although numerous studies have found a positive impact of broadband on economic growth, the estimate of its actual magnitude varies. For example, a 10 percent increase in broadband penetration has been found to increase economic growth from a low of range of 0.24 percent to a high of 1.50 percent (figure 1.2).

While these studies provide important insight into the growth effects of broadband, data collection and further systematic research and analysis in this area are needed, particularly for developing countries. Currently, there is ample anecdotal evidence of the effects of broadband on economic growth, with some cases highlighted in box 1.1. However, these cases provide only limited evidence of the impact that broadband has on the economy as a whole. It is also important to note that investment in broadband or policies fostering its deployment or adoption are unlikely to produce significant GDP gains without complementary investments or policies in other sectors, notably education, innovation, civic participation, and health care. However, even with the implementation of appropriate policies, the impacts of broadband on growth in certain areas may be limited. For example, in seeking to improve health outcomes developing countries may be in less need of high-tech, expensive telemedicine and more in need of low-tech, inexpensive solutions, such as mosquito nets and deworming pills (Kenny 2011).

Figure 1.2   Estimated Minimum and Maximum Impact on GDP of a 10 Percent Increase in Broadband Penetration

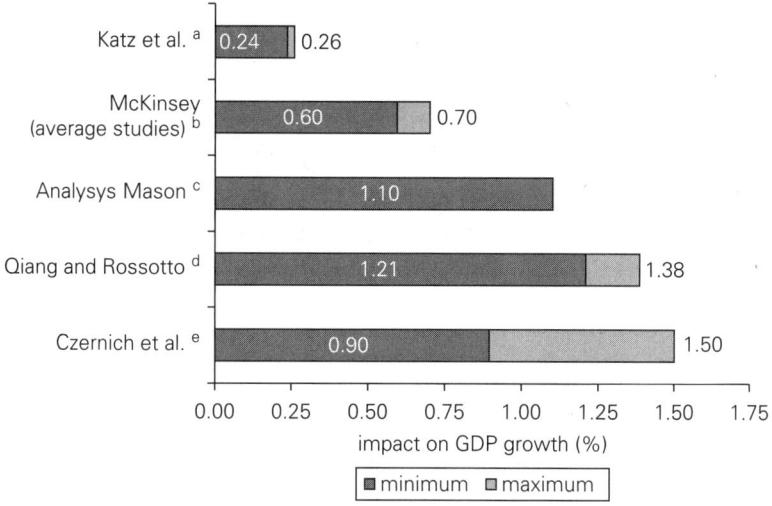

Sources: Czernich et al. 2009; Qiang and Rossotto 2009; Analysys Mason 2010; Beardsley et al. 2010; Katz et al. 2010.

a. Includes only Germany.

b. Average of five country studies: Australia, the Arab Republic of Egypt, Malaysia, New Zealand, and the United Kingdom; various sources for 2003 and 2004 and Qiang and Rossotto for 2009.

c. Limited to mobile broadband impact in India.

d. Various countries; upper range applies to developing countries, and lower range applies to developed countries.

e. Sample of 20 OECD countries.

Box 1.1: Examples of Broadband's Effects on Economic Growth around the World

The following examples highlight how broadband has improved economic outcomes in countries at all levels of development, as well as how different countries are working to improve broadband penetration rates.

Canada. Over the last several years, broadband access studies in Canada have focused on the importance of broadband for economic growth and development, particularly in rural areas. In 2005, for example, Industry Canada commissioned a survey to be conducted in the rural areas of British Columbia regarding subscribers' views of the significance of broadband access. More than 80 percent of all business respondents reported that their businesses would be negatively affected if

(continued)

Box 1.1 *continued*

they did not have broadband access, and over 18 percent stated that they would not be able to operate their businesses without broadband. Additionally, 62 percent of business owners reported that broadband increased productivity to some extent, with a majority stating that broadband increased productivity by over 10 percent.

China. Between 2010 and 2013, China's network operators—China Unicom, China Telecom, and China Mobile—are expected to invest an estimated Y 62 billion (US$9 billion) in the creation of a single wireline broadband access network providing speeds of 1 Mbit/s or more. These investments will be necessary considering that the number of wireline broadband subscribers in China is expected to reach 182 million by 2013, which represents growth of nearly 77 percent between 2010 and 2013. Set against these figures, the impact of broadband on China's GDP is anticipated to be substantial. Dial-up and broadband Internet together is expected to contribute a combined 2.5 percent to GDP growth for every 10 percent increase in penetration.

India. A study released by Analysys Mason in December 2010 on the deployment of wireless broadband in India found that each percentage point increase in mobile broadband penetration could increase India's GDP by 0.11 percent by 2015, which would yield Rs 162 billion (US$3.8 billion). The study

breaks down the impact on GDP based on direct contributions (revenues from services and devices), second-order contributions (revenues or cost savings from increased worker productivity), and ecosystem contributions (revenues from value added and other services enabled by wireless broadband).

South Africa. In July 2010, the South African government issued the Broadband Policy for South Africa. The policy provides that, by 2019, 15 percent of the country's households will have direct access to broadband of at least 256 kbit/s download speed, with broadband reaching within 2 kilometers of the remaining households. A 2010 study by Analysys Mason reviewed the likely direct and indirect effects that the broadband policy might have on South Africa's economy, finding that wireless broadband is expected to increase the country's GDP by 1.8 percent—over R 72 billion (US$9.4 billion)—by 2015. In addition, wireless broadband is expected to create about 28,000 new jobs directly, not including jobs created outside the communications industry. As a result, the direct effect of wireless broadband alone (that is, spending on broadband services and broadband-enabled devices) is expected to increase the GDP of South Africa by 0.71 percent by 2015, or R 28.5 billion (US$3.7 billion). However, the biggest impact on GDP is expected to come from productivity and efficiency gains.

Sources: Zilber, Schneier, and Djwa 2005; Zhao and Ruan 2010; Analysys Mason 2010; South Africa, Department of Communications 2010.

Additionally, despite providing a new educational resource, broadband can also create a new distraction if careful controls are not in place that limit Internet access to nonacademic sites such as Facebook, YouTube, and file-sharing websites (Belo, Ferreiray, and Telangz 2010).

Broadband, Employment, and Job Creation

Broadband enables job creation through three main channels: (1) direct jobs created to deploy the broadband infrastructure, (2) indirect and induced jobs created from this activity, and (3) additional jobs created as a result of broadband network externalities and spillovers (Katz 2009). Each of these channels focuses on a different type of jobs: unskilled, skilled, and highly skilled. Direct jobs relate primarily to civil works and construction of broadband infrastructure, which involve more low-tech positions. Indirect and induced jobs require various levels of skilled workers. However, network-effects (that is, spillover) jobs are mainly high-skill jobs requiring specific technical knowledge and education. Indeed, broadband spillover employment effects are not uniform. Instead, they tend to concentrate in service industries, such as financial services or health care. Broadband can also produce some effects in middle-skill jobs, such as in manufacturing, usually related to the use of ICT and requiring ICT skills.

Numerous studies have estimated the impact of broadband on job creation in specific countries by calculating employment multipliers for each of these job creation categories (table 1.1). While these studies are country specific and cannot be applied directly to other nations, they provide an estimate of the potential employment gains that could result from effective broadband development, which is between 2.5 and 4.0 additional jobs for each broadband job. Some studies have estimated the impact of broadband

Table 1.1 Estimated Broadband Employment Creation Multipliers in Various Countries

Study	Scope	Type I	Type II	Network effects
Crandall, Jackson, and Singer 2003	United States	—	2.17	—
Katz, Zenhäusern, and Suter 2008	Switzerland	1.40	—	—
Atkinson, Castro, and Ezell 2009	United States	—	3.60	1.17
Katz and Suter 2009	United States	1.83	3.43	—
Libenau et al. 2009	United Kingdom	—	2.76	—
Katz et al. 2009	Germany	1.45	1.93	—
Average		1.56	2.78	1.17

Sources: Katz 2009, citing Crandall, Jackson, and Singer 2003; Katz, Zenhäusern, and Suter 2008; Atkinson, Castro, and Ezell 2009; Katz and Suter 2009; Libenau et al. 2009; Katz et al. 2009.

Note: Type I = (direct + indirect) / direct; type II = (direct + indirect + induced) / direct; — = not available.

on the employment creation rate. For instance, Katz (2009) estimated that an increase of about 8 percentage points in broadband penetration in 12 Latin American countries could result in an increase of almost 8 percent on average in their employment rate.[3]

As with broadband's effects on GDP, further data collection and analysis are needed to confirm the positive impact that broadband has on employment growth. Aside from the studies identified in table 1.1, some researchers have reported anecdotal evidence of how broadband development has stimulated the job market, including in the European Union (EU), Brazil, Malaysia, and the United States.

- In the EU, a study estimated that broadband could create more than 2 million jobs throughout Europe by 2015 and result in an increase in GDP of at least €636 billion (Fornefeld, Delaunay, and Elixmann 2008, 6).

- In Brazil, broadband was found to add up to 1.4 percent to the employment growth rate (Broadband Commission for Digital Development 2010, 15).

- In Malaysia, the Malaysian Communications and Multimedia Commission (MCMC) estimated in 2008 that achieving 50 percent broadband penetration by 2010 could create 135,000 new jobs in the country.[4] The MCMC further projected that the number of jobs created would reach 329,000 by 2022, based on a broadband penetration rate of 50 percent.

- Overall, an evaluation of multiple studies showed that, for every 1,000 additional broadband users, approximately 80 new jobs are created (Almqvist 2010).

- In the United States, a nationwide study examined how broadband deployment affects job creation, determining that availability of broadband at the community level adds more than 1 percent to employment growth (Katz and Avila 2010, 3).

Additionally, although broadband is likely to have overall positive effects on job growth, short-term job losses may result from broadband-enabled improvements in productivity due to process optimization and capital-labor substitution. However, countries have confirmed that broadband creates many more jobs than it displaces in the longer term. For example, a study commissioned by the European Commission found a positive impact on employment in 2006, with net creation of 105,000 jobs throughout Europe due to broadband deployment.

Broadband as a General-Purpose Technology

Overall, broadband's importance may be fully realized as it becomes a GPT. Although the notion of broadband as a GPT has been addressed in recent discussions of broadband and development as well as in government-funded stimulus plans, the concept of GPTs was introduced on a more general basis in the 1990s and includes three key characteristics:

- Pervasive use in a wide range of sectors

- Technological dynamism (inherent potential for technical improvements)

- General productivity gains as GPTs evolve, improve, and spread throughout the economy (Bresnahan and Trajtenberg 1995).

In broad terms, GPTs are technologies that enable new and different opportunities across an entire economy, rather than simply addressing one problem or one sector. According to the OECD (2007, 8), GPTs "fundamentally change how and where economic activity is organized." Common examples include electricity, the internal combustion engine, and railways.

Although the initial conception of GPTs did not include the ICT sector, later research has considered ICTs with broadband as the enabling platform, through the lens of the GPT concept. This view of broadband as a potential GPT has also been embraced in publications from, or on behalf of, the World Bank, *info*Dev, and the European Commission as well as in academia (Kim, Kelly, and Raja 2010, 4; *info*Dev 2009, 3; Majumdar, Carare, and Chang 2009, 641).

When taken holistically, broadband as a platform—coupled with services, applications, content, and devices—has the potential to satisfy all three criteria and thus to become a GPT. First, broadband can be used as a key input in nearly all industries. Second, broadband has the potential for technological dynamism through the development of new technologies as well as improvements in the capacity and speed of broadband systems. For example, the average global broadband connection speed at the end of 2010 was slightly below 2 Mbit/s, with the top 20 countries having average speeds of over 7 Mbit/s, which allow services and applications requiring higher bandwidth, such as streaming video, to develop and become accessible to users (Akami 2010, 10). Third, broadband has the potential to enable and engender new organizational methods that result in more general increases in productivity. Global architecture firms, for example, may have offices around the globe, but team members working on a new building design no longer have to be in the same place or even in the same time zone. By using broadband connections to share work products, the team can be completely decentralized.

As broadband's potential as a GPT is realized, it will become an enabler of technology-based innovation and growth throughout the economy by businesses and individuals as well as by academic, government, and other institutions. Businesses and individuals are able to use currently available broadband technologies and services to create entirely new applications and services in areas such as advertising, e-commerce, online video, social networking, and financial services, including online banking and loans (Katz 2010, 9). Innovation in these areas is important for the growth of new markets in developed economies and for the transfer of technology to emerging economies, which can benefit from e-services, particularly mobile health and mobile banking services. Broadband-enabled services also allow the public sector to access new communities and regions as well as to deliver higher-quality services more efficiently and at lower costs, including in online education, telemedicine, and civic participation. The following discussion provides specific examples of how broadband can enable growth in and beyond the ICT sector in both developed and developing countries.

Research and Development: Enabling Product Development and Innovations in Any Sector

Broadband can have a particularly strong impact on research and development (R&D), leading to innovative technologies as well as enabling new ICTs to lead to further innovations (box 1.2). Additionally, broadband may

---

Box 1.2: Examples of Broadband's Potential Impacts on Innovation in R&D and Business Operations

- Enable instant sharing of knowledge and ideas
- Lower the barriers to product and process innovation via faster and less expensive communications
- Accelerate start-ups
- Improve business collaboration
- Enable small businesses to expand their R&D capabilities and collaborate in larger R&D consortia
- Reduce time from idea to final product
- Foster greater networking
- Promote user-led innovation.

Source: OECD 2008.

allow businesses to move more rapidly in the product development cycle from idea to final product. For example, a company could have teams in various locations around the world working on related portions of the same project, using broadband connectivity to provide seamless communication and information sharing.

Increasing broadband penetration may also enable more than just large firms, governments, and academic research institutions to develop innovative products. For example, Apple's iPhone App Store has over 100,000 registered application developers, most of which are small companies.[5] In 2008, those small developers produced five of the top 10 applications sold in the App Store (Dokoupil 2009).

Cloud Computing: Reducing Costs for Businesses

For enterprises of all sizes, the costs of information technology (IT) infrastructure, including hardware, software, and technical support, can be significantly reduced with the adoption of cloud computing technologies. Cloud computing generally allows for instant access to and storage of applications and data via broadband connectivity. Currently, almost every traditional business application has an equivalent application in the cloud, which means that cloud services can effectively replace the more conventional, and typically more expensive, method of accessing and storing applications and data through software installed locally on one's own computer or in-house server (Carr 2008, 72). Additionally, cloud computing reduces or eliminates the need for on-site IT staff since these data processes are handled remotely. According to Zhang, Cheng, and Boutaba (2010), cloud computing has other potential benefits for businesses as well:

- Reduced need for up-front investment, since cloud computing is typically based on a pay-as-you-go pricing model

- Lower operating costs, since the service provider does not need to provision capacities according to the peak load

- Easy access through a variety of broadband-enabled devices

- Lower business risks and maintenance expenses, since business risks (such as hardware failures) and maintenance costs are shifted to infrastructure providers, which often have better expertise and are better equipped to manage these risks.

Harvard Business Review Analytic Services (2011) conducted a global survey of nearly 1,500 businesses and other organizations on their current and planned use of cloud computing as well as the perceived benefits and risks associated with cloud computing services.[6] About 85 percent of respondents stated that their organization would be using cloud computing tools on a moderate or extensive basis over the next three years in order to take advantage of the benefits of cloud computing, including improved speed and flexibility of doing business, lower costs, and new avenues for growth, innovation, and collaboration. Only 7 percent of respondents stated that their business had been using cloud computing for over five years; these early adopters reported that real business value had already been created, including faster time to market, lower operating costs, and easier integration of new operations.

In addition, cloud computing itself can give rise to new business models and create new avenues for generating revenue. For example, Amazon, the largest U.S. online retailer, began offering cloud computing services to businesses and individuals in 2002 because the company had excess computing and storage capacity (Carr 2008, 74–75). In order to accommodate the busiest shopping week of the year in the United States, Amazon had to purchase a much larger amount of capacity than was required the rest of the year. Rather than let the extra capacity go unutilized, Amazon began renting its system to others, thereby becoming a "utility" for computing services.

Despite the promise of cloud computing as a source of substantial cost savings for enterprises, various issues may limit its impact, particularly lack of access to broadband services. Cloud computing requires access to fast, reliable, and affordable broadband in order to achieve the maximum benefits. In addition, cloud computing raises potential network and data security concerns as well as significant concerns about reliability of the technology, lack of interoperability with existing IT systems, and lack of control over the system.

Retail and Services Sectors: Improving Customer Relations

Particularly for the retail and services sectors where customer relations average 50 percent of a company's activities, broadband can improve the ability to reach new customers and maintain contact with existing customers. As such, the ability to send multimedia e-mail or use targeted online advertising to keep and attract customers can increase a company's sales, while using less capital and fewer labor inputs than would be required for postal mailings or door-to-door sales calls. Broadband also enables self-service websites, such as online airline reservations or e-government services, as well as remote services, such as online technical support and video conferencing. For example, broadband is essential for developing countries, particularly India, Mauritius, and China, which are the main off-shore destinations for IT technical support and business process outsourcing.[7]

In addition, sophisticated services, enabled by broadband and the development of ICTs, have become not just an input for trade in goods, but a final export for direct consumption (Mishra, Lundstrom, and Anand 2011, 2). Call centers in Kenya, business-consulting and knowledge-processing offices in Singapore, accountancy services in Sri Lanka, and human resources–processing firms in Abu Dhabi are different forms of this phenomenon. Recent research has found that sophisticated service exports are becoming an economic driver of growing importance in developing countries and may be an additional channel for sustained high growth. The deployment and adoption of broadband also have the potential to provide an additional conduit for economic growth through service exports.

Manufacturing and Industrial Sectors: Improving Supply Chain Management

Broadband allows businesses to manage their supply chains more efficiently by automatically transferring and managing purchase orders, invoices, financial transactions, and other activities. As with any information-based business activity, broadband can enable faster, more secure, and more reliable processing than was previously possible. Broadband connectivity saves processing and transfer time along the supply chain and substantially increases competitiveness by helping businesses to reduce stock levels, optimize the flow of goods, and improve the quality of final products. Since manufacturing and industrial sectors have been the main driver of overall economic growth in developing countries for the last 15 years, broadband is expected to play a vital role in helping developing countries to improve productivity in these sectors and companies to compete effectively in a global market.[8]

Education: Building Human Capital

In order to realize broadband's full potential for economic growth, an educated workforce trained in the use of ICTs is necessary. Additionally, there is a self-reinforcing effect between education and broadband, since broadband can help to improve fundamental educational outcomes, including learning how to use broadband better. For example, a review of 17 impact studies and surveys carried out at the national, European, and international levels by the European Commission found that the services and applications available over broadband networks improve basic educational performance (Balanskat, Blamire, and Kefala 2006, 3). These studies found that broadband and ICTs positively affect learning outcomes in math, science, and language skills. In addition to facilitating basic skills, broadband improves the opportunities for individuals with ICT training, and such individuals generally have a better chance of finding employment as well as higher earning potential (UNCTAD 2009, 57). Bridging the connectivity divide is critical to ensuring that today's students—and tomorrow's high-tech workforce—can take advantage of these benefits.

One way to expand access to broadband and ICTs in rural and remote areas is through the deployment of mobile education labs. These labs, which may simply be vehicles fitted with broadband connectivity, computer equipment, and learning facilities, allow educators to drive to various schools throughout the week (Samudhram 2010). In addition, mobile education labs can provide ICT training for adults to improve digital literacy. As opposed to transporting children in rural areas to where broadband facilities exist or waiting until the network is built out to them, mobile facilities offer a cost-effective way to reach rural populations. The United Nations has noted the success of mobile schools in Mongolia, where 100 mobile "tent" schools have been introduced in 21 provinces, as well as in Bolivia (United Nations 2010). Bolivia has implemented a bilingual education program for three of the most widely used indigenous languages, which has been expanded to include indigenous children in remote areas. In Morocco, the government implemented a program called NAFID@ to help over 100,000 teachers to afford wireline or mobile broadband connections, which has allowed teachers to receive training in the use of ICTs in the classroom as well as to use e-learning programs and online libraries to improve class lessons (Intel 2010).

Health Care Sector: Improving Health and Medical Outcomes

Health-based broadband applications and services are significantly improving health and medical outcomes around the world, particularly for patients in remote areas and those with limited mobility, through e-health and m-health initiatives (WHO 2005). Considering that there are fewer than 27 million doctors and nurses for the more than 6 billion people in the world—and only 1.2 million doctors and nurses in the lowest-income countries—harnessing mobile technologies is a valuable tool for enabling health care practitioners to reach patients. As mobile broadband develops and spreads in developing countries, the benefits are already becoming clear (box 1.3).

Although basic voice and data connections can be useful in improving health and medical care, broadband connectivity is necessary to capture the full potential of e-health services, including telemedicine, which enables real-time audio and video communications between patients and doctors as well as between health care providers. Improvements in telemedicine and other e-health initiatives rely on increasing bandwidth capacity, more storage and processing capabilities, and higher levels of security to protect

Box 1.3: Mobile Health Services in Nigeria

In Nigeria, the government lacked sufficient public health information to allocate health care services efficiently to over 800 villages with no primary health care. A public-private partnership, Project Mailafia, was established to alleviate this situation. Project Mailafia sends teams of mobile health care providers to remote villages, where they treat patients and collect health data that support better public health decision making and resource allocation. Mobile health workers collect the data on ruggedized netbooks and transfer the data to area clinics. The clinics then upload the data to a central database using Worldwide Interoperability for Microwave Access (WiMAX) and Wireless Fidelity (Wi-Fi) technologies.

Source: Intel 2010.

patient information (ITU 2008, 11). As noted in table 1.2, the U.S.-based California Broadband Task Force (2008, 6) estimated that telemedicine will require speeds between 10 and 100 Mbit/s and that high-definition telemedicine will require broadband speeds of over 100 Mbit/s. The current wireline and wireless infrastructure in most countries is insufficient to take advantage of the e-health opportunities in the digital economy. This is particularly important for developing countries, where ensuring access to and adoption of wireline and wireless broadband networks would be particularly useful for including those who have been left out of more traditional health care models.

This happened in Rwanda, where a three-phase e-health project was delayed due to lack of high-speed broadband connectivity (Rwirahira 2009). The first phase of the initiative, which established an electronic data storage system that permitted three hospitals to share patient information, was completed without delay. However, the final two phases, which involved video conferencing and a real-time telemedicine system, were put on hold for a year until a broadband Internet connection could be established to connect the three hospitals with a fiber optic cable network (Ndahiro 2010).

E-Government Applications: Transforming Government Processes and Improving Citizen Participation

E-government covers a broad range of applications that can transform government processes and the ways in which governments connect and interact with businesses and citizens. This allows citizens to participate in society and improves the efficiency, accountability, and effectiveness of government

Table 1.2 Upstream and Downstream Speeds Needed for Various Services and Applications

500 kbit/s to 1 Mbit/s	5 to 10 Mbit/s	100 Mbit/s to 1 Gbit/s
• VoIP • SMS • Basic e-mail • Web browsing (simple sites) • Streaming music (caching) • Low-quality video (highly compressed)	• Telecommuting (converged services) • File sharing (large) • IPTV, SD (multiple channels) • Switched digital video • Video on demand, SD • Broadcast video, SD • Video streaming (2–3 channels) • Video downloading, HD • Low-definition telepresence • Gaming • Medical file sharing (basic) • Remote diagnosis (basic) • Remote education • Building control and management	• Telemedicine, HD • Multiple educational services • Broadcast video, full HD • Full IPTV channel support • Video on demand, HD • Gaming (immersion) • Remote server services for telecommuting

1 to 5 Mbit/s	10 to 100 Mbit/s	1 to 10 Gbit/s
• Web browsing (complex sites) • E-mail (larger attachments) • Remote surveillance • IPTV, SD (1–3 channels) • File sharing (small, medium) • Telecommuting (ordinary) • Digital broadcast video (1 channel) • Streaming music	• Telemedicine • Educational services • Broadcast video, SD and some HD • IPTV, HD • Gaming (complex) • Telecommuting (high-quality video) • High-quality telepresence • Surveillance, HD • Smart, intelligent building control	• Research applications • Telepresence using uncompressed video streams, HD • Live event digital cinema streaming • Telemedicine remote control of scientific or medical instruments • Interactive remote visualization and virtual reality • Movement of terabyte data sets • Remote supercomputing

Source: California Broadband Task Force 2008.

Note: kbit/s = kilobits per second; Mbit/s = megabits per second; Gbit/s = gigabits per second; VoIP = voice over Internet Protocol; SMS = short message service; IPTV = Internet Protocol television; SD = standard definition; HD = high definition.

programs and processes. Broadband is important for e-government, as it provides the foundation for public administration networks that allow processes to flow more smoothly. In turn, e-government can help to drive demand for broadband.

Countries around the world are providing increasing access to online services, including the provision of basic services, the use of multimedia technology to promote two-way exchanges, and the use of technology to facilitate consultation with citizens on public policy issues. Although the Republic of Korea, the United States, and Canada take the top three places with regard to the number of online government services available, several countries have made significant progress over the last two years, including Bahrain, Chile, Colombia, and Singapore. Moreover, the use of mobile phones for e-government services, such as alert messages, applications, and fee payments, is almost as popular in developing countries as it is in developed countries (UNPAN 2010).

What Market Trends Are Fostering Broadband Deployment?

Broadband connectivity is expanding globally. Between 2005 and 2010, the average wireline broadband penetration rate grew 59 percent—from 3.3 to 8 subscribers per 100 inhabitants (ITU-D 2010). The number of mobile broadband subscriptions worldwide is expected to reach the 1 billion mark in 2011, with total mobile subscriptions topping 5 billion.[9] As a result of such growth, the estimated number of wireline broadband subscriptions reached approximately 555 million in 2010, up from 471 million in 2009. A sizable number of these new subscriptions came from Brazil, the Russian Federation, India, and China (known as the BRIC countries), which have collectively doubled their subscriber base in the last four years.[10] Likewise, the number of wireless broadband users has also expanded rapidly. In 2010, the number of third-generation (3G) mobile broadband subscriptions rose to 940 million, an increase from 703 million in 2009. As figure 1.3 shows, the number of wireless broadband subscribers exceeded the number of wireline broadband subscribers for the first time in 2008, and there were an estimated 70 percent more mobile broadband subscribers than wireline broadband subscribers in 2010.

Despite these advances, however, a "digital divide" remains between developed and developing countries; only 4.4 per 100 people in developing countries are broadband subscribers compared to 24.6 in developed countries. In effect, wireline broadband deployments in many developing countries are a decade behind deployments in developed countries. Given the cost and resources required for the deployment of wireline broadband, wireless broadband is more likely to be the broadband solution adopted by users in developing countries, particularly in rural and remote areas.

Figure 1.3 Global Fixed and Mobile Broadband Subscriptions per 100 Inhabitants, 2000–10

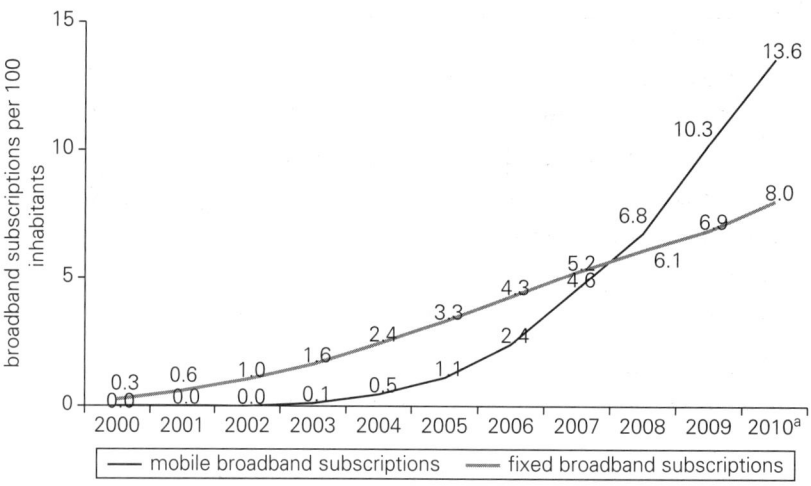

Source: ITU, World Telecommunications Indicators database.

a. Estimated.

Trends in Supply

Developments in the types of technologies and business models used to deploy broadband network infrastructure are allowing operators to supply more people with broadband connections at lower costs. In developed countries, network operators are installing fiber optic cables closer to end users, reaching directly into their neighborhoods, offices, and homes. In developing countries, the spread of high-speed wireless networks promises to gain momentum over the next few years. Wireless broadband is already more prevalent than wireline broadband in many developed and developing countries, but to a much greater extent in developing countries. As noted in table 1.3, the number of wireless broadband subscriptions in Sub-Saharan Africa, for example, is more than eight times the number of wireline subscriptions, suggesting the potential for wireless broadband in areas where traditional wireline infrastructure may be absent.

With the number of wireless broadband subscriptions worldwide expected to reach the 1 billion mark in 2011, developing countries, particularly China and India, are often leading the way. Together, China and India have the top five mobile operators in terms of total number of subscriptions, which is expected to continue as mobile broadband grows.

Broadband Strategies Handbook

Table 1.3 Wireless and Wireline Broadband Subscriptions per 100 Inhabitants, by Region, June 2011

Region	Wireless	Wireline
Sub-Saharan Africa	2.9	0.3
East Asia and Pacific	16.6	10.5
Eastern Europe and Central Asia	14.5	9.2
European Union and Western Europe	45.9	27.6
Latin America and the Caribbean	12.2	7.1
Middle East and North Africa	13.1	2.5
North America	34.0	24.5
South Asia	1.6	0.8
Global	13.6	8.1

Source: World Bank analysis based on data from TeleGeography's GlobalComms database.

Another important trend affecting broadband networks is their ever-increasing speed. In 2010, Akamai, a major Internet content manager, suggested that a global shift was occurring away from narrowband and toward broadband connectivity. Globally, average Internet connection speeds (for users who pass through the company's servers) rose 14 percent year-over-year to 1.9 Mbit/s, and all of the top 10 countries achieved average connection speeds at or above the "high broadband" threshold of 5 Mbit/s (figure 1.4).

In addition to overall growth in wireline and wireless broadband infrastructure, the release of new broadband-enabled devices may also be viewed as a supply-side input. The overall trend for broadband devices is improved capabilities, mobility, and portability. According to the research firm International Data Corporation (IDC), in the third quarter of 2010, global smartphone shipments increased nearly 90 percent from the same quarter of 2009.[11] IDC's analysts predicted in November 2010 that 20 percent of device shipments in 2010 would be smartphones, compared to 15 percent in 2009. The research firm also examined the nascent tablet computing market, noting that shipments in the third quarter of 2010 were 45 percent higher than in the second quarter and forecasting that 2011 shipments would exceed 2010 by more than 160 percent and that 2012 shipments would exceed 2011 by a further 60 percent.[12] In a separate forecast, the IDC predicted that combined shipments of smartphones, tablets, and other application-enabled devices would overtake traditional personal computer (PC) shipments by mid-2011 (Thibodeau 2010). All of these devices are designed to take advantage of broadband connectivity, whether provided by a mobile network or

Figure 1.4   Average Broadband Speed in Top 10 Countries

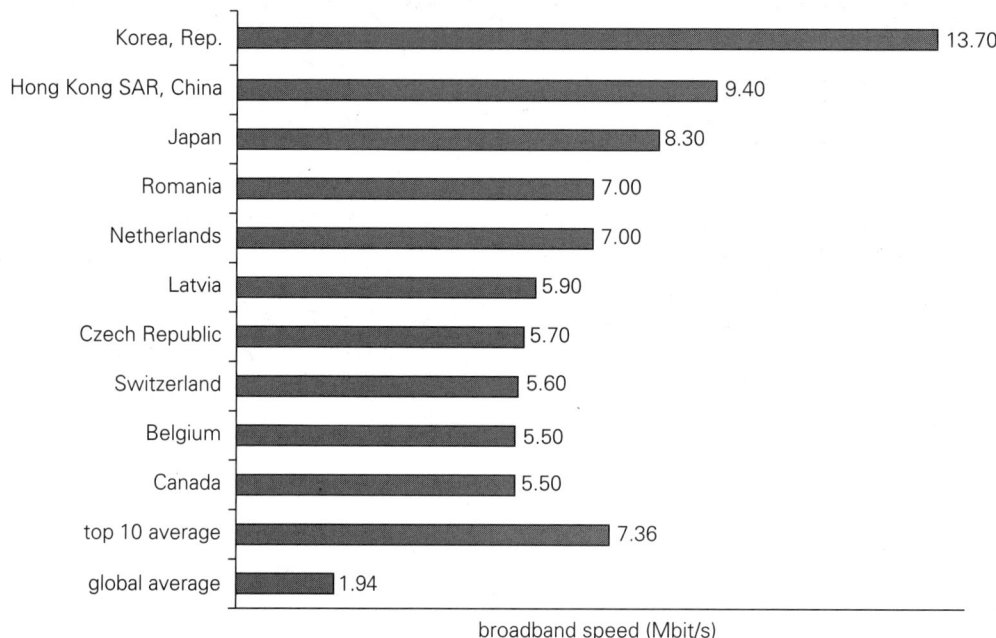

Source: Akamai 2010.

by Wi-Fi distribution of the wired broadband connection in a home, workplace, or Wi-Fi "hotspot."

Trends in Demand

The development of novel or enhanced applications and services enabled by broadband connectivity has served as a key driver of demand for broadband access over the past several years (box 1.4). The availability of broadband networks has facilitated at least a partial migration of existing services from more traditional models to broadband digital networks, including entertainment, banking, education, health care, and shopping, to name a few. While many of these same services saw an initial online presence with dial-up and other narrowband services, the rise of broadband connectivity has facilitated the development of more robust applications and services. From the perspective of the organizations that are using broadband-enabled services to reach consumers, clients, members, and citizens, the efficiency of electronic communications has led to an increasing interest in bringing

Box 1.4: User Trends That Promote Demand for Broadband

As the following broadband-enabled services, applications, and content become an increasingly integral component of daily personal, business, and educational interactions, they are expected to spark further demand for broadband services.

- *Video.* More and higher-quality video and other rich content will continue to drive the demand for higher-capacity broadband services.

- *Apps and cloud-based computing.* Apps are increasingly driving broadband use and development, especially in the wireless broadband context. More robust apps, including productivity applications such as office suites, are being offered by companies such as Google and Microsoft. Some of the benefits of online applications include access to information and documents from multiple locations, decreased processing power requirements for end-user devices, and less responsibility for users to update and maintain applications.

- *Web 2.0.* Web 2.0 applications leverage advances in computing and connectivity to create collaborative, user-centered, and interoperable environments in which users can generate, distribute, and share content in real time. See chapter 6 for more on Web 2.0.

- *Social networking.* As social networking applications have become more sophisticated and diverse, they have also become immensely popular. Each month, Facebook's more than 500 million active users share over 30 billion pieces of content (for example, photos, videos, updates, web links, news stories, and blog posts). YouTube has become the most popular online video-sharing site in the world, and 70 percent of its content is created outside the United States. Broadband access facilitates the use of these social networking applications, which in turn are major drivers of broadband demand. See chapter 6 for more on social networking.

Sources: See Google, http://docs.google.com; Microsoft, http://office.microsoft.com/en-us/web-apps/; Kim, Kelly, and Raja 2010; Facebook, "Statistics, Press Room," http://www.facebook.com/press/info.php?statistics, as of June 2011; YouTube, "YouTube Statistics," http://www.youtube.com/t/press_statistics, as of June 2011.

traditionally offline or nonelectronic services to the Internet or at least augmenting those services with online alternatives.

In addition to the social and personal use of broadband led by the private sector, countries around the world are providing increasing access to online e-government services. Broadband allows business owners to reap substantial benefits. The use of broadband as an input by businesses, both traditional "brick and mortar" and online companies, is expected to be one of the main drivers of broadband adoption and will require network operators to deploy new infrastructure and upgrade existing networks quickly in order to keep pace with demand.

How Can Broadband Development Be Supported?

Despite the rapid growth in demand for broadband and the development of broadband-enabled applications, services, and devices, there are also notable challenges. Whether within a particular economy or across nations or even regions, the more-affluent and better-educated populations generally have had earlier and better access to ICTs than the less-affluent and less-educated populations. With the rise of broadband-enabled services and applications and the increasing migration of many aspects of modern life online, a lack of broadband connectivity can increasingly have a negative impact on social and economic development by excluding those who lack broadband access or do not understand the relevance of broadband-enabled services.

Governments can employ a wide range of strategies and policies to support the development of broadband, such as through market liberalization (for example, opening international gateways to competition) and the allocation and award of new spectrum for wireless broadband (for example, releasing the "digital dividend" spectrum for commercial wireless use once the country's digital television transition is completed). It may be useful for policy makers and stakeholders to view broadband as an ecosystem, as this would encourage the development of coherent, integrated policies that maximize the benefits of broadband across all sectors of the economy and segments of society.

Viewing Broadband as an Ecosystem

So that government policy makers and private sector investors can understand the ways in which broadband networks and services can best be supported, it is useful to have an overarching concept of how to think about broadband from a policy point of view. This section proposes that broadband can be best thought of as an ecosystem of mutually dependent—and reinforcing—components: supply and demand.

Under the ecosystem model (figure 1.5), the supply of broadband network platforms is the first necessary condition (that is, broadband infrastructure must be available). However, demand for broadband is just as important in order to make substantial network investments worthwhile. Also needed is the ability of non-ICT sectors to use and create broadband-enabled services and applications, as this boosts demand and encourages further network deployments. Developing these synergies will largely determine the extent to which broadband affects the economy, serves as an enabling platform, and, ultimately, becomes a GPT that can act as an essential input in driving innovation and growth in all sectors.

The basic elements of supply in the broadband ecosystem consist of four levels: (a) international connectivity, (b) domestic backbones, (c) metropolitan connectivity, and (d) local connectivity (adapted from Kim, Kelly, and Raja 2010). These elements and their importance as the supply-side component of the broadband ecosystem are detailed in subsequent chapters. Chapter 5 discusses each of these components in more detail, while chapters 2 and 3 discuss the policy implications and the legal and regulatory trends, respectively, related to broadband networks.

Without relevant, useful, and innovative advancements in services, applications, and content, there would be little or no demand for broadband. As

Figure 1.5   Broadband Ecosystem and Its Impact on the Economy

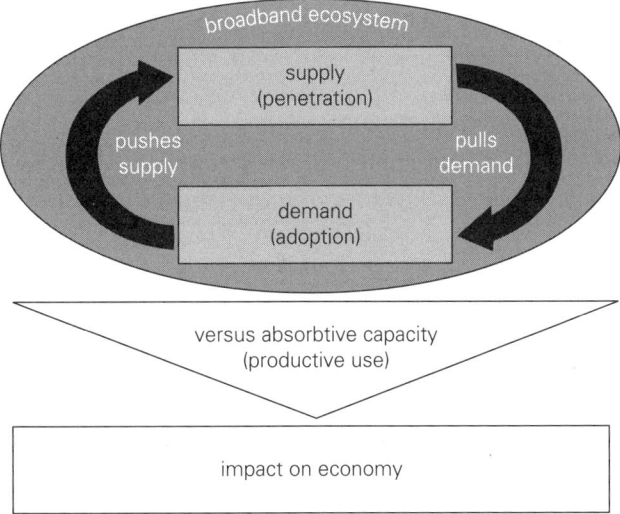

Source: World Bank.

such, the many demand-side components—including services, applications, and content—are essential to promoting a vibrant broadband ecosystem. While a distinction is generally made between services and applications, as technology evolves, services and applications are likely to overlap. For example, mobile banking may be treated as a service or as an application (and maybe even as both), depending on how and what features are offered. In addition, e-government covers an entire range of services and applications that transform government processes and modes of interacting with businesses and citizens (Hanna et al. 2009). The distinction, at least with regard to the ecosystem, may be irrelevant: what is important is that these services and applications drive demand.

Absorptive Capacity of Broadband

Supply and demand are necessary conditions for the promotion of broadband networks and services, but by themselves they are not sufficient to guarantee that broadband can reach its full potential in the economy. For that to happen, broadband users (citizens, businesses, and government) must also have the capacity to understand, learn, and apply the lessons learned about broadband's benefits and capabilities across the economy and society.

Absorptive capacity generally refers to the ability of an organization to recognize the value of new, external information; to assimilate that information; and then to apply it to the organization's benefit. This ability is critical to an organization's innovative capabilities, as new technologies are assimilated by organizations to create, improve, and transform business processes, products, and services (Cohen and Levinthal 1990). As users have the ability to become co-creators of content[13] and as broadband user–led innovation is enabled, this same concept can be extended to include other users of the broadband platform, including citizens (von Hippel 2005). Thus to fully realize the benefits of broadband, the various sectors of the economy and society must have the capacity to acquire, assimilate, transform, and exploit the capabilities enabled by this platform. Under the ecosystem model, absorptive capacity is the mechanism by which the benefits obtained from broadband feed into the greater economy, allowing this technology to unleash its potential as a GPT.

Policy makers can facilitate the capacity to understand and incorporate the many benefits of broadband by developing and implementing policies that are complementary to broadband build-out. In addition, they can encourage the private sector to adopt broadband as an input to drive productivity, growth, innovation, and welfare throughout the economy and society.

As discussed, broadband alone has limited impact as a technological platform, but instead acts as an enabler. As such, it holds the potential to have a significant impact on economic and social progress and to transform the economy.

However, for this potential impact to be unleashed, broadband must be used by businesses, governments, and citizens in a way that increases productivity in the economy.[14] This requires (a) the creation and availability of broadband-enabled services and applications that increase efficiency and productivity and (b) the capacity of businesses, government, and citizens to use broadband-enabled services and applications in a productive and efficient way. These two requirements are critical for achieving the potential economic impact that broadband can produce.

The economy's capacity to absorb broadband depends on how these two requirements are fulfilled in the economy. In a nutshell, a country's absorptive capacity can be thought of as determined by the following:

- The capacity of businesses to create broadband-enabled services and applications and to use these applications and services to make their business processes more productive and efficient

- The capacity of citizens to create and use broadband-enabled services and applications to improve their welfare

- The capacity of government and other institutions (for example, schools) to introduce and accommodate broadband-enabled services to deliver public services more efficiently and transparently to the public.

Components of Broadband Absorptive Capacity

Four components determine the degree to which a country's economy is able to absorb broadband and translate it into economic and social development. These components are (a) the economy's macroeconomic environment, (b) the business environment, (c) the quality of human capital, and (d) the governance structure (figure 1.6). The macroeconomic environment determines the "broadband friendliness" of the economy and whether the economy and its main actors (that is, businesses, government, and citizens) are open to using ICTs. The business environment, which includes access to financing and diffusion of previous technologies, determines the ability of businesses and entrepreneurs to create new broadband-enabled innovations, modify business processes based on these innovations, and update existing products, services, and strategies using broadband and the broadband-enabled environment. The quality of human capital depends on the ability of the labor force, businesses, and academic

Figure 1.6   Examples of the Elements of Absorptive Capacity

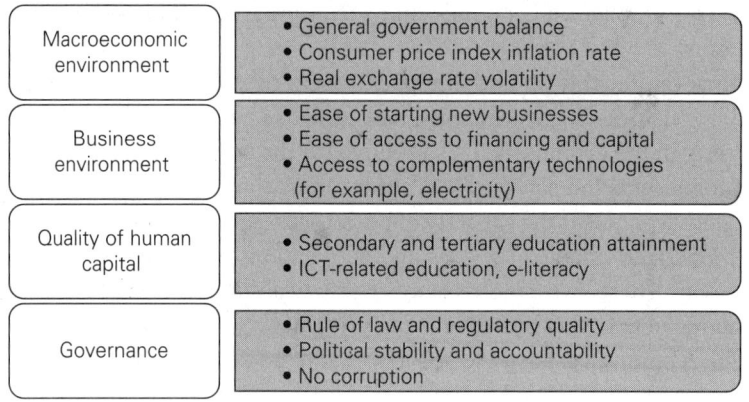

Macroeconomic environment	• General government balance • Consumer price index inflation rate • Real exchange rate volatility
Business environment	• Ease of starting new businesses • Ease of access to financing and capital • Access to complementary technologies (for example, electricity)
Quality of human capital	• Secondary and tertiary education attainment • ICT-related education, e-literacy
Governance	• Rule of law and regulatory quality • Political stability and accountability • No corruption

Source: Partially based on World Bank 2008.

institutions to understand the potential of broadband and adapt their mind-sets to the broadband-enabled environment. Finally, the governance structure determines the degree to which businesses and citizens are permitted to share and access information openly as well as to share broadband-based ideas and innovations. Additionally, governance addresses the security of investment and the cost of creating new broadband-enabled business, services, and products. Governance that promotes the absorptive capacity of broadband generally requires free, open access to information and abidance by the rule of law to protect investments. Although each component of absorptive capacity has a wide range of elements, figure 1.6 provides several examples.

Degree of Broadband Absorptive Capacity

The degree of absorptive capacity in a given economy will determine the amount of broadband-enabled economic development that is possible. Without strong absorptive capacity, the impact of broadband on economic development will be limited or even nonexistent (box 1.5). A country with nationwide broadband coverage and widespread adoption will obtain very little overall economic and social benefit if absorptive capacity is limited. Conversely, a country with relatively limited broadband coverage or adoption can obtain a targeted impact on the economy if there is sufficient absorptive capacity. Moreover, absorptive capacity can be targeted to specific sectors of the economy, which has been the case with the IT and business process outsourcing industry in countries like India. This targeted absorptive capacity can then expand throughout the economy.

Box 1.5: Capacity to Absorb Technology and the Economic Impact of Broadband-Enabled ICTs: The Examples of Italy and Sweden

An economy with a flexible facilitating structure, an entrepreneurial business environment, few technological regulatory restrictions, an ICT-educated workforce, high penetration of complementary technologies (for example, electricity), a business-friendly financing structure, and a responsive public policy structure will experience faster diffusion of broadband-enabled applications and services and a larger economic and social impact. The impact of broadband-enabled ICTs on economic growth will be slower and smaller in an economy that lacks some of these elements or that delays the changes needed to adapt the facilitating structure to broadband-enabled ICTs (for example, by not modifying the regulatory framework to eliminate technological restrictions or to facilitate their diffusion). In relative terms, it can be put as follows: assuming that the maximum and fastest effect on the structure of the economy that a country can obtain from broadband-enabled ICTs is 100 (that is, the potential positive impact from broadband), the degree of absorptive capacity of the economy will determine how much and how fast that 100 value can be actually realized.

Italy and Sweden provide good examples of how this mechanism works. Both countries have relatively similar levels of GDP per capita and an in-depth penetration of previous complementary technologies, such as electricity and telephone lines. However, their absorptive capacity is different (table B1.5.1). Sweden performs better in business environment and human capital and has taken a very active role in modifying the facilitating structure of its economy to allow for faster diffusion of broadband (for example, by establishing a public policy to enable the diffusion of broadband and implementing e-literacy programs).

Many other factors in place explain the ability of Sweden's economy to diffuse broadband-enabled ICTs, but the important point is that Sweden has actively adapted the facilitating structure of its economy to allow broadband to diffuse faster and broader than Italy has. As a result, the economic effects of broadband-enabled ICTs in Sweden have been larger and surfaced faster. For instance, from 1998 to 2007, average annual productivity grew much faster in Sweden than in other peer countries (2.32 percent compared with 0.39 percent in Italy and an average of 1.66 percent among OECD countries). Even though this growth was not due exclusively to broadband, Sweden's policy has transformed the country into a broadband leader, and this transformation has played an important role in its economic growth.

Table B1.5.1 Internet Adoption Proxies in Sweden and Italy, 2007

Proxy	Sweden	Italy
% of population with no Internet skills	22	58
% of enterprises receiving Internet orders	26	4
% of enterprises purchasing on the Internet	72	29

Source: LECG 2009, table citing Commission of the European Communities 2008.

The Role of Governments in Broadband Policy Making

Throughout the rest of this handbook, the concept of broadband as an eco-system, consisting of supply and demand components, is the overarching concept used to frame how policy makers can maximize absorptive capacity and fully realize the potential impact of broadband on economic, social, and policy goals. To this end, as discussed in chapter 2, governments should implement policies that support the supply of broadband networks and services, particularly to economically unviable areas, through a variety of mechanisms such as appropriate market regulation, universal access and service policies, flexible licensing policies, direct infrastructure investments, removal of bottlenecks, and pro-market tax policies. Additionally, governments should seek to stimulate demand and uptake of broadband through the creation of an enabling environment by addressing awareness, affordability, and attractiveness (perceived value) of broadband services (see chapter 6 for more on facilitating demand). In developing these policies, high-level coordination is needed between the ICT ministry and other sector ministries, as is a focus on R&D investments. In addition, the rollout of broadband presents particular concerns for developing countries, as addressed specifically in chapter 7.

It is also essential to ensure that the government possesses the capacity to create and implement effective laws and regulations that give rise to an enabling environment. Chapter 3 discusses this in more detail. This capacity is particularly important in developing countries, where the government must be able to carry out the policies and rules it develops. These policies should emphasize the need for appropriate tools that foster supply and demand and build absorptive capacity.

Notes

1. OECD (2008, 134). Brazil's plan defines broadband as "the provision of telecommunications infrastructure that enables information traffic in a continuous and uninterrupted manner, with sufficient capacity to provide access to data, voice, and video applications that are common or socially relevant to users as determined by the federal government from time to time" (Brazil, Ministério das Comunicações 2009, 24).
2. ITU, "Key Global Telecom Indicators for the World Telecommunication Service Sector," http://www.itu.int/ITU-D/ict/statistics/at_glance/KeyTelecom.html.
3. Countries include Argentina, Brazil, Chile, Colombia, Ecuador, El Salvador, Mexico, Nicaragua, Panama, Peru, Uruguay, and República Bolivariana de Venezuela.
4. MCMC, "Broadband Fact Sheet," Press Release, May 2008, http://www.skmm.gov.my/index.php?c=public&v=art_view&art_id=326.

5. Apple, "Staggering iPhone App Development Statistics to Be Unveiled at Inaugural App Exhibit at Macworld 2010," Press Release, February 2, 2010, http://www.iphoneappquotes.com/press-staggering-iphone-app-develop ment-statistics-unveiled-at-macworld.aspx.

6. The respondents were large companies, with average annual sales for 2010 of US$1.3 billion and average employment of 3,280 employees. However, the respondents were from a wide range of sectors: 17 percent were in the IT sector, 12 percent were in professional services, and 10 percent each were in financial, manufacturing, and government or nonprofit sectors.

7. Gartner, "Gartner Identifies Top 30 Countries for Offshore Services in 2010–2011," Press Release, December 20, 2010, http://www.gartner.com/it/page .jsp?id=1502714.

8. United Nations Industrial Development Organization, "UNIDO Releases Latest International Yearbook of Industrial Statistics," Press Release, March 10, 2010, http://www.unido.org/index.php? id=7881&tx_ttnews[tt_news]=455&cHash =09cad462f0.

9. ABI Research, "One Billion Mobile Broadband Subscriptions in 2011: A Rosy Picture Ahead for Mobile Network Operators," February 2011, http://www .abiresearch.com/press/3607-One+Billion+Mobile+Broadband+Subscriptions+ in+2011:+a+Rosy+Picture+Ahead+for+Mobile+Network+Operators.

10. World Bank analysis based on TeleGeography GlobalComms data (December 2009).

11. IDC, "Worldwide Smartphone Market Grows 89.5% Year-over-Year in Third Quarter as New Devices Launch, Says IDC," Press Release, November 4, 2010, http://www.idc.com/about/viewpressrelease.jsp?containerId=prUS22560610.

12. IDC, "IDC's Worldwide Quarterly Media Tablet and eReader Tracker Makes Its Debut, Projects Nearly 17 Million Media Tablets Shipped Worldwide in 2010," Press Release, January 18, 2011, http://www.idc.com/about/viewpress release.jsp?containerId=prUS22660011&sectionId=null&elementId=null&page Type=SYNOPSIS.

13. See the OECD website, http://www.oecd.org/dataoecd/57/14/38393115.pdf.

14. In particular, the World Bank has defined absorptive capacity in the context of innovation as the quality of the labor force and the business environment (including access to finance) in which firms operate and are able (or unable) to start up, expand, and reap the financial rewards of their new-to-market innovations (see World Bank 2008). Applied to broadband, this concept focuses on broadband-enabled services and applications and expands to the use and creation of these services and applications by businesses, citizens, and govern-ments to modify their behavior and processes in an effort to be more produc-tive and efficient.

References

Akamai. 2010. *The State of the Internet, 4th Quarter, 2010 Report.* Vol. 3, no. 4. Cambridge, MA: Akmai. http://www.akamai.com/dl/whitepapers/Akamai_

state_of_internet_q32010.pdf?curl=/dl/whitepapers/Akamai_state_of_
internet_q32010.pdf&solcheck=1&.

Almqvist, Erik. 2010. "Social Net Benefits of IPTV and BB Infrastructure Investments." Presentation given at the "IPTV World Forum on the Middle East and
Africa," Dubai, November 1. http://www.ericsson.com/campaign/televisionary/
content/pdf/regulation/7f16d52b-d310-4b4f-8a48-1f3fb7fb3c0b.pdf.

Analysys Mason. 2010. "Assessment of Economic Impact of Wireless Broadband in
India." Discussion Document Report for GSMA, Analysis Mason, New Delhi,
December. http://www.analysysmason.com/About-Us/Offices/New-Delhi/
Increase-in-broadband-penetration-by-1-will-contribute-INR162-billion-to-
Indias-GDP-by-2015/.

Atkinson, Robert, Daniel Castro, and Stephen J. Ezell. 2009. *The Digital Road to
Recovery: A Stimulus Plan to Create Jobs, Boost Productivity, and Revitalize
America*. Washington, DC: Information Technology and Innovation Foundation.

Balanskat, Anja, Roger Blamire, and Stella Kefala. 2006. *The ICT Impact Report:
A Review of Studies of ICT Impact on Schools in Europe*. Report for the European
Commission's ICT Cluster. Brussels: European Schoolnet. http://ec.europa.eu/
education/pdf/doc254_en.pdf.

Beardsley, Scott, Luis Enriquez, Sheila Bonini, Sergio Sandoval, and Noëmie Brun.
2010. "Fostering the Economic and Social Benefits of ICT." In *The Global
Information Technology Report 2009–2010: ICT for Sustainability*, ed. Soumitra
Dutta and Irene Mia, 61–70. Geneva: World Economic Forum and INSEAD.

Belo, Rodrigo, Pedro Ferreiray, and Rahul Telangz. 2010. "The Effects of Broadband
in Schools: Evidence from Portugal." Carnegie Mellon University, Pittsburgh,
PA. http://papers.ssrn.com/sol3/papers.cfm?abstract_id=1636584.

Brazil, Ministério das Comunicações. 2009. *Um Plano Nacional para Banda Larga:
O Brasil em alta velocidade* [Brazilian National Broadband Plan], trans. Telecommunications Management Group, Inc. Brasília: Ministério das Comunicações.

Bresnahan, Timothy F., and Manuel Trajtenberg. 1995. "General-Purpose Technologies: Engines of Growth?" *Journal of Econometrics* 65 (1): 83–108.

Broadband Commission for Digital Development. 2010. *Broadband: A Platform for
Progress*. Geneva: International Telecommunication Union.

Buttkereit, Sören, Luis Enriquez, Ferry Grijpink, Suraj Moraje, Wim Torfs, and
Tanja Vaheri-Delmulle. 2009. "Mobile Broadband for the Masses: Regulatory
Levers to Make It Happen." McKinsey and Company, London, February. http://
www.mckinsey.com/clientservice/telecommunications/Mobile_broadband_
for_the_masses.pdf.

California Broadband Task Force. 2008. *The State of Connectivity: Building
Innovation through Broadband; Final Report*. Sacramento: California Broadband
Task Force.

Carr, Nicholas G. 2008. *The Big Switch: Rewiring the World from Edison to Google*.
New York: W. W. Norton.

Cohen, Wesley, and Daniel A. Levinthal. 1990. "Absorptive Capacity: A New
Perspective on Learning and Innovation." *Administrative Science Quarterly* 35
(1): 128–52.

Commission of the European Communities. 2008. "Preparing Europe's Digital Future: i2010 Mid-Term Review." Commission Staff Working Document, Commission of the European Communities, Brussels.

Crandall, Robert W., Charles L. Jackson, and Hal J. Singer. 2003. *The Effects of Ubiquitous Broadband Adoption on Investment, Jobs, and the U.S. Economy.* Washington, DC: Criterion Economics.

Czernich, Nina, Oliver Falk, Tobias Kretschmer, and Ludger Woessmann. 2009. "Broadband Infrastructure and Economic Growth." CESifo Working Paper 2861, CESifo, Munich, December. http://www.cesifo-group.de/portal/page/portal/ DocBase_Content/WP/WP-CESifo_Working_Papers/wp-cesifo-2009/ wp-cesifo-2009-12/cesifo1_wp2861.pdf.

Dokoupil, Tony. 2009. "Striking It Rich: Is There an App for That?" *Newsweek,* October 6. http://www.newsweek.com/2009/10/05/striking-it-rich-is-there -an-app-for-that.html.

Fornefeld, Martin, Gilles Delaunay, and Dieter Elixmann. 2008. *The Impact of Broadband on Growth and Productivity*. Study on behalf of the European Commission. Dusseldorf, Germany: MICUS Management Consulting.

Friedrich, Roman, Karim Sabbagh, Bahjat El-Darwiche, and Milind Singh. 2009. "Digital Highways: The Role of Government in 21stCentury Infrastructure." Booz & Company, New York. http://www.booz.com/media/uploads/ Digital_ Highways_Role_of_Government.pdf.

Hanna, Nagy K., Christine Zhen-Wei Qiang, Kaoru Kimura, and Siou Chew Kuek. 2009. "National E-Government Institutions: Functions, Models, and Trends." In *Information and Communications for Development 2009*, ch. 6. Washington, DC: World Bank. http://siteresources.worldbank.org/EXTIC4D/Resources/5870635 -1242066347456/IC4D_2009_Chapter6.pdf.

Harvard Business Review Analytic Services. 2011. "How the Cloud Looks from the Top: Achieving Competitive Advantage in the Age of Cloud Computing." Report commissioned by Microsoft, Seattle, May. http://www.microsoft.com/en-us/ cloud/tools-resources/whitepaper.aspx?resourceId=Achieving_Competitive_ Advantage&fbid=nSj309bhGW3.

*info*Dev. 2009. "What Role Should Governments Play in Broadband Development?" Paper prepared for *info*Dev and Organisation for Economic Co-operation and Development workshop on "Policy Coherence in ICT for Development," Paris, September 10–11.

Intel. 2010. "Realizing the Benefits of Broadband." Intel, Santa Clara, CA. http:// www.intel.com/Assets/PDF/Article/WA-323857001.pdf.

ITU (International Telecommunication Union). 2008. "Implementing e-Health in Developing Countries: Guidance and Principles." ITU, Geneva. http://www.itu .int/ITU-D/cyb/app/docs/e-Health_prefinal_15092008.PDF.

———. 2009. *Manual for Measuring ICT Access and Use by Households and Individuals: 2009 Edition*. Geneva: ITU.

ITU-D (International Telecommunication Union–Digital). 2010. "The World in 2010: ICT Facts and Figures." ITU-D, Geneva. http://www.itu.int/ITU-D/ict/ material/FactsFigures2010.pdf.

Katz, Raúl. 2009. "Estimating Broadband Demand and Its Economic Impact in Latin America." Paper prepared for the third ACORN-REDECOM conference, Mexico City, May 22–23. http://unpan1.un.org/intradoc/groups/public/documents/gaid/unpan036761.pdf.

——. 2010. "The Impact of the National Broadband Plan on Jobs: A Quantification Framework." Paper prepared for "The National Broadband Plan," a roundtable discussion, New York University Law School, April 19.

Katz, Raúl, and Javier G. Avila. 2010. "The Impact of Broadband Policy on the Economy." Paper prepared for the fourth ACORN-REDECOM conference, Brasília, May 14–15.

Katz, Raúl, and Stephan Suter. 2009. "Estimating the Economic Impact of the Broadband Stimulus Plan." Working Paper, Columbia Institute for Tele-information, New York, February.

Katz, Raúl, Stephan Vaterlaus, Patrick Zenhäusern, and Stephan Suter. 2009. "The Impact of Broadband on Jobs and the German Economy." Working Paper, Columbia Institute for Tele-information, New York.

——. 2010. "The Impact of Broadband on Jobs and the German Economy." *Intereconomics: Review of European Economic Policy* 45 (1): 26–34. http://www.polynomics.ch/dokumente/Polynomics_Broadband_Brochure_E.pdf.

Katz, Raúl, Patrick Zenhäusern, and Stephan Suter. 2008. "An Evaluation of Socio-Economic Impact of a Fiber Network in Switzerland." Polynomics and Telecom Advisory Services, Olten, Switzerland.

Kenny, Charles. 2011. "No Need for Speed." *Foreign Policy*, May 16. http://www.foreignpolicy.com/articles/2011/05/16/no_need_for_speed?page=full.

Kim, Yongsoo, Tim Kelly, and Siddhartha Raja. 2010. "Building Broadband: Strategies and Policies for the Developing World." Global Information and Communication Technologies Department, World Bank, Washington, DC, January. http://siteresources.worldbank.org/EXTINFORMATIONAND COMMUNICATIONANDTECHNOLOGIES/Resources/282822-1208273252769/Building_broadband.pdf.

LECG. 2009. *Economic Impact of Broadband: An Empirical Study.* London: LECG. http://www.connectivityscorecard.org/images/uploads/media/Report_BroadbandStudy_LECG_March6.pdfLECG.

Libenau, Jonathan, Robert Atkinson, Patrik Kärrberg, Daniel Castro, and Stephen Ezell. 2009. *The UK's Digital Road to Recovery.* London: LSE Enterprise; Washington, DC: Information Technology and Innovation Foundation.

Majumdar, Sumit K., Octavian Carare, and Hsihui Chang. 2009. "Broadband Adoption and Firm Productivity: Evaluating the Benefits of General-Purpose Technology." *Industrial and Corporate Change* 19 (3): 641–74. http://icc.oxfordjournals.org/content/19/3/641.full.pdf+html.

Mishra, Saurabh, Susanna Lundstrom, and Rahul Anand. 2011. "Service Export Sophistication and Economic Growth." Policy Research Working Paper 5606, World Bank, Washington, DC. http://www-wds.worldbank.org/servlet/WDSContentServer/WDSP/IB/2011/03/22/000158349_20110322141057/Rendered/PDF/WPS5606.pdf.

Ndahiro, Moses. 2010. "RDB Forging Ahead with E-Health Development." *New Times,* September 20. http://www.newtimes.co.rw/news.php.

OECD (Organisation for Economic Co-operation and Development). 2007. "Broadband and the Economy." Ministerial Background Report DSTI/ICCP/IE (2007)/Final. Report prepared for the OECD Ministerial Meeting on the Future of the Internet Economy, Seoul, June 17–18.

——. 2008. "Broadband Growth and Policies in OECD Countries." Report prepared for the OECD Ministerial Meeting on the Future of the Internet Economy, Seoul, June 17–18.

Qiang, Christine Zhen-Wei, and Carlo M. Rossotto. 2009. "Economic Impacts of Broadband." In *Information and Communications for Development: Extending Reach and Increasing Impact*, ch. 3. Washington, DC: World Bank, Global Information and Communication Technologies Department.

Rwirahira, Rodrigue. 2009. "Rwanda: Better Health Care by Using ICT in Medicine." *All Africa*, August 28. http://allafrica.com/stories/200908280497.html.

Samudhram, Ananda. 2010. "Building ICT Literate Human Capital in the Third World: A Cost-Effective, Environmentally Friendly Option." Paper prepared for the Ascilite conference, Sydney, December 5–8. http://www.ascilite.org.au/conferences/sydney10/Ascilite%20conference%20proceedings%202010/Samudhram-poster.pdf.

South Africa, Department of Communications. 2010. "Broadband Policy for South Africa." *Government Gazette* 541 (33377, July 13): 1–24. http://blogs.timeslive.co.za/vault/2010/07/14/south-africas-new-broadband-policy/.

Thibodeau, Patrick. 2010. "In Historic Shift, Smartphones, Tablets to Overtake PCs." *Computerworld*, December 6. http://www.computerworld.com/s/article/9199918/In_historic_shift_smartphones_tablets_to_overtake_PCs.

UNCTAD (United Nations Conference on Trade and Development). 2009. *Trends and Outlook in Turbulent Times*. Information Economy Report. Geneva: UNCTAD.

United Nations. 2010. "Millennium Development Goals: Goal 2 Achieve Universal Primary Education." United Nations Summit, New York, September 20–22. http://www.un.org/millenniumgoals/pdf/MDG_FS_2_EN.pdf.

UNPAN (United Nations Public Administration Programme). 2010 "Leveraging eGovernment at a Time of Financial and Economic Crisis." UNPAN, New York. http://www2.unpan.org/egovkb/global_reports/10report.htm.

von Hippel, Eric. 2005. *Democratizing Innovation*. Cambridge, MA: MIT Press.

WHO (World Health Organization). 2005. "Global Survey for eHealth." WHO, Geneva. http://www.who.int/goe/data/en/.

World Bank. 2008. *Global Economic Prospects 2008: Technology Diffusion in the Developing World*. Washington, DC: World Bank. http://econ.worldbank.org/WBSITE/EXTERNAL/EXTDEC/EXTDECPROSPECTS/GEPEXT/EXTGEP2008/0,,menuPK:4503385~pagePK:64167702~piPK:64167676~theSitePK:4503324,00.html.

Zhang, Qi, Lu Cheng, and Raouf Boutaba. 2010. "Cloud Computing: State-of-the-Art and Research Challenges." *Journal of Internet Services and Applications*

1 (May): 7–18. http://www.springerlink.com/content/n2646591h5447777/fulltext.pdf.

Zhao, Judy, and Levi Ruan. 2010. "Broadband in China: Accelerate Development to Serve the Public." Value Partners, Milan, March. http://www.valuepartners.com/VP_pubbl_pdf/PDF_Comunicati/Media%20e%20Eventi/2010/value-partners-PR_100301_BroadbandInChinaZhaoRuan.pdf.

Zilber, Julie, David Schneier, and Philip Djwa. 2005. "You Snooze, You Lose: The Economic Impact of Broadband in the Peace River and South Similkameen Regions." Report prepared for Industry Canada, Ottawa. http://www.7thfloormedia.com/news/71.

CHAPTER 2

Policy Approaches to Promoting Broadband Development

The development of broadband networks and services over the last decade or so has been largely focused in developed countries. In that time, private sector investment, coupled with enabling polices put in place through liberalization and regulatory reform, has driven the building of broadband networks and the adoption of broadband services throughout the developed world. But as more economic and social activity has moved onto broadband networks in recent years, developing countries have been implementing their own broadband plans and initiatives to realize the benefits that broadband can bring to a country and its citizens.

The development of strategies and policies to promote broadband, however, is not an easy task. Policy makers are quickly realizing that it might be more difficult to promote broadband than other types of services, such as mobile and wireline telephony. The usefulness of a mobile or wireline telephone is typically obvious to consumers regardless of income or education level, and, coupled with relatively low prices, such intuitive services have grown rapidly. But the same cannot necessarily be said of broadband, especially if the opportunity to try it is undermined by high prices. Using broadband services requires access to a computer or smartphone and some way to pay for using the network through a subscription (and often some form of term contract), a pay-as-you-go approach, or prepaid services. In the absence of access through the workplace, school, or community centers, this can make ownership relatively costly for individual users (even with falling prices for hardware and subscriptions). In addition, understanding the benefits of broadband, and having the skills to make use of the available services, requires some level of digital literacy as well as basic literacy (that is, the ability to read and write).

As they consider how best to promote broadband, policy makers and analysts have come to view broadband as an ecosystem with supply and demand considerations. On the supply side, the building of networks to carry broadband services is the top priority. But simply having a network available does not guarantee that broadband services will automatically be used. It will also be necessary for government policy and private sector investment to focus on driving demand for broadband services, whether by putting more services online or educating users about the benefits of broadband and the skills needed to use the new services effectively. Those countries with the best success in broadband development have focused on developing holistic policies to support both sides of the broadband supply and demand equation.

This chapter identifies the issues that policy makers must address as they seek to create an enabling environment for broadband and examine what policies and regulatory approaches may be effective in encouraging broadband development. It is designed to introduce the issues, policies, and strategies that are discussed in more detail in subsequent chapters. Those chapters analyze the issues extensively and provide many examples of how different countries have approached broadband development. They deal with the technologies that make broadband possible, how broadband networks and services can be universalized, how demand for broadband can be stimulated, and what changes to laws and regulations can help broadband to reach its greatest potential.

The Public Sector's Evolving Role in Broadband

The public sector has played two roles in promoting the growth of information and communication technologies (ICT): (a) making markets more competitive, efficient, accountable, and transparent and (b) ensuring equitable access for all. This has enabled the private sector to lead the rollout of and investment in ICT. This same approach should be pursued with broadband development. The role of government should be to enable, facilitate, and complement market development, rather than to substitute government decisions for market forces and public sector investment for private investment.

Due to broadband's importance, however, there have been calls to view broadband as a public good in order to ensure affordable universal access and spread the benefits across the full range of economic sectors.[1] Based at least partially on a public goods analysis, some countries have taken more direct action to promote broadband development, establishing initiatives and strategies where the government intervenes more directly to promote, oversee, and universalize their broadband markets. This was particularly the case in the wake of the economic crisis of 2008, as many governments came to see broadband networks and services as a way to preserve and enhance their economies. In 2009, for example, countries with different economic philosophies included broadband in their economic stimulus plans (for example, Australia), which indicated that they were no longer averse to making strategic investments. By 2011, however, such policies were being increasingly called into question as government debt levels rose, in some cases dramatically, forcing austerity programs and corresponding cuts in government spending on a wide range of priorities, including broadband.

Defining the Challenges: Barriers to Broadband Growth

As policy makers and regulators consider approaches to stimulating and promoting broadband development, they need to recognize the full scope of the challenges that must be addressed. These challenges tend to be multilayered and involve stimulating the supply of broadband infrastructure and encouraging demand for broadband applications and services, as discussed in chapter 6. On the supply side, the problem is not as simple as just building more networks; as operators roll out their broadband business plans, issues of cost, service quality (data speeds), and technology choice will also play important roles in decisions about how best to bring access to a nation's citizens. Even then, just building more networks or providing access to all will not guarantee success. Governments may need to support broadband development by encouraging demand for broadband in those limited instances where the private sector does not generate useful and relevant applications, services, and content. In sum, governments must think of broadband as an ecosystem, holistically, with supply and demand components, if they are to maximize their chances for broadband development success.

Supply: Reaching Unserved and Underserved Users

As one considers policies and strategies to promote broadband development, one important goal is to ensure that access is available to the widest possible user base. This means that networks need to be built out to reach as many people as possible. But facilitating broadband supply presents at least two significant issues. First, some areas in virtually every country have no meaningful access to broadband services at all. This problem is most pronounced in developing countries, which have seen less investment in the construction of networks outside metropolitan areas. This situation has improved in recent years with the spread of wireless networks, but some areas still lack network coverage. Second, some areas have networks in place, but these networks are not capable of supporting broadband speeds and services. These areas will need to be upgraded, either through the construction of high-speed wireline networks or through advanced wireless networks (3G [third-generation mobile telecommunications systems] or 4G [fourth-generation mobile telecommunications systems] services). In many developing countries, where wireless penetration can far exceed wireline penetration, upgraded wireless networks capable of providing true broadband speeds are expected to be the main supplier of broadband services.

Demand: Lowering the Barriers to Adoption

Improving the availability of broadband networks only addresses one impediment linked to broadband development. Even with networks in place and accessible, there are likely to be barriers due to lack of demand. This problem involves people who have access to broadband network(s), but are unable or unwilling to obtain service. Addressing lack of demand is important because low adoption rates will leave networks underutilized. This has at least two implications. First, from a network externalities standpoint, fewer users reduce the economic and social utility of the networks. Where relatively few people can communicate online, the network externalities are reduced since there is a smaller number of potential customers for businesses to serve. This further means that there may be fewer local businesses and consumers offering broadband-enabled services and applications, such as video streaming services (for example, Hulu+), voice and video communications (for example, Skype), and download services for a variety of applications such as software and e-books.

Second, low adoption and use will undermine the business case of any network—even those built with public funds. Fewer users mean that networks are correspondingly higher cost or that their costs are spread over a smaller user base, making them relatively more expensive to build and operate. Thus it is important for the overall goal of improving broadband development for governments to focus on developing policies that not only facilitate and encourage the building of broadband networks, but also ensure that as many people as possible can and do use them. Barriers to adoption vary and will likely not be the same in all countries, but some broad categories are identifiable. In studies conducted to identify barriers to Internet and broadband adoption, the primary reasons respondents cite for not subscribing to broadband services can be grouped into four main categories: (1) broadband is not relevant; (2) equipment or service is too expensive; (3) individuals lack training in or are not comfortable using broadband Internet services; and (4) broadband is not available (Pew Internet and American Life Project 2010; EUROSTAT 2009). This is not to say that demand inhibitors are exactly the same in all countries. The factors seen as impediments to adoption in some countries may be less of a factor in other countries, due to different social and cultural histories and experiences as well as different socioeconomic conditions (Hernandez, Leza, and Ballot-Lena 2010, 4). Figure 2.1, which presents survey data collected from nonadopters of Internet services in Brazil and the United States, shows how some factors are more important than others.[2] Respondents in the United States, for example, see digital literacy as a much bigger problem than respondents in Brazil, who consider high

Figure 2.1 Reasons Given for Not Adopting the Internet in Brazil and Broadband in the United States

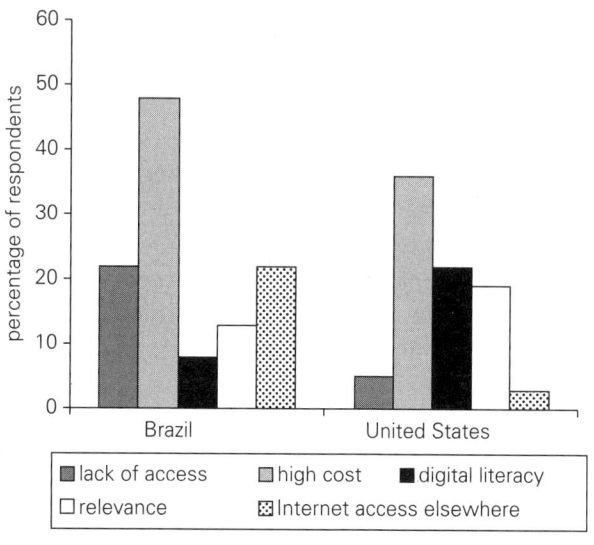

Sources: Brazil, Núcleo de Informação e Coordenação 2009; United States, FCC 2009.

cost to be a larger issue. Therefore, each country must analyze and address the demand-reducing factors on a case-by-case basis and tailor solutions to the individual problems.

Developing Country-Specific Solutions

No "one-size-fits-all" approach will guarantee greater broadband deployment and adoption in every country. Political and economic conditions vary, and each country is endowed with different technological resources. Some countries have a relatively well-developed wireline telephone network that could support broadband deployment, while others have widely deployed cable television networks that might be able to provide a measure of facilities-based competition from the start. In yet other countries, various regulatory, political, economic, or other barriers to entry may prevent potential competitors from offering broadband services or building broadband networks.

This variance makes it unwise to propose a uniform solution to promote broadband development. In some cases, the challenge will be to create incentives so that widespread networks can be used to offer broadband services. In other countries, the main challenge may be to find ways to educate potential users about the benefits of broadband and train them to use broadband applications and services. As a result, each country will have its own unique circumstances that will drive policy and investment decisions. However, the key objective for governments is to pursue policies that will create an enabling environment that will foster broadband development.

Important lessons can be learned from those countries that have pursued broadband development policies (box 2.1).[3] The focus in those countries has been on improving the incentives and climate for private investment, a policy that even highly resource-constrained countries might be able to follow (and many have successfully attained with mobile telephony). Many of the policies and programs that have been developed support private sector investments and call for specific, limited, and well-justified public funding interventions only in exceptional circumstances. In particular, governments that are trying to promote the growth of underdeveloped markets should avoid policies and regulations that may reduce private sector investment.

Government funding or policy should not have the effect of "crowding out" private sector investment. For example, governments can encourage private investments in many cases without direct subsidies, such as by developing passive infrastructure—ducting, towers, and cable conduits—which can significantly cut costs and create minimal market distortions

Box 2.1: Public Sector's Role in Fostering Broadband Development: Key Lessons

- Government should focus on maximizing competition, including removing entry barriers and improving the incentives and climate for private investment.

- Government should provide for specific, limited, and well-justified public funding interventions only in exceptional circumstances (for example, where governments are trying to promote growth of underdeveloped markets).

- Government funding or policy should not compete with or displace private sector investment.

- Government should maintain a level playing field for competition by avoiding favor-ing one company (or type of company, for example, telephony vs. cable) over another.

- Subsidized networks should be open access (that is, they should offer capacity or access to all market participants in a non-discriminatory way).

- Government may need to regulate domi-nant providers to avoid market concentra-tion or other adverse impacts on overall market competition.

- Government should eliminate barriers to content creation and refrain from block-ing access to· content, including social networking sites, or restricting local con-tent creation.

Source: Telecommunications Management Group, Inc.

(OECD 2008; Qiang 2009). Public investments should be considered only when no or insufficient private investments are expected for a significant period. Furthermore, to maintain a level playing field for competition even with public investments, governments should seek to avoid favoring one company (or type of company, for example, telephony vs. cable) over another. For example, if and when governments intervene to increase net-work availability, it may be necessary to ensure that subsidized networks are open access, meaning that network operators offer capacity or access to all market participants in a nondiscriminatory way. Nonetheless, there may be cases where a dominant provider may need to be appropriately regulated to avoid market concentration or other adverse impacts on over-all market competition.

Developing countries in particular will also need to identify ways to leverage limited resources to maximize impact, prioritizing programs based on demand and market evolution, rather than shying away from policy reform altogether. For most developing countries, the most effective approach to promoting broadband development is likely to involve a mix of approaches and policies that rely on private sector investment, coupled with regulatory reform that will promote efficient and competitive markets (which will also increase private sector investment). Direct government intervention should be limited to those cases where markets may not function efficiently (for example, providing service to high-cost areas) or where larger social goals are clearly identified (for example, providing digital literacy training). The basic principle remains the same: governments should only intervene based on sound economic principles, where the benefits of intervention outweigh the costs. For example, particularly at the initial stage of broadband market development, there may be a need for aggressive government policies to generate demand, expand networks, and reach underserved areas and communities.

How to Do It: Implementing Policies and Strategies to Enhance Broadband Development

Governments have various ways to promote the development of broadband networks and services in their countries. In most cases, the most effective government strategies are those that seek to harness the power of private sector investment to spur broadband growth. This handbook examines four broad categories of government action in this regard: (a) legal and regulatory policies and reform, (b) universal access policies, (c) support for private sector broadband network build-out, and (d) policies to stimulate demand and spur adoption. These areas are discussed in more detail in chapters 3, 4, 5, and 6, respectively.

General Approaches to Promoting Broadband

As policy makers seek ways to promote the development of broadband in their countries, certain general lessons can be learned from those countries with more developed broadband networks and services. This section briefly describes the general elements that governments should be aware of as policies and strategies are created.

Establish Specific Plans and Policies
Based on an evaluation of the supply and demand challenges that exist in a country, the next step is developing specific policies and strategies to address those challenges. This will entail setting concrete, measurable objectives for improving the supply of broadband through infrastructure build-out as well as promoting demand for broadband services and applications. Setting specific plans or policies will provide a clear sense of direction that will encourage investment as well as provide a blueprint for long-term action.

A good plan should aim to promote efficiency and equity, facilitate demand, and help to support the social and economic goals of the country. The most successful plans will start with a clear vision of what broadband development should be and contain well-articulated goals that can be used to develop specific strategies to achieve success. Such frameworks can launch or revise ambitious national broadband visions, including definitions of broadband, service goals (including national and rural coverage), transmission capacity, service quality, and demand-side issues such as education and skills development. The government of the Republic of Korea, for example, was one of the early broadband leaders. It has developed six plans since the mid-1980s that have helped to shape broadband policy in the country. The Korea example shows that policy approaches can effectively move beyond network rollout and include research, manufacturing promotion, user awareness, and digital literacy. It also highlights the possibilities for sector growth based on long-term interventions focused predominantly on opportunity generation rather than on direct public investment.

For many countries, the development of an extensive national broadband plan or strategy is an important step toward elaborating more specific broadband development policies. The countries highlighted in table 2.1 have national broadband strategies containing specific broadband development goals.

As table 2.1 shows, however, countries differ in their approach to setting targets and goals. Some focus on improving access, while others set specific targets for data transfer speeds.

But policies and programs to spur broadband development have not been confined to developed countries. Other countries have also sought to develop national broadband strategies, as shown in box 2.2.

Allow Ample Opportunity for Stakeholder Input on Plans and Policies
The development of broadband plans should involve the participation of all relevant stakeholders, both public and private. As such, governments should provide for a public consultation process that allows ample opportunities to obtain input from the private sector, consumers, and other relevant stakeholders. Given the complexity, varied issues, and broad impact of broadband, these transparent discussions are an important part of bringing stakeholders to the table in an open, objective, and neutral manner so as to maximize cooperation between the public and private sectors. Such services make it much easier for all parties, but particularly ordinary citizens, to learn about and comment on the issues being considered. A variety of mechanisms can be used to foster stakeholder input—presentation of filings

Table 2.1 Publicly Stated Policy Goals for Broadband Service Delivery and Adoption in Selected Countries

Country	Goal for service delivery, access, and adoption
Brazil	Access to broadband for 50 of every 100 households
Finland	Legal right of all citizens to 1 megabit per second (Mbit/s) access at affordable levels by 2010; by year-end 2015, 99% of all permanent residences to have access, within 2 kilometers, to an optical fiber or cable network delivering 100 Mbit/s service
France	By 2012, ubiquitous access to 512 kilobits per second (kbit/s) service at monthly rates at or below €35
Germany	75% of households with high-speed broadband access at transmission rates of at least 50 Mbit/s by 2014
Malaysia	By year-end 2010, broadband penetration rate for households of 75%
Morocco	One out of three households connected by 2013
South Africa	Broadband penetration rate for households of at least 15% by 2019
Sweden	By 2010, near ubiquitous access to 2 Mbit/s service; 40% of households with access to 100 Mbit/s connections by 2015 and 90% by 2020
United Kingdom	By 2012, 2 Mbit/s service to all households
United States	By 2020, 100 million households with access to actual (not advertised) speeds of 100 Mbit/s and universal connections with actual speeds of at least 4 Mbit/s download and 1 Mbit/s upload

Source: Rob Frieden for the World Bank and Telecommunications Management Group, Inc.

by stakeholders, workshops, hearings, and inputs made through an online comment mechanism on a regulatory website or blog. Moreover, as e-government services have expanded, the effectiveness of public consultations has grown as well. The broadband development process will benefit from the broader range of perspectives that can now be presented to regulators and policy makers. Consultations and discussions are also proven mechanisms for helping regulators and ministries to understand the varying challenges and potential opportunities that are part of the reform process, for increasing capacity and knowledge, and for exchanging ideas in an open, transparent setting.

Recognize and Take into Account That Implementation of the Plan Will Take Time and Persistence

In many cases, the success of programs that have increased broadband adoption has simply been the result of longevity. Some countries prioritized broadband in the 1990s or early 2000s and have been promoting broadband for quite a number of years, giving them a meaningful head start

Box 2.2: Broadband Strategies in Middle-Income Countries

Chile was the first Latin American country to announce a national broadband strategy. The strategy identifies ICT as a priority for economic development. Chile has also planned and implemented ICT policies from both the supply and demand sides. On the supply side, the government has authorized four Worldwide Interoperability for Microwave Access (WiMAX) operators as regional providers, and the regulator plans to award additional spectrum to a new 3G operator. The demand-side strategy has included programs for e-literacy, e-government, and ICT diffusion. For example, almost all taxes are filed electronically, and government e-procurement more than doubled the volume of transactions processed between 2005 and 2008. The government has also promoted broadband use by municipalities. By 2008, almost all municipalities had Internet access, and 80 percent had websites. In May 2010, Chile's wireline broadband penetration was 10.66 percent, while mobile broadband penetration was less than half that, but growing at a much faster rate.

Turkey's government recognizes the importance of a vibrant telecommunications market and is keen to promote the spread of broadband. For instance, many educational institutions have been given broadband access. The Information Society Strategy for 2006–10 aims to develop regulation for effective competition and to expand broadband access. Targets include extending broadband coverage to 95 percent of the population by 2010 and reducing tariffs to 2 percent of per capita income. The regulator has also considered issuing licenses for the operation of broadband fixed wireless access networks in the 2.4 gigahertz (GHz) and 3.5 GHz bands. In June 2010, Turkey had penetration rates of 9 percent for wireline broadband and 4 percent for mobile broadband.

Malaysia developed its Information, Communications, and Multimedia Services (My-ICMS) 886 Strategy in 2006, setting several goals for broadband services. One was to increase broadband penetration to 25 percent of households by the end of 2006 and 75 percent by the end of 2010. Although these targets were not met, the results have been impressive: the household broadband penetration rate in the country topped 53 percent in October 2010. Now the government is focusing on WiMAX, 3G, and fiber to the home (FTTH) platforms to boost broadband adoption. To that end, the government is funding a fiber optic network that will connect about 2.2 million urban households by 2012. The network will be rolled out by Telekom Malaysia under a public-private partnership, where the government will invest RM 2.4 billion (US$700 million) in the project over 10 years, with Telekom Malaysia covering the remaining costs. The partnership is expected to cost a total of RM 11.3 billion (US$3.28 billion).

Sources: Kim, Kelly, and Raja 2010; Cisco, "Broadband Barometer for Chile," Press Release, February 8, 2011, http://newsroom.cisco.com/dlls/2011/prod_020811.html; "Broadband Penetration Target for 2010 Exceeded, says Muhyiddin," *Malaysian Insider*, October 27, 2010, http://www.themalaysianinsider.com/Malaysia/article/broadband-penetration-target-for-2010-exceeded-says-muhyiddin/.

over other countries. For example, in 2000, Sweden enacted an information technology (IT) bill, which established the pillars of its ICT strategy as "competencies, confidence, and access." Sustained, focused efforts with continual updates over a number of years contribute to the long-term success of any broadband strategy. Conversely, seeking a "one-shot" solution that can be achieved with minimal time and resources is not likely to produce the best long-term outcome.

Develop Research Mechanisms to Track Progress of the Plan

As broadband technologies and applications evolve over time, the various segments of the broadband market will change as well. Further, notions of digital literacy and underserved populations will also be in flux. Various agencies and organizations are already tracking various parts of the broadband equation. To keep up with this dynamic and ever-changing sector, governments may wish to create an ongoing, multiyear, broadband-specific research program that tracks population use, ongoing barriers, and levels of digital literacy (box 2.3). This program could complement the ministry's or regulator's efforts to encourage the supply-side parameters of broadband

---

Box 2.3: General Elements for Governments to Consider When Creating Policies and Strategies

- Establish specific plans and policies that define broadband development and contain concrete, measurable objectives that can be used to develop specific strategies to achieve success

- Ensure that plans address mechanisms for improving the supply of broadband through infrastructure build-out as well as for promoting demand for broadband services and applications

- Allow ample opportunity for stakeholder input in developing the plan

- Be realistic when establishing objectives: recognize and take into account that implementation of a plan will take time and persistence

- Focus on long-term success by developing sustained, focused efforts (with continual updates) over a number of years

- Avoid seeking a "one-shot" solution that can be achieved with minimal time and resources, as this approach is not likely to produce the best outcome

- Consider developing an ongoing, multiyear broadband-specific research program for tracking use, ongoing barriers, and levels of digital literacy and for determining whether objectives are being met or modifications are needed

- Assign one coordinating agency responsibility for implementation of the plan

Source: Telecommunications Management Group, Inc.

(for example, network build-out, speeds, and capabilities). The program could be housed within the agency responsible for broadband development or could be run out of one of the existing government agencies that perform such research. The ongoing issues of measurement and assessment, including international benchmarking, are discussed in more detail later in this chapter.

Pilot projects can play an important role in ongoing research and development (R&D) efforts related to broadband deployment. Such projects can help to demonstrate the viability of a new technology or service, but, even more important, they may help to identify those policies and strategies that do not work very well. Pilot projects may be a cost-effective approach to broadband development, as they allow concepts, plans, and methods to be tested on a small scale before committing larger amounts of resources. In the United Kingdom, for example, Broadband Delivery U.K., a unit of the government, gives out grants (supplemented with private funds) for pilot projects to build or upgrade broadband networks in rural areas. Once the upgrades have been completed, Internet service providers (ISPs) gain access to the infrastructure, which may use any technology, on a wholesale basis.

Provide a National Focal Point for Broadband and Develop Broadband Capacity

To optimize the benefits of broadband, governments need a comprehensive national-level focus on promoting broadband use, a clearinghouse for successful projects, and a consistent evaluation of what works and what does not. An important part of establishing and maintaining that focus over time will be developing capacity-building programs for government officials to provide education on how broadband can provide benefits across many sectors of the economy. Such programs, in turn, can help to shape the design of effective broadband development strategies throughout all levels of government—from local training programs to national network regulatory regimes.

Numerous countries have established agencies or special offices specifically to oversee broadband development issues. In Sweden, for example, the IT Policy Strategy Group recommended the creation of an internal strategic coordination function to oversee "holistic" IT policy development and implementation. This internal coordination function was also envisioned to improve coordination between central government, local authorities, county councils, and the business sector. The United Kingdom now has a minister of digital inclusion. Brazil has appointed a digital inclusion secretary housed within the Ministry of Communications, which will be in charge of the National Broadband Plan as well as of all digital inclusion projects that are currently being carried out by various branches of the federal government.

Often, broadband development efforts are overseen by the ministry responsible for communications or the regulator. In many cases, this responsibility is exercised in conjunction with a comprehensive broadband development plan. In Singapore, for example, the government developed and is actively pursuing its Intelligent Nation 2015 master plan, which is designed to transform Singapore into "an intelligent nation and a global city, powered by info-communications" (IDA 2009). As part of that plan, a next-generation nationwide broadband network (NBN) is being developed to bring fiber to homes and businesses across the whole territory. A wireless broadband network is also part of the strategy. All of these efforts are being overseen by the Info-communications Development Authority of Singapore (IDA), which is providing government leadership in the development of these networks. In India, the Ministry of Communications and Information Technology established an advisory group with members from telecommunications companies, industry associations, and various government departments (including health, education, and rural development) to help to guide India's plan for a national fiber network that is envisioned to reach all villages and towns with more than 500 people. India's approach is particularly noteworthy because it recognizes not only the importance of a central focal point, but also the cross-cutting impact of broadband on various sectors of the economy and the need for a coordinated approach that involves all relevant agencies.

The decision regarding whether to set up such an agency or office will depend on the local situation in each country and will need to take into account existing laws and institutional responsibilities as well as the ability of the government to provide adequate funding for such an activity. For developing countries with limited financial and human resources, devoting a whole agency or branch of government to broadband development may seem ambitious. Nevertheless, given the importance of broadband development and its potential role as a general-purpose technology (GPT) capable of supporting advances in many sectors of any economy, developing such human resource capacities will be critically important.

The issues surrounding the development of effective broadband policies are extremely complex and cover a wide range of disciplines, including engineering, law, and economics, among others. This will require governments to build capacity so that trained, knowledgeable professionals can guide the implementation of a country's broadband plan from concept through construction and adoption. Without such leadership, even the best laid plans may fail through inattention and neglect.

Develop Policies for Both Sides of the Broadband Coin: Supply and Demand

The experience in high-penetration countries shows that successful broadband diffusion requires that both supply- and demand-side factors be addressed (figure 2.2). While supply-side policies focus on promoting the build-out of the network infrastructure over which broadband applications and services can be delivered, the main goal of demand-side policies is to enhance the awareness and adoption of broadband services so that more people will make use of them.

The interaction of both supply- and demand-side factors is crucial to achieving the highest penetration and adoption of broadband. However, these factors do not always appear naturally, as market failures may hinder their development. For instance, broadband diffusion can be limited if the market is not able to reach the required critical mass that leads to a sustainable growth cycle. More important, even if both types of factors (that is, supply and demand side) are present in an economy, they will not reach their full potential if they are not coordinated, as lack of coordination may result in slow supply of broadband infrastructure or in poor demand and uptake

Figure 2.2   Framework for Government Intervention to Facilitate Broadband Development

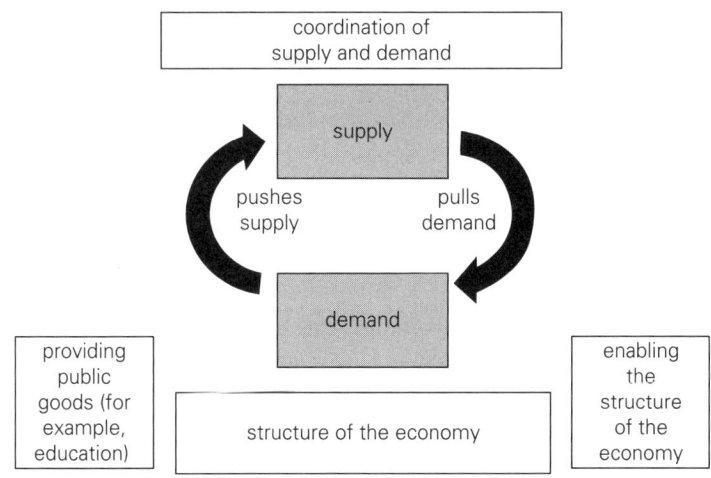

Source: World Bank 2010.

once networks are available. For this reason, countries with high rates of broadband penetration have comprehensive broadband policies that coordinate both supply- and demand-side actions.

In assessing the strategic options for improving broadband build-out and adoption (supply and demand), one must remember that many factors are involved and that no two countries have followed identical routes. Nevertheless, it is possible to recognize common elements in national broadband success stories (table 2.2). In reality, most countries will use a mix of policies, with supply-side policies generally focusing on how to stimulate private sector investment in networks, especially in the early years, and demand-side policies being more long term and focusing on how governments can help to drive broadband demand and adoption.

Build Infrastructure: Promoting the Supply of Broadband

Most developing countries have not yet seen their broadband markets penetrate more than a very small segment of their population. Hence, the government's role is even more important in promoting and accelerating growth of the broadband market. Promoting the build-out of broadband networks throughout a country will likely require governments to pursue multiple strategies, depending on local circumstances. As each country has its own unique history, regulatory structure, economic conditions, social goals and expectations, and political processes, the path a country follows to improve broadband networks and services will necessarily have to reflect its specific advantages and disadvantages.

Nevertheless, some general policy approaches may be applicable across the world. It is generally accepted that the private sector should be the primary driver of broadband development in most cases. Particularly when government debt is high and resources are limited, sufficient public money may not be available for broadband infrastructure spending. Consequently, policy makers and regulators must consider how best to attract and encourage private sector involvement and investment in broadband. This, in turn, will require governments to conduct an honest evaluation of the extent to which their country represents—or can be made into—a profitable market opportunity for private sector investors and operators. Questions to be answered may include the following: Are companies willing to invest? If not, why not? Will such companies drive the broadband market forward on their own, or will they need help? What government strategies, policies, and regulations can foster and support private sector initiatives, and what policies may hold back investment? Many countries have taken this approach: they have attempted to facilitate and, where possible, accelerate, broadband

Table 2.2 Elements of Broadband Strategies in Selected Countries

Strategy	Brazil	Colombia	Finland	France	Japan	Oman	Singapore	South Africa	United States
Establish open-access wholesale networks	✓						✓		
Encourage private sector investment	✓	✓	✓	✓	✓	✓	✓	✓	✓
Include broadband under universal service definition	✓			✓		✓[a]	✓	✓	✓
Encourage demand for broadband services	✓	✓	✓	✓	✓		✓	✓	✓
Promote, improve, and expand public-private partnerships	✓	✓		✓	✓		✓	✓	✓
Subsidize local (citywide), regional, or national ventures		✓		✓		✓	✓	✓	✓
Promote facilities-based resale competition		✓		✓	✓		✓	✓	✓
Mandate local loop unbundling (LLU)	a	✓	✓	✓	✓	✓	✓	b	✓

Source: Rob Frieden for the World Bank and Telecommunications Management Group, Inc. For a list of weblinks to each of these national broadband plans, see appendix A in this volume.

a. Article 155 of Brazil's Telecommunications Law, as well as various other regulatory instruments issued by Anatel, notably Order no. 172 of May 12, 2004, issued by the Superintendent of Public Services of Anatel, requires wireline providers to unbundle. However, because wireline network unbundling prices are high, in practice, unbundling does not really occur in Brazil. Anatel has identified as a short-term priority the need to review its policies with regard to LLU as well as to adopt a pricing model for network use so that LLU can be mandated.

b. Under consideration as of 2011.

rollout through regulatory measures rather than more direct forms of intervention such as investment.

In the context of a private sector–led approach to broadband development, it is recognized that allowing competition to flourish will usually lead to greater deployment and efficiencies in network build-out. Competition in broadband supply is crucial for reducing prices, improving quality of service, and improving customer service (ITU 2003). It has a positive effect on market growth, as it expands access, increases affordability, and augments the value proposition. Conversely, lack of access to infrastructure and high prices can act as strong barriers to broadband diffusion. If no broadband infrastructure is available, consumers cannot access the service. Even if a network is available, it will be of little use for consumers if the service is not affordable. The government, therefore, should place a priority on developing enabling policies that will facilitate competition throughout the supply chain to encourage deployment and lower consumer prices.

However, in certain instances, competition and market forces will not be sufficient for broadband to develop. In those cases—due to factors such as geography or low population density, for example—private sector players will be unwilling to invest capital where they perceive that they will get a low (or no) return on their investment. For these areas, it will be necessary for the government to intervene more directly to ensure that un- and underserved areas and populations are able to get access to broadband networks and services.

Use Competition to Promote Market Growth

A key lesson from countries surveyed in Kim, Kelly, and Raja (2010) is that competition is critical to successful broadband market promotion. Each country studied used different mechanisms to spur competition and promote broadband market growth. Some focused primarily on fostering facilities-based competition, while others focused more generally on increasing the level of competition at the service level. The presence of established, competitive telecommunications operators in many countries also contributed to broadband market development.

In the long term, liberalization and promotion of competition among facilities are the best ways to guarantee lower costs. For example, the initiation of the Southern and East Africa Cable System (SEACOM) network that links Kenya, Madagascar, Mozambique, South Africa, and Tanzania resulted in Kenya Data Networks (KDN), a Kenyan data services provider, announcing that it would reduce its Internet prices by up to 90 percent.[5] However, liberalization may be difficult in some developing countries, particularly those that have small populations, are geographically isolated, or are small island developing states (SIDSs)[6] with limited access to multiple sources for connectivity. Specific countries may exhibit features that make developing competitive markets in certain segments of the supply chain particularly difficult.

Develop Enabling Policies to Eliminate Bottlenecks in the Broadband Supply Chain

Broadband networks are not simple systems; they consist of multiple components, all of which must work together in order for broadband services to be delivered to end users in the most efficient and effective way possible. The technologies that make up the "broadband supply chain" are discussed in detail in chapter 5, while the legal and regulatory issues associated with each level of the supply chain are addressed in chapter 3.

In order to be most effective, competition must be present throughout the different levels of the broadband supply chain (figure 2.3). If not, bottlenecks will arise and the benefits of broadband diffusion will be severely reduced. As such, it is important to develop enabling policies to eliminate bottlenecks across the broadband supply chain. For instance, if domestic and local levels are competitive but access to international connectivity is limited or too expensive because there is only one provider of submarine cable, broadband prices will remain high and diffusion will not achieve its

Figure 2.3   Addressing Bottlenecks in Broadband Networks: Policies on the Supply Side

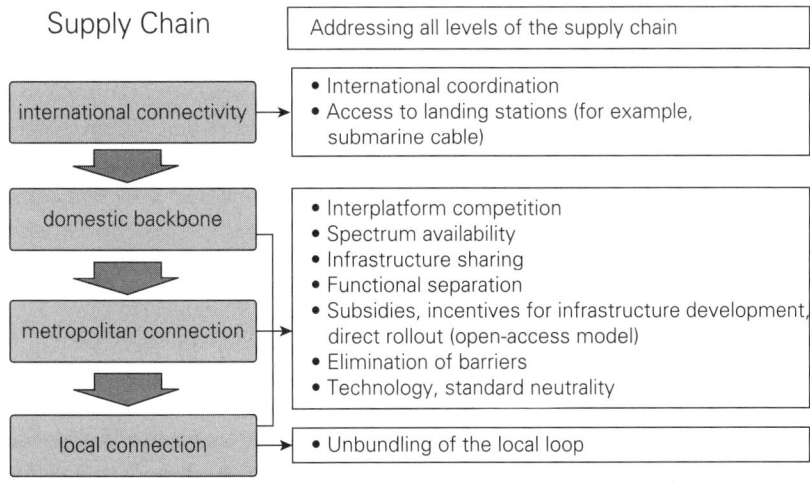

Source: Adapted from World Bank 2010, 56; World Bank 2009.

potential. The same can happen if all levels of the supply chain are competitive but local connectivity is limited to a single operator.

Recognizing the role of competition, high-penetration countries tend to address competition issues throughout the supply chain. However, the particular conditions in each country may lead to the creation of different bottlenecks and require different policy approaches applicable to their specific broadband market. As a result, not all countries have identified the same bottlenecks in the supply chain, nor have they adopted the same policies to ensure competition in these markets. Governments can foster competition in each level of the supply chain through various public policy options. Additional information on these issues can be found in chapters 3 and 5.

Promote Effective Competition and Encourage Investment

Some issues involved in promoting competition do not apply to one particular part of the broadband supply chain, but rather involve the interaction between different levels or policies that may be applied in a complementary fashion across levels. Furthermore, the dominant service providers in a country often operate at several levels in the supply chain. Thus, policies may be needed to ensure that they do not use their dominance in one market segment to affect other levels of the supply chain. For example, policies that foster open access to network infrastructure can be implemented at all levels, and interconnection agreements are needed between operators at all levels as well (see chapter 3 for more detailed discussion of these issues).

Access to Infrastructure

Network operators and service providers wishing to enter the downstream market (that is, building access networks and offering services to customers) must either build their own backbone network or access the network of another operator. The terms under which operators can obtain access to the backbone networks of other operators will have a significant impact on the success of their business and will influence whether effective competition in the downstream market develops. At the same time, the demand created by these downstream operators will affect the financial viability of the backbone networks, since they are the entities that generate traffic and revenues on those networks. Thus, by promoting effective competition in the downstream market, governments will help to stimulate backbone network development.

The role of the regulator is crucial, since it often defines and enforces the terms of access. The decision about whether to regulate directly the terms of access to infrastructure has a major effect on the investment incentives.

In Europe, for example, where the incumbent operator historically dominated both the local loop and the backbone markets, the priority for regulators was to provide access to these operators' networks for companies entering the markets, since this was seen as crucial to the development of competition. Subsequently, as competition has grown, regulators have developed systems for determining which operators should be regulated and how, and these systems are based on a well-established framework of general competition regulation. In developing countries (for example, most countries in Sub-Saharan Africa), such frameworks often do not exist. Regulators therefore need to develop alternative sets of guidelines to govern how access to the infrastructure of private operators in competitive markets is regulated. This involves a trade-off between supporting the development of competition in the downstream market and maintaining the incentives to invest in upstream infrastructure. In areas of a country where public support is provided for backbone infrastructure, this trade-off is relatively straightforward, since one of the conditions of public support typically includes the provision of wholesale services on regulated terms. In other areas of the country and in other parts of the infrastructure, the trade-off may be more difficult to determine.

Infrastructure Sharing

Many governments have sought to encourage deployment of networks and improve the overall competitive situation by allowing or, in more limited instances, even requiring infrastructure sharing. In most cases, infrastructure sharing has been instituted in areas where having competing physical infrastructures was not considered economically viable (such as in rural or remote areas) or where the construction of competing infrastructures could prove unacceptable for social or political reasons (too much disruption from repeated construction projects). By sharing network infrastructure, builders of networks can significantly reduce costs and make investment in them more commercially viable. This is particularly relevant for fiber optic networks in rural areas, where the revenues generated by such networks are low. In some cases, operators have a commercial incentive to enter into these sharing arrangements. For example, in Nigeria, where there has been extensive fiber optic cable network rollout, operators have entered into a variety of network-sharing agreements aimed at reducing costs and improving the quality of supply. In addition, operators may also be required to install multiple fibers in their cables, even if they only need one. These additional "dark" (unused) fibers may not be used initially, but may be held in reserve for future use by an existing operator or new entrant. This may be a very cost-efficient way to manage fiber optic networks because installation (and the associated civil works costs) only needs to be done once as opposed to multiple rounds of digging to install multiple fibers.

Including broadband in land use planning efforts may also promote build-out and reduce costs. For example, requiring all new housing and building developments to include broadband infrastructure, particularly fiber cables, alongside other utility requirements, including electricity and water, can help to lower long-term costs by ensuring that broadband infrastructure is laid at the outset, which avoids the higher costs associated with retrofitting.

With wireless networks, particularly in low-density areas where the economics may not support multiple competing infrastructures, carriers can share cell towers and some backhaul facilities as a way of reducing network build-out costs and bringing competition to such areas more quickly. Such arrangements have slowly been gaining acceptance in both developing and developed countries, particularly as carriers seek to manage costs as they expand their networks or upgrade their services to support higher-speed broadband.[7] However, in some cases, it may be necessary to overcome resistance from incumbents or dominant operators, since they are likely to accrue relatively little benefit from sharing with competitors.

Access to Rights-of-Way

Most of the cost of constructing wireline networks lies in the civil works. By lowering the barriers to and costs of accessing the rights-of-way associated with public infrastructure (for example, roads, railways, pipelines, or electricity transmission lines), governments can significantly increase incentives for private investment in broadband networks at all levels of the supply chain. Such incentives can be achieved in several ways, but primarily by (a) making rights-of-way readily available to network developers at low cost, (b) simplifying the legal process and limiting the fees that local authorities can charge for granting rights-of-way, and (c) providing direct access to existing infrastructure owned by the government through state-owned enterprises (for example, a railway company partnering with one or more operators to build a fiber optic cable network along the railway lines). Such access can also be valuable to wireless operators as they seek to locate towers to expand services.

Accounting and Functional Separation

In those countries where bottlenecks persist in the supply chain and especially where the historic monopoly provider still retains a dominant position in the backbone, middle-mile, or local access segments, governments have intervened even more directly. To bring added transparency to the operations of a dominant provider, regulators have sometimes required the provider to separate the accounting for different parts of its business—keeping wholesale and retail accounts separate, for example. This better enables stakeholders to identify unfair discrimination against nonaffiliated providers and can help to ensure that competition takes place on fair and equal terms.

One of the most severe remedies imposed by regulators is functional separation. Functional separation requires the incumbent operator to establish a new business division—separate from its other divisions—to manage the network and provide wholesale services to all retail service providers on a nondiscriminatory basis. Functional separation is not the same as (and is less severe than) structural separation or the spinoff and sale of network operations: the incumbent operator maintains ownership of the network division, but it must be independent from the operator's retail and commercial divisions. In many cases, other regulatory obligations are used as a complement to functional separation, such as local loop unbundling (LLU) or bitstream obligations. As (fiber) broadband networks are being deployed, governments have begun to consider whether similar obligations should be placed on those new or upgraded networks. Finally, as a last resort, full structural separation may be warranted if the government does not believe that anticompetitive conduct—either by an incumbent or by a new broadband or fiber optic network operator—can be otherwise controlled. This entails the creation of a totally separate entity—for example, to build and manage the network's physical infrastructure. The various types of separation policies and examples are discussed in chapter 3.

Table 2.3 presents an overview of some of the policies that can be used to promote the supply of broadband. For a more in-depth view of the various policies and programs for promoting the build-out and uptake of broadband, see appendix B to this volume.

Encourage the Adoption of Broadband: Promoting Demand

Countries are beginning to view broadband promotion not only as a problem of supply of broadband (access to networks), but also as a problem of demand for it (adoption by businesses, government, and households). As a result, demand facilitation is becoming an important part of broadband development strategies and policies. Chapter 6 discusses demand-side policies in more detail.

Most of the experiences to date in stimulating demand for broadband applications and services come from developed countries. Similar policies may work more or less well in developing countries, where economic and

Table 2.3 Checklist of Policies to Promote the Supply of Broadband Networks

Goal	Policy
Promote competition and investment	• Implement policies or regulations to create conditions to attract private investment in broadband networks
	• Implement technology- and service-neutral rules or policies giving operators greater flexibility
	• Promote effective competition for international gateways and possible policies for service-based competition for gateway operators to provide access to their facilities on a wholesale nondiscriminatory basis
	• Develop policies to facilitate interplatform competition
Encourage government coordination	• Adopt common technical standards and facilitate the development of international, regional, and national backbones
	• Incorporate broadband planning into land use and city planning efforts
Allocate and assign spectrum	• Assign additional spectrum to allow new and existing companies to provide bandwidth-intensive broadband services
	• Allow operators to engage in spectrum trading
Promote effective competition and encourage investment	• Encourage multiple providers to share physical networks (wireline and wireless), which can be more efficient, especially in low-density areas
Facilitate access to rights-of-way	• Facilitate access to public rights-of-way to ease the construction of both long-distance (backbone) and local connections
	• Develop policies that provide open access to government-sponsored and dominant-operator networks to enable greater competition in downstream markets
Facilitate open access to critical infrastructure	• Develop policies that provide open access to government-sponsored and dominant-operator networks to enable greater competition in downstream markets
	• Consider implementation of LLU if necessary to facilitate competition

Source: Telecommunications Management Group, Inc.

social conditions differ; the ability to adapt the lessons learned and successful policies to local needs will be critical. This particularly applies to policies that are focused on demand-side issues, where culture and socioeconomic status are important variables. For example, with the first availability of broadband services, demand (measured by subscriber growth, for example) may be initially very high—reflecting pent-up demand among users who previously had no broadband access. In such cases, governments may decide that there is no need to stimulate demand. In Kenya, for example, at the end of September 2010, broadband subscriptions increased to 84,726 subscribers from 18,626 in the previous quarter (a growth rate of over 450 percent) without any specific attempt by the government at demand-side stimulation (CCK 2011).

As time passes, however, growth in demand is expected to slow down as the potential pool of users evolves from motivated early adopters to potential users who do not necessarily understand all that broadband has to offer and may be concerned with the potential threats to privacy and data security. This is when government policies to stimulate demand may have the most beneficial impact. By educating users through digital literacy programs, governments can help to drive adoption to a broader user base and educate them at the same time. Such programs may become increasingly important as adoption rates rise in order to avoid the social and economic inequities associated with broadband "haves" and "have nots." One important issue that policy makers should consider as they address broadband demand development is the opportunity cost of using (limited) public moneys for broadband demand programs as opposed to other worthy public uses. In some cases, governments have decided that stimulating broadband demand was important enough for reaching national economic and social goals. This may not be the case in all countries, particularly in countries with the fewest resources to spare.

The role of government in stimulating demand will vary by country. In some countries, with populations that are more technically literate, there may be less need for direct government intervention. The appeal of social networking and video streaming as an entertainment source may be more self-evident than more mundane uses such as e-government or multimedia mail. In such cases, demand will be driven by attractive offerings made available by private sector developers. In other cases, however, basic illiteracy, lack of understanding of what the Internet can do, or costs may require governments to step in to fill out and aggregate demand, particularly among at-risk groups. Policies to support digital inclusion will be an important leveler to ensure that broadband can bring benefits to all segments of the population.

Efforts to increase demand typically fall into three categories: awareness, affordability, and attractiveness. In order to drive broadband adoption and use, policies must address these three categories, especially targeting those populations that are generally less likely to adopt and use broadband Internet services. Mechanisms to address awareness include improving digital literacy and encouraging the use of broadband in education and by small and medium enterprises (SMEs). Efforts to address affordability focus on the costs of both hardware and services, and attractiveness initiatives include promotion of services, applications, and local content as well as delivery of government services over the Internet (e-government). E-literacy and e-skills, in particular, are vital for broadband diffusion to succeed. Recognizing this, governments with high penetration and adoption rates have been very active in trying to raise e-literacy. The three main barriers to broadband adoption are discussed in further detail in chapter 6.

Table 2.4 provides an overview of some of the policies that governments can use to promote demand for broadband applications and services, including programs to provide users with a place where they can obtain access.

Consider Other Sectors of the Economy and Society

As policy makers and regulators consider policies and strategies to promote broadband development in their countries, they need to consider the issues in the broader context of larger economic and social goals.

Table 2.4 Checklist of Policies to Promote Demand for Broadband

Focus	Policy
Infrastructure	• Connect schools to broadband networks
	• Make government an anchor tenant
	• Expand access to underserved communities with universal service fund support
	• Construct community access centers
	• Consider expanding universal service to include broadband
Services, applications, and content	• Undertake government-led demand aggregation
	• Provide e-government applications
	• Promote creation of digital content
	• Implement reasonable intellectual property protections
	• Ensure nondiscriminatory access
Users	• Provide low-cost user devices in education
	• Develop digital literacy programs for citizens
	• Address content and security concerns
	• Facilitate affordability of broadband devices
	• Monitor service quality
	• Support secure e-transactions
	• Provide training to small and medium enterprises

Source: Telecommunications Management Group, Inc.

Broadband applications and services are increasingly intersecting with virtually every other major sector of the economy—including education, health, banking, the environment and climate change, and cyber-security—and broadband policy makers and regulators will likely need to coordinate their efforts with their counterparts in other areas of the government to achieve larger policy objectives, whether they be social, economic, or political.

Tackling such cross-sector goals will require close coordination among various regulators so that policies and approaches support each other. It will also require policy approaches and regulatory frameworks that are broad enough to allow policy makers to consider the relevant interrelated issues as well as a high degree of committed leadership at the most senior levels to ensure that all parts of government work together to promote the development of broadband as part of the more general goals of promoting social and economic growth. Despite increasing recognition of the importance of broadband and its impact on the policies and implementation of programs in other sectors, most countries' laws do not typically address the jurisdictional issues related to other sectors of the economy vis-à-vis broadband. As a result, it will be increasingly important for governments to adopt provisions outlining the cooperative arrangements between the broadband regulator and other government agencies. For agencies that are not used to working together—and that come to the same issues with vastly different points of view—such guidelines or arrangements will be crucial to ensuring that policies and decisions are mutually supportive of both broadband development and sector-specific goals and programs. This section briefly describes how broadband development policies interact with policies in other key sectors of a country's economy. Specific examples of such collaboration and the ways broadband can support other sectors of the economy are found in chapter 1. To view how applications and services developed within these other sectors can help to drive demand for broadband, see chapter 6.

Financing Broadband Development

In the past 20 years, markets have liberalized, competition has increased, and the private sector has been the primary vehicle for financing telecommunications projects, especially in profitable areas. Nonetheless, in many developing countries, significant barriers to entry persist and legacy-dominant carriers continue to control markets and distort competition. Thus, the government's primary role has been twofold: to develop policies that support and encourage private sector investment, while also seeking more effective ways to regulate dominant carriers and promote competition.

Today, most countries emphasize competition and a significant role for private sector investment to spur the growth of their broadband markets. In developed countries and some developing countries, the majority of the private investment likely comes from within the country itself. In the least financially endowed countries, however, private investment may also come from foreign sources. Governments seeking to promote broadband development in their countries should bear in mind that investors and companies around the world may be looking for opportunities to invest in good projects wherever they are located. Thus, attracting foreign private investment—through appropriate incentives, a clear regulatory and legal environment, and a good development plan—may be important components of a broadband strategy.

Where governments choose to finance broadband networks, they should avoid replacing private investment or substituting for the normal operation of market mechanisms. Rather, governments should facilitate and support private sector investment and be capable of developing, promoting, and implementing timely policies based on a thorough understanding of the market. For this reason, an essential element in effectively deploying broadband is the ability to find an appropriate financing model in which government oversight and intervention are focused mainly on funding and financing only those initiatives targeted at addressing actual or expected market failures in the availability of a broadband network and at driving the early adoption of broadband services.

In addition to private sector investment and direct funding by governments, several other options exist for countries to finance broadband deployment, including government grants or subsidies to both private and public entities and to partnerships where private funding is matched by government. The following sections briefly address the main ways that governments can support the financing of broadband development.

Government Support to Enhance Private Investment

As stated by the 2004 report of the Task Force on Financial Mechanisms for ICT for Development, the engine of ICT development and finance over the past two decades has been private sector investment, including foreign direct investment by an increasingly diverse and competitive array of multinational and regional ICT sector corporations (Task Force on Financial Mechanisms for ICT for Development 2004, 22). Companies target and provide service to profitable, high-revenue customers, neighborhoods, and regions to the detriment of those that are less commercially viable. This is the result of the tendency to see profitability and return on investment as drivers of private investment.

In addition to the purely economic decisions involved, private investment also depends heavily on the regulatory climate. The government's challenge is to put in place the necessary policy measures and regulatory framework to allow and encourage the deployment and financing of broadband networks as widely as possible and thus ensure not only that high-value users receive high-quality services, but also that the benefits of broadband are spread throughout all populations and areas.

The Organisation for Economic Co-operation and Development (OECD), based on a survey of broadband policies in member states, identified particular policy initiatives that may promote broadband investments, including policies to undertake the following:

- Improve access to passive infrastructure (conduit, poles, and ducts) and coordinate civil works as an effective means to encourage investment
- Ensure access to rights-of-way in a fair and nondiscriminatory manner
- Encourage and promote the installation of open access to passive infrastructure when public works are undertaken
- Allow municipalities or utilities to enter telecommunications markets; where market distortion is a concern, policy makers could limit municipal participation to basic investments (such as the provision of dark fiber networks under open-access rules)
- Provide greater access to spectrum (which is a significant market barrier to wireless broadband provision) and adopt more market mechanisms to promote more efficient spectrum use.

Many countries have used these policies to spur the build-out of broadband networks. In Korea, for example, thanks to greater market liberalization over the past decade, several new service providers entered the telecommunications market and began to fund and deploy fiber-based networks. Many advanced broadband networks are now available, and the country has an impressive number of users.

In Africa, wireless broadband licenses have been granted by governments since 2004, allowing mobile operators to roll out networks capable of supporting high-speed data. Although uptake was initially slow, several factors have led to a growing number of African operators, boosting investments for 3G or 4G, including (a) more affordable international and backhaul capacity, (b) increasing competition in the mobile sector, (c) greater demand for more advanced services (for example, through the launch of e-health and e-education projects relying on mobile as well as other technologies), (d) slower growth in voice subscribers and revenues, and (e) the lack of wireline networks on the continent.[8]

In some cases, private investors may also look to multilateral investment banks to assist in financing, particularly where investment proposals are perceived by potential investors as higher-risk transactions or where difficult liquidity conditions and uncertain economic prospects are seen as additional risk factors. Such conditions decrease the possibility of private financing and raise the costs of financing. In such cases, investment banks have become involved in broadband projects. The European Investment Bank, for example, is already lending an average of €2 billion each year to support broadband projects; it also develops and finances pilot projects and innovative funding schemes.

Fiscal Support to Facilitate Broadband

In some cases, regulatory reform and private sector investment still will not permit a government to reach its broadband development goals. In those cases, policy makers may turn to fiscal support to fill broadband development gaps. Fiscal support can be directed to a company or end users and can be provided in various forms, including cash subsidies, in-kind grants, tax incentives, capital contributions, risk assumption, or other fiscal resources (Irwin 2003).

Economic Justification of Fiscal Support

Fiscal resources are limited and face competing demands from many sectors. As a result, policy makers considering providing more direct support for broadband development must carefully analyze the expected costs and benefits of providing that support. First, a persuasive case must be made that the benefits of supporting broadband development are likely to outweigh the cost to be incurred by all participating private and public sector entities, as seen from the viewpoint of the economy as a whole. Fiscal support should not be provided for components of the broadband strategy that will leave the economy worse off than without it. Second, if a component is desirable for the economy overall, it is essential to determine how much fiscal support should be provided.

For example, the government of Australia committed in its 2008–09 budget to base its spending on infrastructure projects on rigorous cost-benefit analysis to ensure the highest economic and social returns to the nation over the long term. The national broadband network is an open-access wholesale-only network that was expected at the time to be capable of delivering fiber-based coverage at 100 Mbit/s to 93 percent of all premises and fixed wireless and satellite coverage at 12 Mbit/s to the rest. The total construction and initial maintenance cost was estimated at $A 35.7 billion, including $A 27.1 billion of equity contributed by the Australian government. A study in 2009 also estimated that the cost of the network would exceed its benefits by between $A 14 billion to $A 20 billion (present value in 2009). The study concluded that the investment should not be undertaken if the total cost would exceed $A 17 billion, even accounting for rising demand for broadband-enabled services and the negative outcome that the typical end user would not have access to more than 20 Mbit/s (Ergas and Robinson 2009).

Fiscal support often involves the direct use of government money. Subsidizing investment requires cash outlays up-front that will never be recovered. Subsidizing use involves payments during a long time, possibly for the lifetime of the strategy. Investing equity in public-private partnerships (PPPs) involves cash contributions up-front that may be recovered in the long run (for example, as dividends) to the extent that the ventures are commercially successful. Long-term debt financing comprises cash outlays that may be recovered over the years, provided the beneficiaries do not default on repayment obligations.

Fiscal support that does not involve direct use of government money also has a cost. Giving investors free use of spectrum for last-mile access has an opportunity cost related to the revenues that the government could obtain from the sale of spectrum licenses for profitable business use. Preferential taxation (for example, income tax holidays, custom duty exemptions) implies fiscal revenues forgone. On-lending international development loans and credits reduce the funding available from these sources for other initiatives in the same country.[9] Regulatory risk (for example, changes in the pricing rules) can be mitigated through government guarantees, which create contingent liabilities. The government can pick up part of the commercial risk of uncertain market outlook for new investments by committing to future purchases, which may result in obligations unrelated to actual need.

Estimating costs and benefits. In order to determine whether to move ahead with some form of fiscal support for broadband development, the costs and benefits must be determined. Economic costs and benefits of a component of the broadband strategy are valued to reflect real scarcities of goods and services. Financial analysis values costs and benefits at market prices. Both economic and financial analyses compare the situations with and without the component. Sunk costs are not taken into account.

The principles for estimating economic and financial costs and benefits are well known, but applying these principles in practice is subject to assumptions about market and technology development. This can be a challenge, especially when some players (for example, incumbent operators) have more detailed information and analytical capabilities than others (for example, government authorities, new entrants). To some extent, this limitation can be overcome by using the calculus of costs and benefits to provide guidance on fiscal support, but relying primarily on market mechanisms (for example, minimum subsidy auctions) to reach the final decisions on support awards.

When costs and benefits can be measured in monetary terms, economic costs and benefits can be derived from financial costs and benefits. Transfers from one part of the economy to another, such as sales taxes or customs duties, are excluded from the cost stream. Prices that are distorted by market interventions, such as unskilled labor, foreign exchange, capital, and the radio spectrum, are adjusted to reflect their real scarcity in the economy. External costs (for example, business losses resulting from digging up streets to install fiber) should be quantified, to the greatest extent possible.

Benefits can be harder to calculate. Starting from the financial analysis of network and service providers, economic benefits can be estimated by adding consumer and producer surpluses to the revenue streams. For example, U.S. consumers have been increasingly willing to spend more money for fixed broadband connectivity than they are actually paying. This resulted in a consumer surplus of about US$32 billion in 2008, up 58 percent from about US$20 billion in 2005. Higher speed is expected to add a further US$6 billion for existing customers. The study underestimated the wider economic impact of broadband, as it excluded business users and wireless access (Dutz, Orzag, and Willig 2009).

Comparing costs and benefits. The net present value (NPV) of the expected benefits is the discounted monetary value of benefits minus costs over time. For the government, which values costs and benefits to reflect real scarcities in the economy, an economic NPV > 0 means the project would have a positive effect on the country's welfare. For a private company, which values costs and benefits at market prices, NPV > 0 means the project could be commercially viable. This analysis can be applied to the broadband strategy as a whole as well as to each major separable component.

Projects that have negative economic NPV should not be supported. Projects that have positive financial NPV do not need support. Components that have positive economic NPV but negative financial NPV would be good for the economy, but are unlikely to be undertaken as a business. Fiscal support of these components would be justified, up to a maximum support equal to the absolute value of the (negative) financial NPV. This is the amount of support that would make the component just viable commercially. Support above this level would not be justified.

Types of Fiscal Support

Private investment. Where government does decide to provide some type of fiscal support, the re-creation of monopolies with public support is a fundamental concern to many governments around the world, as is avoiding contributing to established carriers' dominance and displacing private investment. The European Union (EU) supports the construction of broadband infrastructure and Internet take-up through both rural development and structural funds and has clarified the application of state aid rules on the use of public funds for broadband deployment.

The 2009 European Commission's Community Guidelines for the Application of State Aid Rules in Relation to Rapid Deployment of Broadband Network were drafted specifically to address concerns relating to public support, and they contain safeguards to ensure that any broadband infrastructure funded with public money does not favor existing operators, including provisions that a company receiving public money must provide effective open access to its competitors to allow them to compete in an equal, nondiscriminatory way (European Commission 2009). Although the state aid guidelines focus on the role of public authorities in fostering the deployment of such networks in unprofitable areas (that is, areas where private operators do not have the commercial incentives to invest), they clearly note that state aid should not replace or "crowd out" private investment. Instead, public funds should complement private operators' investments and thereby achieve higher and faster broadband coverage. Box 2.4 provides an overview of the EU experience with the state aid guidelines.

In the United Kingdom, for example, the government set a goal in 2009 of ensuring 100 percent access to next-generation broadband and planned to support the rollout of fiber-based broadband and other next-generation technologies via a tax on telephone lines (United Kingdom, Department for Business Innovation and Skills 2009). Since then, BT (formerly British Telecom) has started to roll out fiber broadband to most of the United Kingdom by 2015. However, BT has made clear that without some form of public sector support it will not provide fiber coverage beyond around two-thirds of households. It noted that such support could come from national funding or regional funding combined with local partnerships to deploy networks in specific areas (Lomas 2010). Within this context, BT has announced plans

Box 2.4: Experience in the European Union with State Aid for Financing Broadband

In the context of market reform, good practice in financing universal access projects using public financing other than funds in international jurisdictions includes the practice of setting out rules or guidelines on the provision of public funding for universal service and access. The EU state aid guidelines for funding broadband assist in bringing universal access and service through the presence of clear rules that do the following:

- Facilitate next-generation access (NGA) and broadband investments from public funds in order to bring broadband connectivity to underserved areas

- Enable the rapid deployment of broadband and especially NGA networks, thus avoiding the creation of a new digital divide
- Due to the conditions laid down for the granting of state aid (such as open access, open tenders), allow the maintenance of competition, thus helping to ensure better and more broadband services.

Historically, funding decisions could be made on a case-by-case basis in the EU. However, in light of the significant level of investments, stakeholders require a level of certainty—hence the need for the guidelines.

Source: Box taken from European Commission and ITU 2011.

to roll out superfast fiber broadband to unprofitable areas with the help of European funding. The European Regional Development Fund's Convergence Program is investing £53.5 million, or just over 40 percent of the total funding, with BT providing £78.5 million (France 2011).

Direct government intervention. Market-based investments should be the mainstay for broadband deployment, but some degree of direct government funding may be required to enable and complement the market, particularly in areas that are not considered economically viable by private operators. The form of this more direct intervention will vary from country to country. In many countries, subsidies are used to underpin private sector investment.

Some governments have effectively used subsidies and other financial incentives to spur broadband deployment. Canada, Germany, Greece, Korea, Malaysia, Portugal, Singapore, the United Kingdom, and the United States have all announced plans to undertake and are implementing substantial direct government funding for network infrastructure development. In some countries (for example, Canada, Finland, Germany, Portugal, the United Kingdom, and the United States), measures to expand broadband access and to bolster connection speeds have been included in the country's planned economic stimulus packages. Most of these plans seek to speed up existing links to build faster wireline and wireless next-generation networks. Countries are spending public funding for rolling out high-speed networks to areas that are underserved or unserved by commercial ISPs. In other countries, however, the debate over public financing is not over how much to contribute to broadband efforts, but rather how to cut budgets in line with the economic realities of 2011. In such a context, funding for broadband may assume lesser importance compared to other, more important, social and economic goals. Consequently, the focus on finding private sector–led solutions is likely to increase.

A few governments are pushing the build-out of broadband networks through direct investment by a government-backed company specifically tasked with building new networks. In most, if not all, cases, these government-led efforts will deliver only wholesale services that service providers can then use to offer retail services. In April 2009, for example, Australian Prime Minister Kevin Rudd announced that the government would commit $A 43 billion (US$30 billion) to building a national broadband network across Australia, with wireline services reaching 93 percent of the population and wireless or satellite broadband networks serving the other 7 percent. In March 2011, Qatar announced a similar plan to build a fiber to the home network to reach 95 percent of the population by 2015, with a government-backed company focusing on supplying the passive infrastructure for the network.[10] In 2007, the Rwandan government awarded a US$7.7 million contract to Korea Telecom to construct a wireless broadband network that was the first of its kind in Africa (Malakata 2009). By 2012, the government plans to make broadband Internet access available to more than 4 million Rwandans through the new wireless network and the Kigali Metropolitan Network project.

Public-private partnership models. Apart from implementing policies and regulations to ensure competition (between networks or services), the public sector can promote broadband development by sharing financial, technical, or operational risks with the private sector. Indeed, experience has shown that, in some cases, private sector–only development, or direct government or subsidy funding, may not be sufficient to reach the most remote areas, provide the most bandwidth-intensive services, or provide ongoing public funding, even with "smart subsidies."[11] Within this context, many countries are now adopting approaches that combine public and private sector skills and resources, as well as combining public financing with some form of matching funding from private investors. This approach helps to reduce investment risk, while also recognizing that market participation is essential to the financial sustainability of projects. PPPs are also increasingly being considered as a solution for ICT development, including for broadband backbones and the supply of transmission bandwidth sufficient to catalyze advanced broadband applications.

In Africa, for example, much attention has been given in recent years to the funding and financing of projects aimed at bringing more affordable broadband connectivity to the continent by means of submarine cables, regional fiber optic backbones, and satellites. Such projects have generally been financed through a mixture of public and private sector funding. In 2010, for example, Alcatel-Lucent was selected by Africa Coast to Europe (ACE), a consortium of 20 operators and governments, to build a submarine cable network that will link 23 African countries to Europe.[12] The system will be over 17,000 kilometers long and capable of transmitting data between the African continent and Europe at initial speeds of 40 Gbit/s.

In Finland, the main objective of the December 2008 plan for 2009–15 is to ensure that more than 99 percent of the population in permanent places of residence, as well as businesses and public administration offices, are no farther than 2 kilometers from a 100 Mbit/s fiber optic or cable network. The government expects telecommunications operators to increase the rate of coverage to 94 percent by 2015, depending on market conditions, while public finances will be used to extend services to sparsely populated areas where commercial projects may not be viable, bringing coverage to the target of 99 percent. The plan stipulates that, where public financial intervention is required, it should be in the form of public-private partnerships, with federal funding only being allocated to projects deemed not viable for 100 percent private investment. The plan limits such interventions, providing that the federal subsidy amount cannot exceed one-third of the total project cost, with additional EU and municipal support capped at another one-third, thereby requiring private participants to invest at least one-third of the cost.

Spain has relied greatly on inputs from the private sector through PPPs. Of the public funds used, €31 million were structural funds and €53 million were zero-interest public credits. Operators invested about €280 million. The funded projects use asymmetric digital subscriber line (ADSL), WiMAX, and satellite technologies depending on geography, rollout dates, and available technologies. The government set the minimum download speed at 256 kbit/s, and prices were capped at a "reasonable fee."

Malaysia's 2006 MyICMS Strategy also set out several goals for broadband services as well as strategies to achieve them.[13] The government is funding a fiber optic network under a public-private partnership with Telekom Malaysia that is aimed at connecting about 2.2 million urban households by 2012. Under the terms of the agreement, the government committed to investing RM 2.4 billion (US$700 million) in the project over 10 years, with Telekom Malaysia committing to covering the remaining costs.

Local efforts, bottom-up networks. There are also some interesting examples of how local efforts or bottom-up networks have resulted in the financing of broadband deployment. Module 4, Universal Access and Service, of the *info*Dev ICT Regulation Toolkit, for example, notes that the emergence of municipal broadband networks provides an additional source of financing from local governments for ICT service development.[14] The toolkit highlights the Pirai municipal network in Brazil as a successful initiative that was based on the needs of the municipal authority and included e-government, education, and public access, with a range of application support and development activities. The project established numerous broadband access nodes that allowed all local government offices and most public schools, libraries, and general public access points to be connected. Initially, all financing was provided by the municipal government. A commercial enterprise was later established, but continues to be funded and supported by the municipality.

Local governments have also been instrumental in driving broadband deployments (box 2.5). A 2010 study notes that in some European countries (for example, the Netherlands and Italy) municipal involvement is the result of incumbent carriers' reluctance to deploy networks in areas with less chance of investment return, increasing demand for broadband services, and a perception that broadband networks may serve to reduce the digital divide and stimulate economic growth (Nucciarelli, Sadowski, and Achard 2010). This, the study claims, has led some European municipalities to become more directly involved in broadband network development.

Universal service funds for broadband. In the past, many countries defined their universal service funds (USFs) in a way that gave priority to providing voice telephony (traditionally provided over wireline) services to unserved or underserved regions. Recently, however, some countries have revised their definitions and the scope of the funds to include broadband, mobile telephony, or Internet access. For example, the EU and the United States are adding resources to existing rural development funds or USFs to accommodate broadband. Some countries have turned or are considering turning broadband provision into a universal service obligation and are reforming their universal service policies. Other countries are contracting commercial providers to build the network with service obligations through a competitive bidding process (for example, France, Ireland, Japan, and Singapore). Chapter 4 discusses these issues in more detail.

Comparing Alternative Instruments

Not all fiscal support instruments are equally effective. They differ primarily in terms of accuracy and also regarding transparency, targeting, cost, and sustainability.

Table 2.5 illustrates which instruments of fiscal support can help to overcome each type of obstacle to broadband development ahead of or beyond the market (that is, their effectiveness in addressing specific impediments to broadband development).[15] For example, subsidizing investment is particularly effective at reducing investors' costs and also can help to overcome financial market failures. Alternatively, subsidizing the use of broadband is an effective way to increase revenues by making service affordable to people who otherwise would not buy the service; however, it can also enhance competition among firms in the provision of services and reduce commercial risk by building up demand that otherwise would materialize at some point in the future as incomes rise and costs decline. The choice of instrument can be further narrowed down by considering the transparency of the instrument's cost and its ability to target specified categories of beneficiaries effectively.

Table 2.5 Effectiveness of Fiscal Support for Broadband Development

Objective	Subsidy of investment	Subsidy use and devices	Rights-of-way, spectrum	Preferential taxation	Equity investment	Long-term loans	On-lending international loans	Partial risk guarantees
Reduce costs	X		X	X				
Increase revenues		X						
Facilitate competition		✓	✓					
Improve business environment								
Address financial market failures	✓				X	X	X	
Reduce regulatory and political risk					✓		✓	X
Reduce commercial risk		✓						

Source: Telecommunications Management Group, Inc., adapted from Irwin 2003.
Key: X = instrument is very effective; ✓ = instrument has additional effects

Measurement, Monitoring, and Evaluation: Checking Progress

Why Measure Performance?

Policy makers seeking to promote broadband development need mechanisms to ensure that their objectives are being achieved and to identify whether corrections and refinements to policies and programs are needed. In short, they need to measure progress through regular monitoring in order to identify successes and failures. Different countries will adopt broadband strategies with different objectives, which will affect the appropriate indicators to monitor. It is best to build the indicators most appropriate for the selected objectives into the design of the programs from the beginning and to allocate the necessary resources for data collection and analysis from the start. Broadband indicators are also needed for analysis, for example, to examine trends and the link between broadband adoption and social and economic development. They are also important for monitoring license compliance in areas such as coverage and quality. As a result, the specific indicators appropriate for a particular country, the frequency of data collection and reporting, the geographic unit of analysis, and so forth will also differ from country to country. Consequently, the following section provides a range of options rather than a single prescription for countries to consider when looking at measurement issues.

What to Measure?

The broadband indicators likely to be of the most interest to policy makers are availability, demand, quality, and pricing (figure 2.4). These indicators relate to local retail access rather than to wholesale and backbone markets.[16] Additional indicators also may be useful for monitoring and analysis, including monetary-based statistics such as broadband revenues. The Partnership on Measuring ICTs for Development, a coalition of intergovernmental agencies, has produced a methodological manual identifying core ICT statistics including several broadband indicators.[17] This manual provides a useful list of key broadband indicators based on definitions with international consensus.

Availability (Supply)

Availability refers to the ability to access wireline and wireless broadband networks and services. Different modes of providing broadband exist; therefore, different indicators of availability are needed for each of the modes. In

Figure 2.4   Categories of Broadband Indicators

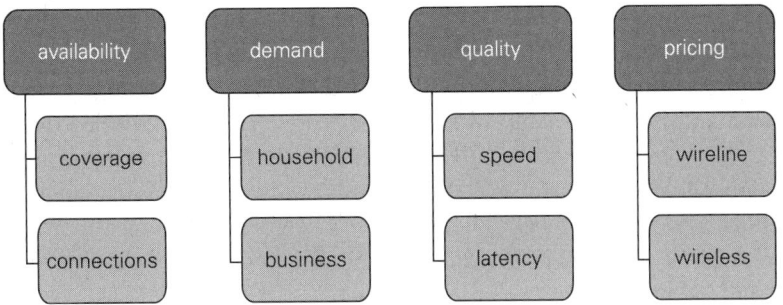

Source: Telecommunications Management Group, Inc.

the case of wireline systems, availability can be measured by the percentage of households passed. This is a conventional measure in the cable industry that can be extended to fiber and digital subscriber line (DSL) as well. The indicator reflects the number of copper (telephone), coaxial (cable television), or fiber optic lines accessible by a premise, regardless of whether users actually subscribe to the broadband service.[18] It may also be useful to distinguish between the type of technology, such as DSL, cable modem, and fiber to the premises (FTTP). This provides an idea of the relative importance of each to broadband development as well as the degree of intermodal competition between technologies. It may also be useful to provide a breakdown of subscriptions by speed ranges and geographic area. These considerations are becoming increasingly important as countries seek to deploy minimum-speed broadband services to unserved and underserved populations.

In the case of wireless, the obvious indicator of availability is signal coverage. This can be measured in terms of population or area. The International Telecommunication Union (ITU) has developed a definition for wireless broadband coverage in the form of 3G or 4G network coverage, although the data are not reported for most countries. Parallel definitions for fixed wireless, satellite, and wireline coverage do not exist within the ITU definitions. However, several countries in the OECD report these data using definitions developed either by national governments or by industry organizations (OECD 2009). They may be adapted by countries wishing to develop comprehensive coverage indicators.

Adoption (Demand)

While supply-side indicators give a general idea of high-speed Internet availability, they do not reflect concrete adoption or usage. Measuring the uptake or adoption of wireline and wireless technologies, however, is significantly more difficult than measuring the supply. While coverage measures the theoretical ability to access broadband services, the number of subscribed connections measures actual demand for the service. Subscriptions should be minimally broken down by wireline and wireless broadband and preferably by additional categories to allow for deeper analysis. A growing number of countries are measuring broadband access by households and businesses through surveys typically carried out by the national statistical offices. These demand-side surveys typically include various indicators of use, which can illuminate factors contributing to broadband take-up.

Determining the number of wireless broadband subscriptions presents several methodological challenges. Although it is useful to distinguish between different types and modes of wireless broadband delivery such as mobile, fixed wireless, and satellite, the line between fixed and mobile broadband is not always clear. For example, in some countries there is a legal rather than a technical restraint on nationwide roaming for some wireless broadband networks. Even with this restriction, users can move with their mobile handset or data card within a limited area, so the distinction between fixed and mobile is not so clear. Another consideration is that the use of wireless broadband on laptops via data cards is different from the use via mobile handsets, and countries define wireless broadband differently. Some countries only consider the former to be mobile broadband and consequently include it in their overall broadband counts, while smartphone broadband use can go uncounted, which could lead to misleading results.

Conversely, another major issue is that users may have the theoretical ability to access mobile broadband services if they have an appropriate handset, regardless of whether they are using it or not. Counting this theoretical availability can significantly overstate the take-up of wireless broadband services in a country. Therefore, it is important to distinguish between active and inactive data subscriptions. The OECD has defined active wireless subscriptions as access to the Internet in the previous three months or the use of a separate data subscription (OECD 2010). However, even activity is a blurred concept since some countries count access to any high-speed service such as video chat, mobile television, and so forth, and users may not be accessing the Internet.

Regulators in some countries publish broadband subscription data, highlighting trends and making comparisons. The Turkish Information Communications and Technology Authority, for example, contrasts the availability of different broadband subscriptions in Turkey with that in the EU and also provides a breakdown of speeds over ADSL, the most prevalent wireline broadband technology in the country (figures 2.5 and 2.6).

There is no international indicator for the percentage of the population that uses broadband, although some countries conduct surveys to determine the percentage of Internet users. This would be a useful supplementary indicator for monitoring and evaluating broadband markets.

Quality

In order to use or fully utilize certain applications, certain performance parameters must be met by the broadband connection. Two of the most important are latency (the amount of time it takes for a packet to travel between sender and receiver) and speed, which can be monitored for both

Figure 2.5   Wireline Broadband (ADSL) Penetration in the European Union and Turkey, by Technology, 2010

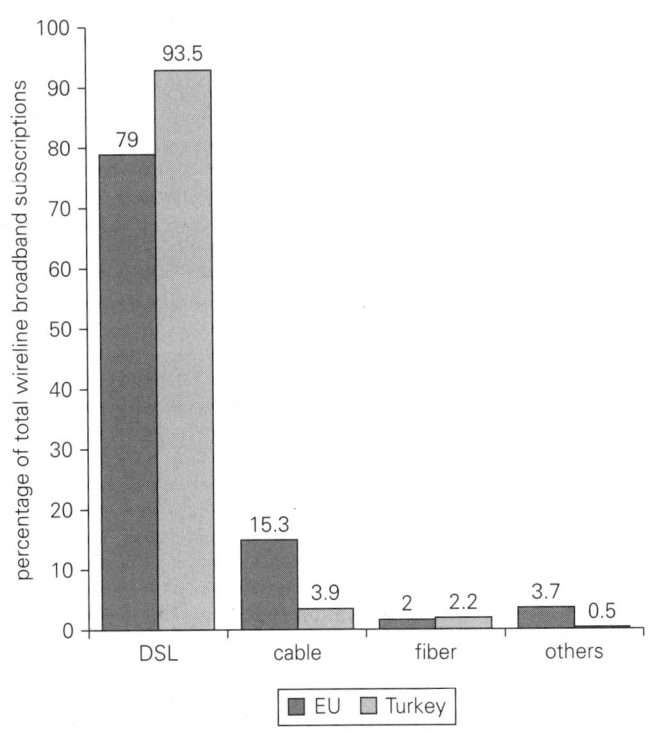

Source: ICTA 2010.

Note: Data for EU refer to January 2010.

Figure 2.6   Wireline Broadband (ADSL) Penetration in Turkey, by Speed, 2010

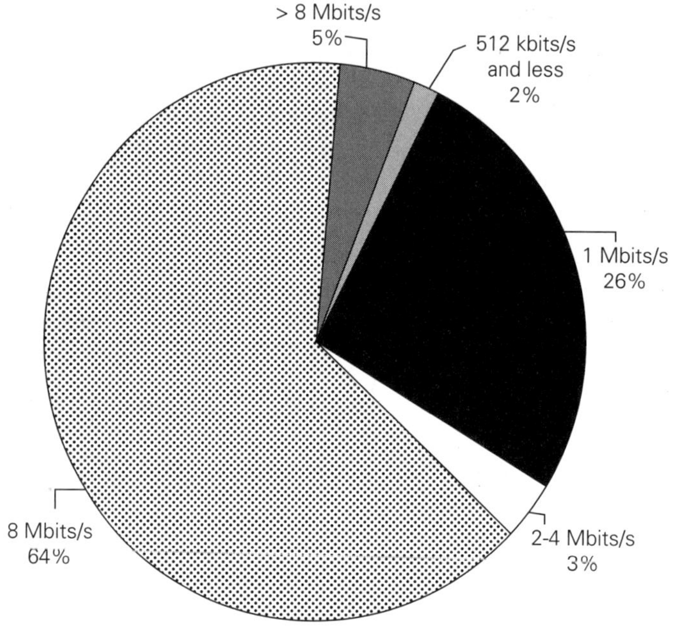

Source: ICTA 2010.

fixed and wireless networks. Other broadband performance metrics include signal quality, availability ("uptime"), complaint ratios, and service activation and restoration times. Technical means exist to measure these aspects at various points in the link between the end user and the server providing the application. Such information is important both to policy makers, who can use it to ensure that the broadband networks and services being supplied are up to industry standards, and to consumers, who can use it to decide which service will provide them with the highest quality. Many consumer complaints hinge on differences between advertised and actual speeds.[19]

In Bahrain, for example, the Telecommunications Regulatory Authority publishes quarterly reports measuring average download and upload speeds and domain name system (DNS) and latency times (figure 2.7; see Bahrain, Telecommunications Regulatory Authority 2011). In the absence of regular monitoring, some regulators publish links on their websites to third-party applications for measuring speed and other aspects of quality.[20]

Because differences exist in performance inside the ISP domain (the user and the server are within the ISP's system), the national domain (the

Figure 2.7   Average Download Speed and Ping Time in Bahrain, January–March 2011

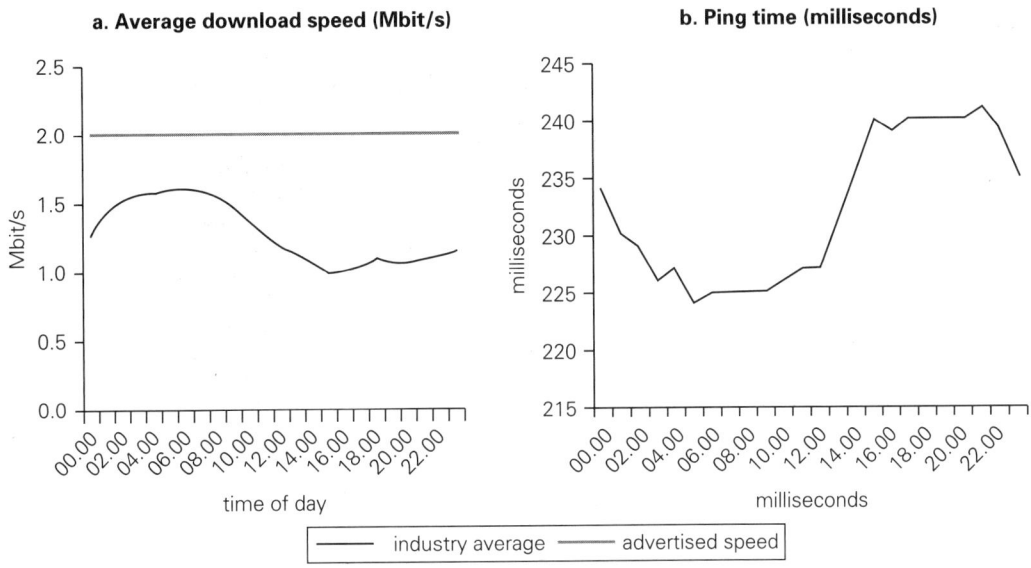

a. Average download speed (Mbit/s)   **b. Ping time (milliseconds)**

industry average ——— advertised speed

Source: Bahrain, Telecommunications Regulatory Authority 2011.

Note: Ping time measures latency by taking the average time for the data to make the round trip to servers located in Bahrain, Europe, and the United States.

user and the server are on different systems, but within the national territory), and the international domain (the user and server are in different countries), measuring performance in each domain yields diagnostic information useful for regulators, operators, and consumers. For example, the Info-communications Development Authority of Singapore establishes different latency parameters depending on whether Internet traffic is national or international (IDA 2011).

Pricing

A government that launches a broadband initiative using public resources will want the service to be affordable to the intended beneficiaries. One could argue that prices need not be monitored in the case of purely private supply, where no public resources have been expended. However, when broadband is seen as an essential public utility or where prices are high due to market failure, governments may want to monitor pricing. Concerns about this issue have prompted countries such as India and the United States to include "affordable" broadband access as a key factor or goal in their broadband initiatives (United States, FCC n.d., 10; TRAI 2010).

Competitive broadband markets typically have multiple tariffs with varying levels of bandwidth, data download caps, and discounts. This presents methodological challenges in terms of compiling comparative broadband tariff indicators across technologies. Baskets of monthly services are often used as a common measure of price trends that factor in caps and speeds. The key components include the monthly price of broadband service, the corresponding speed, and, if applicable, the cap and prices for exceeding the cap. Capped versus unlimited packages pose comparison problems, but can be mitigated somewhat by comparing price per advertised Mbit/s.

An example comparing wireline and mobile broadband monthly prices for selected economies is shown in table 2.6. The example illustrates the various ways of looking at broadband pricing and highlights comparability issues. One notable aspect is the difference between entry-level prices, speeds, and affordability (in terms of price as a percentage of per capita income). For example, although an entry-level fixed broadband package in Turkey costs almost twice as much as in Brazil, the Turkish tariff is a slightly better value since the download speed is twice as fast as in Brazil. Similarly, although the entry-level price for fixed broadband in Brazil is more than twice that in Vietnam, it is much more affordable in Brazil than in Vietnam (although the value of the Vietnamese package is 10 times more).

Mobile broadband pricing is a bit more difficult to compare, since some operators do not guarantee advertised speeds. Instead, prices tend to vary by the volume of data downloaded per month. Nevertheless, as table 2.6 shows, the price of mobile broadband is the same as or lower than the price of wireline broadband (except in Brazil). It is important to note that mobile broadband is more often capped than wireline and that real mobile broadband speeds are also lower in many cases.

How to Measure?

Broadband indicators affect many parties. Government agencies responsible for broadband policy should consult internationally comparable indicators and identify those most suitable for monitoring and evaluation. Best practice suggests that national regulatory agencies should compile broadband statistics, such as number of subscriptions, solicited from operators. This arises out of their mandate to regulate and monitor the sector. Ideally, policy makers consult and cooperate with national statistical agencies that have the technical skills to produce demand-side statistics through household and enterprise surveys, asking about broadband possession or use of different ICTs by households and businesses (or by

Table 2.6 Monthly Prices for Wireline and Mobile Broadband in Selected Countries, 2011

Indicator	Brazil	Kenya	Morocco	Sri Lanka	Turkey	Vietnam
Fixed broadband basket, unlimited (US$)	16.99	39.36[c]	11.86	14.18	30.10	7.93
Speed (Mbit/s)	0.512	0.256	1.000	0.512	1.000	2.56
US$ per Mbit/s	33	154	12	28	30	3
% GDP per capita	1.9	28.4	4.4	7.0	3.5	8.1
Mobile broadband basket, 1 GB, (US$)	51.27	26.24	11.86	4.34	19.93	6.34
Speed (Mbit/s)	1.0[a]	7.2[b]	1.8[a]	7.2[b]	7.2[b]	3.6[a]
US$ per Mbit/s	51	3.7	7	1	2.8	2
% GDP per capita	5.7	18.9	4.4	2.1	2.3	6.5
Annual average exchange rate, 2010 (LCU per US$1)	1.7536	76.1926	8.3507	112.7960	1.5054	18,919.1000
GDP per capita (US$)	10,816	1,662	3,249	2,435	10,399	1,174
Fixed-broadband basket, 1 GB (LCU)	29.8	2,9990	99.00	1,600.00	45.31	150,000.00
Mobile broadband basket, 1 GB, (LCU)	89.9	1,999.0	99.0	490.0	30.0	120,000.0

Sources: Adapted from Telefonica, VIVO (Brazil); Orange, Safricom (Kenya); Maroc Telecom (Morocco); SLT, Dialog (Sri Lanka); TTNET, Turkcell (Turkey); VNN, MobiFone (Vietnam).

Note: GB = gigabyte; LCU = local currency unit.

a. Advertised download speed.

b. Theoretical download speed.

c. Includes 30 minutes of on-net calls. For fixed broadband, least expensive uncapped plan providing download speed of at least 256 kilobits per second (kbp/s). For mobile broadband, least expensive plan offering 1 GB per month of download and download speed of at least 256 kbit/s.

individuals). Broadband operators play a key role, both as providers and as consumers of the data.

The entities best positioned to provide supply-side data are the suppliers of the relevant services. It is common for provisions mandating the reporting of data to the government or the regulatory agency to be included in statutes governing the industry or in licenses or concession contracts. Irrespective of legal provisions, the principal challenge will be to ensure regular and timely reporting of the required indicators based on adherence to agreed-upon standard definitions and procedures.

Table 2.7   Sources of Official Broadband Statistics

Source	Site	Note	Link
ITU	ICT Data and Statistics Division	Worldwide scope; fixed and mobile broadband subscriptions; fixed broadband tariffs	http://www.itu.int/ITU-D/ict/statistics/
EUROSTAT	Information Society	EU members and sometimes other countries; household and enterprise broadband penetration	http://epp.eurostat.ec.europa.eu/portal/page/portal/information_society/introduction
OECD	Broadband Portal	OECD member data, including broadband indicators covering penetration, usage, coverage prices, services, and speeds	http://www.oecd.org/document/54/0,3746,en_2649_33703_38690102_1_1_1_1,00.html

Source: Telecommunications Management Group, Inc.

Most governments do not monitor their country's broadband development in a vacuum. They typically need data from other countries to put their nation's high-speed market evolution in perspective and benchmark it with that of other countries. Brazil, for instance, compared its broadband penetration and forecast evolution to those of Argentina, Chile, China, Mexico, and Turkey.

Several international sources harmonize and disseminate statistics for different countries. The ITU has been the traditional repository of supply-side data on telecommunications and now on ICTs, including some demand-side data. Similarly, the OECD collects and disseminates a number of broadband indicators for its member countries, as does EUROSTAT, the statistical arm of the EU. All of these organizations make the data available on dedicated websites (table 2.7). The Economic Commission for Latin America and the Caribbean recently launched a broadband indicator site for its members.[21] In addition, several private sector entities publish broadband statistics on mobile broadband subscriptions as well as average download speeds and other quality metrics.[22]

Notes

1. In a technical sense, public goods are nonrivalrous (that is, one person's use does not diminish another person's ability to use it) and nonexcludable (people cannot be stopped from using it). Examples include free over-the-air radio and television and national defense. However, some argue that broadband is not a pure public good, as broadband access is excludable, as demonstrated by the unevenness of broadband deployment, even within the same country. Some

may also argue that broadband is not a public good since it is also rivalrous—one person's use can diminish another's use if the network is congested. See Atkinson (2010).

2. Brazil, Núcleo de Informação e Coordenação (2009, 14). The total percentage of respondents is more than 100 because some respondents provided more than one reason for nonadoption. The total percentage of respondents is less than 100 because, for purposes of comparison, not all factors addressed in the study are included in this figure (U.S. FCC 2009, 30).

3. The experiences of the countries surveyed in Kim, Kelly, and Raja (2010), for example, may provide good approaches that could be adapted for use in many countries.

4. Oman in 2009 short-listed firms competing to become the sultanate's first universal service provider.

5. "Kenya Data Networks Cuts Internet Rates by 90 Percent," *telecompaper*, August 5, 2009, http://www.telecompaper.com/news/kenya-data-networks-cuts-internet-rates-by-90-percent.

6. For SIDSs, obtaining submarine cable connectivity has been a mixture of geography, history, and luck. Investment in a submarine cable depends on traffic, which is itself a function of the number of people and the intensity of use. SIDSs have very small populations and modest levels of teledensity and Internet usage, making it challenging to obtain submarine cable connectivity. See Sutherland (2009, 8).

7. See, for example, TeleGeography, "Telia and Telenor Share Danish Networks," June 11, 2011, http://www.telegeography.com/products/commsupdate/articles/2011/06/14/telia-and-telenor-share-danish-networks/; TeleGeography, "Safaricom and Telkom Poised to Ink Tower Sharing Deal," June 14, 2011, http://www.telegeography.com/products/commsupdate/articles/2011/06/14/safaricom-and-telkom-poised-to-ink-tower-sharing-deal/.

8. "Connecting Africa: Continent-Wide Mobile Broadband Rollout Intensifies," *Oxford Analytica*, July 2010, http://www.forbes.com/2010/07/28/africa-mobile-broadband-business-oxford-analytica.html.

9. International development organizations, such as the World Bank and regional development banks, typically have an overall funding envelope for a particular country at a given time, which involves trade-offs among competing eligible initiatives.

10. "Qatar's Government Establishes Q.NBN to Accelerate Rollout of Nationwide Broadband Fiber to the Home (FTTH) Network," *Zawya*, March 27, 2011, https://www.zawya.com/Story.cfm/sidZAWYA20110327102507/Q.NBN%20To%20Build%20Qatar%20National%20Broadband%20Network.

11. *info*Dev and ITU, "ICT Regulation Toolkit, Module 4: Universal Access and Service," sec. 5.4.1, Public-Private Partnerships, http://www.ictregulationtoolkit.org/en/Section.3288.html.

12. The consortium is headed by France Telecom-Orange and includes Baharicom Development Company, Benin Telecoms, Cable Consortium of Liberia, Orange Cameroun, Companhia Santomense de Telecomunicações, Côte d'Ivoire Telecom Expresso Telecom Group, Gambia Telecommunications Company,

International Mauritania Telecom, Office Congolais des Postes et Telecommunication, Orange Guinea, Orange Mali, Orange Niger, PT Comunicações, Equatorial Guinea, Gabon, Sierra Leone Cable, Société des Télécommunications de Guinée, and Sonatel; see "20 Operators Team with Alcatel-Lucent to Bring Fast, Lower-Cost Broadband Connectivity in Africa with a New 17,000 Km Submarine System," *PR Newswire*, June 2010, http://www.prnewswire.com/news-releases/20-operators-team-with-alcatel-lucent-to-bring-fast-lower-cost-broadband-connectivity-in-africa-with-a-new-17000-km-submarine-system-95852004.html.

13. See the MCMC website, "MyICMS," http://www.skmm.gov.my/index.php?c=public&v=art_view&art_id=62.

14. *Info*Dev and ITU, "ICT Regulation Toolkit, Module 4: Universal Access and Service," http://www.ictregulationtoolkit.org/en/Section.3126.html.

15. Some instruments can actually compound the obstacles. For example, granting tax holidays or custom duty exemptions weakens the business climate by discriminating among economic activities and increasing the cost of tax administration and compliance.

16. Although the deployment of national backbones is an important goal of some broadband plans, the indicators to measure developments in these areas have not been identified or defined by the international statistical community, and the data are not widely available. Nevertheless, perusal of plans from some countries can help to identify relevant indicators. For example, India's proposed broadband plan calls for the construction of a national fiber optic backbone throughout the country. Deployment might be measured by indicators such as the number of localities served by the national fiber optic backbone and kilometers of fiber backbone in the network. See TeleGeography, "India's National Broadband Policy to Be Sent for Cabinet Approval Shortly," March 31, 2011, http://www.telegeography.com/products/commsupdate/articles/2011/03/31/indias-national-broadband-policy-to-be-sent-for-cabinet-approval-shortly/.

17. The Partnership on Measuring ICTs for Development aims to develop further initiatives regarding the availability and measurement of ICT indicators at the regional and international levels. It provides an open framework for developing a coherent and structured approach to advancing the development of ICT indicators globally, particularly in developing countries. Partners include EUROSTAT, ITU, the OECD, United Nations Conference on Trade and Development, United Nations Educational, Scientific, and Cultural Organization's Institute for Statistics, the United Nations regional commissions (Economic Commission for Latin America and the Caribbean, Economic and Social Commission for Western Asia, Economic and Social Commission for Asia and the Pacific, and Economic Commission for Africa), United Nations Department of Economic and Social Affairs, and the World Bank. See http://www.itu.int/ITU-D/ict/partnership/index.html. See also UNCTAD (2010). In addition, the ITU has identified and defined other broadband-related statistics. See ITU (2010).

18. For more on issues related to measuring broadband coverage, see OECD (2009).

19. "Average Broadband Speed Is Still Less Than Half Advertised Speed," *Ofcom*, March 2, 2011, http://media.ofcom.org.uk/2011/03/02/average-broadband-speed-is-still-less-than-half-advertised-speed/.

20. For example, the Federal Communications Commission in the United States has a broadband webpage, where consumers can test their speed, latency, and jitter. See http://www.broadband.gov/qualitytest/about/.

21. ECLAC, "ECLAC Launched Regional Broadband Observatory," Press Release, May 27, 2011.

22. GSM World, "Market Data and Analysis," http://www.gsmworld.com/newsroom/market-data/market_data_and_analysis.htm. Ookla's Net Index provides average download speeds for 170 economies; see http://www.netindex.com/download/allcountries/. Akamai compiles performance data for a number of economies. Also see Akamai's network performance comparison, http://www.akamai.com/html/technology/dataviz2.html.

References

Atkinson, Robert. 2010. "Network Policy and Economic." Paper presented at the Information Technology and Innovation Foundation's "Telecommunications Policy Research Conference," Washington, DC, October. http://www.itif.org/files/2010-network-policy.pdf.

Bahrain, Telecommunications Regulatory Authority. 2011. "TRA Bahrain Broadband Analysis Report 01 Feb 2011–31 Mar 2011." Telecommunications Regulatory Authority, Manama, April 3. http://www.tra.org.bh/en/marketQuality.asp.

Brazil, Núcleo de Informação e Coordenação. 2009. *Análise dos resultados da TIC domicílios*. Saõ Paulo: Núcleo de Informação e Coordenação.

CCK (Communications Commission of Kenya). 2011. *Quarterly Sector Statistics Report: 1st Quarter July–Sept 2010/2011*. Nairobi: CCK.

Dutz, Mark, Jonathan Orzag, and Robert Willig. 2009. "The Substantial Consumer Benefits of Broadband Connectivity for U.S. Households." Internet Innovation Alliance, Washington, DC. http://internetinnovation.org.

Ergas, Henry, and Alex Robinson. 2009. "The Social Losses from Inefficient Infrastructure Projects: Recent Australian Experience." Paper presented at the Productivity Commission Roundtable, "Strengthening Evidence-Based Policy in the Australian Federation." http://ssrn.com/abstract=1465226.

European Commission. 2009. "Communication from the Commission: Community Guidelines for the Application of State Aid Rules in Relation to Rapid Deployment of Broadband Network." European Commission, Brussels, September 30. http://eur-lex.europa.eu/LexUriServ/LexUriServ.do?uri=OJ:C:2009:235:0007:0025:EN:PDF.

European Commission and ITU (International Telecommunication Union). 2011. "SADC Toolkit on Universal Access Funding and Universal Service Fund Implementation." European Commission, Brussels; ITU, Geneva. http://www.itu.int/ITU-D/projects/ITU_EC_ACP/hipssa/events/2011/SA2.2.html.

EUROSTAT (Statistical Office of the European Communities). 2009. "Information Society Statistics at Regional Level." European Commission, Brussels, March. http://epp.eurostat.ec.europa.eu/statistics_explained/index.php/Information_society_statistics_at_regional_level#Publications.

France, Paul. 2011. "BT Brings Fibre to First Cornwall Homes." *cable*, March 28. http://www.cable.co.uk/news/bt-brings-fibre-broadband-to-first-cornwall-homes-800478200/.

Hernandez, Janet, Daniel Leza, and Kari Ballot-Lena. 2010. "ICT Regulation in the Digital Economy." GSR Discussion Paper, ITU, Telecommunications Management Group, Inc., Geneva. http://www.itu.int/ITU-D/treg/Events/Seminars/GSR/GSR10/documents/GSR10-ppt2.pdf.

ICTA (Information and Communication Technology Agency of Sri Lanka). 2010. *Annual Report 2010.* Colombo: ICTA.

IDA (Info-communications Development Authority of Singapore). 2009. "What Is Next Gen NBN?" IDA, Mapletree Business City. http://www.ida.gov.sg/Infrastructure/20090717105113.aspx.

———. 2011. "Quality of Service." IDA, Mapletree Business City. http://www.ida.gov.sg/Policies%20and%20Regulation/20060424141236.aspx.

Irwin, Timothy. 2003. "Public Money for Private Infrastructure." Working Paper 10, World Bank, Washington, DC.

ITU (International Telecommunication Union). 2003. *Birth of Broadband.* Geneva: ITU.

———. 2010. "Definitions of World Telecommunication/ICT Indicators." ITU, Geneva. http://www.itu.int/ITU-D/ict/handbook.html.

Kim, Yongsoo, Tim Kelly, and Siddhartha Raja. 2010. "Building Broadband: Strategies and Policies for the Developing World." World Bank, Washington, DC, June. http://www.infodev.org/en/Publication.1045.html.

Lomas, Natasha. 2010. "BT: Fibre Broadband Coming to Two-Thirds of UK by 2015." *silicon.com*, May 13. http://www.silicon.com/technology/networks/2010/05/13/bt-fibre-broadband-coming-to-two-thirds-of-uk-by-2015-39745802/.

Malakata, Michael. 2009. "Rwanda's Mobile Broadband Is Africa's First." *Computerworld*, December. http://news.idg.no/cw/art.cfm?id=D6F3D422-1A64-67EA-E4FF70C29D8BDB9D.

Nucciarelli, Alberto, Bert M. Sadowski, and Paola O. Achard. 2010. "Emerging Models of Public-Private Interplay for European Broadband Access: Evidence from the Netherlands and Italy." *Telecommunications Policy* 34 (9): 513–27.

OECD (Organisation for Economic Co-operation and Development). 2008. *Broadband Growth and Policies in OECD Countries.* Paris: OECD. http://www.oecd.org/document/1/0,3343,en_2649_34223_40931201_1_1_1_1,00.html.

———. 2009. "Indicators of Broadband Coverage." DSTI/ICCP/CISP(2009)3/FINAL, OECD, Committee for Information, Computer, and Communication Policy, December 10. http://www.oecd.org/dataoecd/41/39/44381795.pdf.

———. 2010. "Wireless Broadband Indicator Methodology." OECD Digital Economy Paper 169, OECD, Paris. http://www.oecd-ilibrary.org/science-and-technology/wireless-broadband-indicator-methodology_5kmh7b6sw2d4-en.

Pew Internet and American Life Project. 2010. "Home Broadband Survey." Pew Internet and American Life Project, Washington, DC. http://www.pewinternet.org/Reports/2010/Home-Broadband-2010/Summary-of-Findings.aspx.

Qiang, Christine. 2009. *Broadband Infrastructure Investment in Stimulus Packages: Relevance for Developing Countries.* Washington, DC: World Bank. http://siteresources.worldbank.org/EXTINFORMATIONANDCOMMUNICATIONANDTECHNOLOGIES/Resources/282822-1208273252769/Broadband_Investment_in_Stimulus_Packages.pdf .

Sutherland, Ewan. 2009. "Telecommunications in Small Island Developing States." Paper presented at the 37th "Research Conference on Communication, Information, and Internet Policy," George Mason University School of Law, Arlington, VA.

Task Force on Financial Mechanisms for ICT for Development. 2004. *A Review of Trends and an Analysis of Gaps and Promising Practices.* Geneva: ITU, December.

TRAI (Telecommunications Regulatory Authority of India). 2010. "Consultation Paper on National Broadband." TRAI, New Delhi.

United Kingdom, Department for Business Innovation and Skills. 2009. *Digital Britain: Final Report.* London: Department for Business Innovation Skills. http://webarchive.nationalarchives.gov.uk/+/http://www.culture.gov.uk/images/publications/digitalbritain-finalreport-jun09.pdf.

UNCTAD (United Nations Conference on Trade and Development). 2010. "Partnership on Measuring ICT for Development: Core ICT Indicators." UNCTAD, Geneva. http://new.unctad.org/upload/docs/ICT_CORE-2010.pdf.

United States, FCC (Federal Communications Commission). 2009. *Broadband Adoption and Use in America.* Washington, DC: FCC.

———. n.d. *National Broadband Plan: Connecting America.* Washington, DC: FCC.

World Bank. 2009. "Advancing the Development of Backbone Networks in Sub-Saharan Africa." In *Information and Communication for Development: Extending Reach and Increasing Impact*, ch. 4. Washington, DC: World Bank.

———. 2010. "Strategic Options for Broadband Development in the Arab Republic of Egypt." World Bank, Washington, DC.

Vos, Esme. 2009. "Groningen, Netherlands Deploys Municipal Wireless Network." *MuniWireless*, April 15. http://www.muniwireless.com/2009/04/15/groningen-deploys-muni-wireless-network.

CHAPTER 3

Law and Regulation for a Broadband World

Throughout this handbook, we refer to the two primary components necessary to promote broadband development—one related to supply (the availability of and access to broadband networks, services, applications, and devices) and the other related to demand (the adoption and use of broadband). As the world moves to a converged information and communication technology (ICT) environment, countries are revisiting their traditional legal and regulatory frameworks and crafting new laws and regulations to address some of the supply and demand issues associated with developing broadband networks and services.

On the supply side, certain key legal and regulatory issues are being considered, such as determining how legal and regulatory licensing frameworks may facilitate voice, video, and data offerings. Other issues are related to spectrum management reforms, Internet interconnection, and infrastructure access policies. On the demand side, legal and regulatory issues are also arising. As more of our social, political, and economic transactions occur online, it becomes critical to ensure user trust and confidence. Policy makers are therefore considering measures to ensure users' privacy and rights online.

At this time, the legal and regulatory responses to address many of these issues are still being debated around the world. As broadband expands and its full potential is realized, a clearer picture may emerge. This chapter discusses the key policies and regulatory approaches that are being considered and implemented by policy makers and regulatory authorities to address some of these issues.

Licensing and Authorization Frameworks

Technological convergence in the telecommunications and broadcasting markets is hastened by the growth of broadband networks, since the higher speeds and larger capacities of broadband create new opportunities for operators to offer an array of services, including voice, data, and video. For example, two of the largest broadband network operators in the world, Comcast and Time Warner, began as cable television (TV) operators, but now derive substantial revenues from Internet and voice services, as well as from pay TV, particularly through their "triple-play" packages (Raja 2010). Broadband also supports the expansion of markets and competition as well as helping to reduce prices, improve the efficiency of service provision, and increase the variety of offerings for subscribers. To facilitate the supply of emerging wireline and mobile broadband networks, an enabling licensing framework is necessary.

Convergence and the distributed nature of networks and communications have unleashed a disruptive force across traditionally segregated industries that demands new, flexible, enabling responses (Benkler 2006). Traditional, service-specific regulatory frameworks have typically required separate licenses for wireline, wireless, and broadcasting networks as well as for different types of services. In many instances, operators have been prohibited from offering services outside their traditional, rigidly defined industry—even though new digital broadband technologies make this easily possible. For example, Internet Protocol television (IPTV) was restricted in the Republic of Korea until the IPTV Business Act of 2008 permitted telecommunications operators to offer television programs in real time over their broadband networks.[1] Within a year of enabling this converged technology and licensing three IPTV operators, Korea had more than 1 million IPTV subscribers.[2]

As this and similar cases demonstrate, distinctions between types of network infrastructure are becoming increasingly impractical in a converged environment. Thus policy makers and regulators in both the developed and developing worlds are enacting reforms to transform legacy regulatory regimes so that they can effectively address converged networks and services. These efforts generally have two key elements: (a) the introduction of the principles of technology and service neutrality and (b) the establishment of greater flexibility in key aspects of licensing and authorization frameworks, particularly the authorization of a wide range of networks and services under a single license. At the same time, there is expected to be greater reliance on broad competition law and regulation, as the historic restrictions contained in licenses and authorizations are progressively reduced.

Technology and Service Neutrality

Technology neutrality is based on the premise that service providers and network operators should be allowed to use the technology that best meets the needs of their network and the demands of their customers; such choices should not be dictated by governments. In the licensing context, technology neutrality means that different technologies capable of providing similar or substitute services should be licensed and regulated in a similar way.[3] In the broadband context, this means that broadband service providers abide by similar licensing processes and conditions regardless of whether they deliver services via wireless, digital subscriber line (DSL), fiber, cable modem, or other technology. However, a licensing framework that is generally considered technology neutral does not have to treat all providers in exactly the same way; it may treat certain broadband technologies or services differently. For example, the promotion of nascent services (for example, voice over Internet Protocol, or VoIP) using a light-handed regulatory approach may warrant departure from technology neutrality, at least on a temporary basis, to promote the development of those technologies. This also may be the case for wireless vs. wireline broadband technologies due to the need for separate spectrum authorizations and other spectrum-related matters, such as capacity constraints and interference.

Service neutrality is based on the similar premise that network operators should be allowed to provide whatever services their technology and infrastructure can deliver. In the past, due to the limitations of technology, networks were "purpose built." As information and communications became increasingly digitized, however, it became possible for different networks to support similar or substitute services. Thus, both cable and telecommunications networks can now support a wide range of voice, data, and video services. More relevant for developing countries, mobile service providers are increasingly able to offer such services as well. Given this convergence, constraining network operators' services based on old conceptions of technology is no longer appropriate. Adoption of more liberal licensing regimes allows companies to provide a wide range of services under a single license or authorization, which thereby enables the operator to take "cues from the market as to which services are most in demand or most cost-effective" (ITU 2004).

For example, Botswana, Ghana, Kenya, South Africa, Tanzania, and Uganda have already implemented technology- and service-neutral licensing frameworks. In Tanzania, the Electronic and Postal Communications Act, 2010, specifically incorporates both principles into the converged licensing framework, providing that "a licensee is authorized to provide any electronic communication service" (that is, service neutrality) and allowing the licensee to "use any technology for the provision of electronic communication services" (that is, technology neutrality).[4]

Together, technology and service neutrality recognize and facilitate technological convergence and promote new and innovative services and applications by reducing the number of licenses that an operator must obtain and expanding the variety and breadth of services an operator may provide. Neutrality may also contribute to reducing unnecessary or even contradictory regulatory obligations, such as different reporting standards and requirements provided under service-specific regimes. However, a country's licensing regime often requires substantial reforms from traditional service-specific licensing to a more unified licensing framework capable of accommodating technology and service neutrality.

New Authorization Options and Their Implications for Broadband

In light of the regulatory implications that flow from convergence and the transition to a next-generation network (NGN) environment, regulators have begun to adopt more unified frameworks based generally on one of the following approaches: (a) unified or general authorization or (b) multiservice authorization (ITU-D 2009a). Establishing some form of converged licensing framework that includes technology and service neutrality can be a key step for developing countries to foster the supply of broadband, increase investment, and improve the uptake of broadband.

Unified or General Authorizations

In principle, these authorizations are technology and service neutral, allowing licensees to provide all forms of services under the umbrella of a single authorization and permitting them to use any type of communications infrastructure and technology capable of delivering the desired service (figure 3.1). This is the most flexible approach, and it typically permits any number of operators to be authorized, except where scarce resources, such

Figure 3.1   General Elements of a Unified and General Authorization Framework

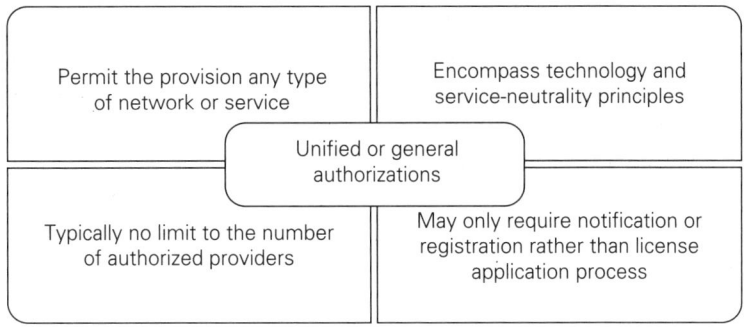

Permit the provision any type of network or service

Encompass technology and service-neutrality principles

Unified or general authorizations

Typically no limit to the number of authorized providers

May only require notification or registration rather than license application process

Source: Telecommunications Management Group, Inc.

as spectrum, are involved. In addition, this type of framework may only require registration or notification in order for the operator to begin offering services. The general authorization regime established by the European Union (EU) Authorization Directive in 2002, as amended in 2009, characterizes this type of framework (European Union 2002). Under that regime, a provider may offer any type of electronic communications network or service with a simple notification to the relevant national regulator. No license application or approval process is generally required.

Multiservice Authorizations

A multiservice licensing framework allows operators to offer a wide range of services under a single authorization and may also permit certain categories of licensees to use any type of communications infrastructure and technology capable of delivering the licensed services. However, the multiservice authorization framework is generally not as flexible or as streamlined as a general authorization approach: (a) there are multiple license categories rather than a single license category; (b) the various license categories may limit the number and types of services that may be provided; (c) licensees may be required to hold multiple licenses; and (d) rules may bar licensees from holding more than one type of license, which may stifle convergence if, for example, a telecommunications licensee is not permitted to hold a broadcasting license and therefore cannot offer video services.

Singapore has adopted a simplified variation of the multiservice licensing framework, which is based on two main types of licenses: facilities-based operator (FBO) and services-based operator (SBO) (IDA 2011a, 2011b; figure 3.2).

Figure 3.2 Example of Multiservice Licensing Framework in Singapore

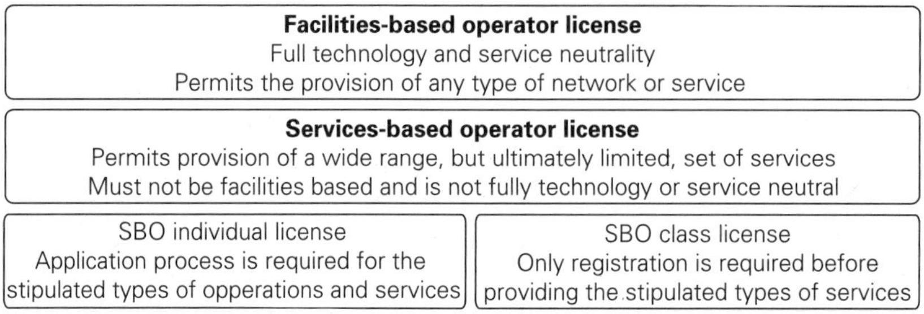

Facilities-based operator license
Full technology and service neutrality
Permits the provision of any type of network or service

Services-based operator license
Permits provision of a wide range, but ultimately limited, set of services
Must not be facilities based and is not fully technology or service neutral

SBO individual license	SBO class license
Application process is required for the stipulated types of opperations and services	Only registration is required before providing the stipulated types of services

Sources: IDA 2011a, 2011b.

Spectrum Management to Foster Broadband

In the past, as new technologies and services developed, legal and regulatory frameworks often evolved in a piecemeal fashion, with regulators often charging different fees, using different assignment mechanisms, and imposing different conditions on the various types of spectrum authorizations or licenses. However, these practices do not facilitate converged service offerings or maximize the value and use of spectrum, since new technologies enable multiple services and applications to be provided over one network, allow multiple services to be provided using the same spectrum, and enable the spectrum to be used more efficiently and intensively.

As a result, policy makers and regulators are looking to replace narrowly defined technical and service rules with more flexible assignments that allow providers to match their network and service. In today's broadband environment, access to spectrum is particularly relevant, given the anticipated likelihood that for many countries, particularly developing ones, wireless will be the primary vehicle for deploying broadband networks. For example, in Morocco, third-generation (3G) mobile broadband connections surpassed asymmetric DSL (ADSL) wireline connections in September 2009 and represented over 76 percent of the total Internet connections in the country as of March 2011 (Morocco, ANRT 2011). As a result of this trend, regulatory authorities and policy makers in many countries are looking at legal and regulatory reforms as necessary to facilitate the supply of wireless broadband services and the build-out of networks. Such policies include spectrum allocation and licensing, license terms and conditions (for example, coverage obligations), license renewals, and procedures to reclaim and reuse spectrum (for example, the transition from analog to digital television).

Spectrum Licensing Regimes

The process for licensing spectrum use typically depends on a country's general licensing regime for electronic communications services. The traditional approach that developed in many countries was to issue a service-specific license to cover both the network or service and the spectrum in a single document. For instance, prior to adoption of the unified access service license in India, the cellular license was one of 12 service-specific licenses under the traditional framework (TRAI 2003, 8). As is typical under a traditional licensing framework, tying the network, service, and spectrum license together may limit or eliminate the ability of licensees and consumers to capture the benefits of convergence.

More recently, converged licensing frameworks have developed that generally involve two authorizations: one covering the networks and services to be provided and another covering the spectrum.[5] Additionally, the network or service and spectrum licenses may be issued separately. If issued separately, the licenses to provide networks and services should be granted or registered simultaneously with the spectrum licenses to ensure regulatory certainty.

Regardless of the initial procedures for issuing spectrum licenses, spectrum licensing regimes for commercial services should be as flexible as possible, since limiting the flexibility of spectrum licenses can diminish the value of the broadband service and ultimately undermine the service provider's investment incentives (Kim, Kelly, and Raja 2010, 47). Regulators can introduce flexibility through rules that are technology and service neutral, allocating certain frequency bands for unlicensed or license-exempt use and using market-based assignment mechanisms, including spectrum trading.

Flexible-Use Technical and Service Rules

A key tool for promoting wireless broadband development is for governments to allow flexible use of spectrum, particularly through technical and service rules that enable wireless providers to offer any type of broadband service or application, including voice, video, and data. Flexible-use rules may be applied to both current and future commercial assignments to maximize the benefits of technological evolution and development of advanced services. For example, the EU's 1987 Global System for Mobile Communications (GSM) Directive reserved the 900 megahertz (MHz) band (890–915 MHz/935–960 MHz) for GSM networks and services only; however, this was revised in 2009 to permit greater flexibility in choice of technology and encourage the growth of mobile broadband in this band (European Union 2009c).

When considering adopting flexible-use rules for existing licenses, however, regulators should evaluate the potential competitive implications of such liberalization and the possible safeguards that would need to be put in place to address them. This includes determining whether this policy would place certain providers at a competitive advantage vis-à-vis their rivals or whether operators should be allowed to retain all or part of the liberalized spectrum. In case existing providers are allowed to retain the spectrum, the regulator should consider the possible mechanisms to control for potential windfalls (for example, regulatory obligations and fees). If some spectrum is to be released back into the market, the regulator should also consider the manner and timetable in which the assignment of such spectrum will take place. Regulators will also need to address the impact that flexible-use rules for broadband spectrum licensing will have on processes in their pipeline, including the assignment of various spectrum bands in a single process or the adoption of caps to facilitate new entry or make it possible for an operator to obtain an even blend of spectrum across different bands.

Cognitive radio technologies (CRTs) are also expected to lead to a significant increase in the flexible use of spectrum. A cognitive radio is able to sense and understand its local radio environment and to identify temporarily vacant spectrum in which to operate. At present, most attention relating to CRT is placed on opportunistic or unlicensed use (that is, identifying "unused" portions of spectrum using CRT and sharing the spectrum dynamically with existing users), but it is expected that in the future licensed operators may use CRTs to improve the management of their spectrum assignments. This represents significant opportunities to optimize the use of spectrum for the provision of bandwidth-intensive wireless broadband services and applications.

Spectrum Allocation and Assignment

As the deployment and adoption of wireless broadband increases, additional spectrum is widely expected to be needed to accommodate the demand of bandwidth-hungry broadband services, including video and data. For example, the average smartphone user generated 10 times the amount of traffic as the average non–smartphone user between 2009 and 2010. And the number of smartphone users is expected to grow substantially—it is anticipated that most people in the world will use mobile devices as their primary connection to the Internet by 2020 (GSA 2010).

Beyond introducing converged, flexible licensing frameworks, countries are also looking at the way (a) spectrum bands are planned and harmonized and (b) spectrum blocks are configured, assigned, and transferred. Wherever possible, a key initial step in promoting commercial wireless broadband networks and services is for regulatory authorities to adopt internationally harmonized band plans when considering the allocation and assignment of spectrum. This approach facilitates the commercial launch of broadband services by allowing providers to take advantage of scale economies in network equipment and devices, thus reducing the costs of deployment and, ultimately, the prices for consumers. International harmonization also facilitates the ability to offer roaming services.

Wireless broadband also requires additional bandwidth to be made available to keep pace with the high data rates needed to support bandwidth-hungry services and applications, such as video. The specific amount of spectrum will vary by country, depending on the current assignments and the expected growth in the demand for data services and traffic. Nevertheless, to deliver new data-intensive services and applications in a technically efficient and cost-effective manner and at the desired level of quality, providers will need to obtain additional spectrum. This is especially the case in large, densely populated urban areas in both developed and developing countries. For example, the U.S. National Broadband Plan seeks to make 500 MHz of spectrum available for broadband use by 2020, of which 300 MHz between 225 MHz and 3.7 gigahertz (GHz) should be made available by 2015.

In addition, the size of the spectrum blocks awarded to licensees may need to be revisited. For example, scalable, new International Mobile Telecommunications–Advanced (IMT-Advanced) technologies are best suited for wider blocks of contiguous spectrum, ranging from 2×15 or 2×20 MHz for paired spectrum and a minimum of 20 MHz for unpaired spectrum. Therefore, regulators are increasingly designating larger spectrum blocks for the provision of wireless broadband services. Recent assignments for IMT in countries such as Brazil, Chile, Costa Rica, Colombia, Denmark, the Netherlands, Norway, and Mexico highlight this approach.

Also relevant is the method of awarding spectrum. Increasingly, countries are using market mechanisms to assign spectrum use rights, particularly through auctions. Competitive award methods are generally viewed as more open, nondiscriminatory, and transparent than other assignment processes, such as administrative proceedings, and they provide an opportunity for new entrants. Auctions are also more economically efficient, since those willing to pay the highest price place the most value on spectrum, while the winning bids provide additional revenues to governments. For example, India's 2010 auction of 3G spectrum garnered over US$14.5 billion for the government (Kinetz 2010).

Spectrum trading (also known as "secondary markets") is another method that facilitates aggregation of spectrum to meet future data traffic demand requirements by permitting existing licensees to transfer all or a part of their spectrum assignments to third parties with little or no government involvement in the process. Implemented in Australia, New Zealand, and the United States, spectrum trading has allowed late entrants to the mobile market to obtain spectrum rights, which can reduce constraints on new entrants with regard to the timing of their market entry. In the absence of spectrum trading, potential entrants and existing operators seeking to build out their networks further must wait for the government to award new spectrum assignments. Ultimately, spectrum trading provides the opportunity for secondary markets to emerge that can improve the rollout of new services, increase the potential for competitive service provision, and encourage investments in the sector (for an in-depth study of spectrum management and reform in developing countries, see Wellenius and Neto 2008).

Although placing greater emphasis on market forces and spectrum trading offers many advantages over the traditional models of spectrum management, ineffective regulatory environments may allow incumbent or dominant operators to control key, high-value spectrum bands. This could result in spectrum hoarding and concentration of the wireless broadband market. As such, there is a trade-off between operators having sufficient spectrum and monopolizing the available spectrum. Many countries seek to mitigate this through build-out obligations, while others impose spectrum caps or set aside spectrum blocks for new entrants. However, at least one study argues that spectrum caps in Latin America may hinder the development of mobile broadband.[6] Overall, making as much spectrum available as possible through transparent and nondiscriminatory procedures is a key step toward ensuring that operators are able to meet future wireless broadband demands.

Spectrum License Renewal

As spectrum licenses granted in the 1990s and early 2000s reach the end of their initial terms, license renewal policies will become an increasingly relevant regulatory issue to fostering investment in wireless broadband. In establishing renewal policies, policy makers and regulators should strive to promote investors' confidence and provide incentives for long-term investment while preserving the flexibility of the regulatory process to accommodate market and policy developments (Guermazi and Neto 2005, 2). Legal certainty is of utmost importance to create an environment conducive to investment and the technological upgrades required to deploy wireless broadband services.

Principle and Procedure for Renewal

While legal regimes vary, most frameworks have adopted a system based on the "presumption of renewal" or "renewal expectancy." Under a presumption of renewal, the licensing authority must renew a license as long as the licensee has fulfilled its obligations and has not violated the law or the terms of its license. In general, renewal expectancy provisions seek to provide regulators with the flexibility to review and adjust license conditions in response to technological developments and market conditions, while providing the regulatory certainty necessary for licensees to continue investments. In Canada, for example, the licensing framework provides a high expectation of renewal unless a breach of license condition has occurred, a fundamental reallocation of spectrum to a new service is required, or an overriding policy need arises.[7] Similarly, in Antigua and Barbuda, there are both a renewal expectancy for the same period as the original license and a requirement for the regulator to provide 180 days written notice of its intention not to renew. An appeals process to the regulator is also established.

Other countries rely on automatic renewals. For example, Portugal's Decree-Law no. 151-A/2000, regarding the use of radio communications, automatically renews licenses every five years unless the regulator provides at least 60 days written notice to the licensee stating the reasons for nonrenewal.[8] In the Dominican Republic, automatic renewal is warranted in the absence of a negative finding from the regulator. Some countries, such as Australia, provide less long-term certainty to incumbents, opting instead for a legal presumption that, when a spectrum license expires, the license will be reassigned via a price-based method (for example, auction), unless it is in the public interest to do otherwise.[9]

Change in License Conditions and Review of License Fees

Renewal expectation, however, does not necessarily imply that licenses will be renewed under the same terms as the original license. In setting the terms and conditions of license renewal, regulators must strike the right balance between giving certainty to operators and investors and ensuring that license conditions reflect current policy objectives, respond to technological and market developments, and consider the consumers' needs. If an appropriate balance is not struck, proposed changes to licenses and the review of associated fees in particular can become highly controversial.

For example, France's regulator, ARCEP (Autorité de Régulation des Communications Électronique et des Postes), initiated a public consultation in 2003 on the renewal of GSM licenses, which were set to expire in 2006 and 2009 (France, ARCEP 2003). ARCEP originally announced that it would charge a 5 percent progressive tax on annual turnover, but licensees protested that this amount would harm investment and the development of services. The regulator issued its decision in March 2004 after comments from licensees demonstrated that the high annual fees would negatively affect investment and the market generally (France, ARCEP 2004b). Although the government set out new licensing fees, as well as higher coverage obligations and quality of service levels, the annual fees were substantially less onerous (France, ARCEP 2004a). The new fees required GSM licensees to pay €25 million annually and 1 percent of annual turnover (France, Ministry of the Economy, Finance, and Industry 2004). A similar controversy surrounded the renewal process for mobile licenses in Bangladesh, where a proposal to extract large renewal fees from existing licensees and the imposition of additional obligations created significant opposition from service providers, apparently causing the government to abandon the idea.[10]

License-Exempt (Unlicensed) Spectrum

In an effort to provide maximum flexibility for innovation and lower entry costs for some types of ubiquitous wireless devices, policy makers and regulators in many countries have set aside certain bands exclusively for license-exempt (also known as unlicensed) uses. In other bands, license-exempt devices and licensed services share frequencies. Many commonly used wireless devices, such as cordless phones, garage door openers, and smart meters for water and gas metering, depend on unlicensed spectrum. In addition, municipal wireless networks also use unlicensed spectrum to create mesh networks that cover downtown areas or even entire cities.[11]

Wireless Fidelity (Wi-Fi) is perhaps the most well-known and widespread example of unlicensed use. Many countries have opened the 2.4 and 5.8 GHz spectrum bands for unlicensed use, allowing for the tremendous growth of Wi-Fi devices. According to ABI Research, consumer devices with Wi-Fi functionality surpassed 770 million units in 2010, an increase of nearly 33 percent compared to 2009.[12] Over half of all Wi-Fi devices are mobile handsets and laptop computers; however, a wide and expanding range of equipment is equipped with Wi-Fi, including cameras, fax machines, and printers. Furthermore, in many countries there has been significant development of "Wi-Fi hotspots" in cafés, libraries, universities, and other public areas where users can access the Internet for free or at low cost.

An important emerging use for Wi-Fi is as a complement to commercial wireless networks. As wireless broadband services spread, the demand placed on mobile network capacity is increasing exponentially, putting significant strain on available resources. The combination of licensed and unlicensed spectrum usage—Wi-Fi in particular—is becoming a key complement of the wireless broadband experience, allowing users to offload their traffic from mobile operators' networks in certain circumstances, thereby reducing potential congestion and enhancing broadband access.

Technical and service rules for unlicensed spectrum typically specify that unlicensed devices must operate at low power and may not cause harmful interference to a licensed user. In addition, unlicensed devices must generally accept interference from licensed users and other unlicensed devices. Although interference and economic issues may make it difficult or impossible to replace all spectrum licenses with unlicensed use, opening bands to unlicensed devices can support broadband development through the growth of new technologies, efficient use of spectrum, and the entry of new network, service, and applications providers.

Spectrum Refarming and the Digital Dividend

In order to maximize the ability to offer wireless broadband, particularly where spectrum is intensively used, many countries are engaging in spectrum refarming, whereby existing spectrum users are moved out of a band to allow for new broadband uses. The refarming process is often lengthy and costly, since it typically involves negotiations with existing private and public spectrum holders and potential licensees and may also include compensation for the existing licensees to change spectrum bands. As such, it is important to conduct a thorough spectrum inventory to identify unused or underutilized spectrum as well as heavily used bands before implementing a refarming process. In many developing countries, refarming may be less necessary in the near future since available spectrum may be sufficient and more easily allocated for wireless broadband services.

One of the most promising and active areas of spectrum refarming is the result of the transition from analog to digital television. As countries around the world prepare for or complete the transition to digital terrestrial television (DTT), they are examining procedures for reallocating the spectrum that becomes available as broadcasters vacate the 700 MHz or 800 MHz bands, depending on the region. This freed-up spectrum, which is widely known as the "digital dividend," offers excellent propagation characteristics for mobile broadband services by providing an ideal balance between transmission capacity and distance coverage. This means that the digital dividend spectrum is well suited to providing mobile services to rural areas as well as to providing effective in-building performance in urban areas. For countries where rural coverage is an important policy goal, this is a notable advantage.

However, given the various timelines for the DTT transition—some countries have completed the transition, while others are planning for the analog switch-off (ASO) between 2011 and 2020)—many countries are only beginning to consider rules and timeframes for refarming digital dividend spectrum. Many countries are waiting to award digital dividend spectrum until after the ASO is completed and the spectrum is no longer encumbered by broadcasters. However, some countries, such as the United States, Colombia, and Peru, have awarded or are planning to award the digital dividend ahead of their ASO dates. Regardless of the approach, considerable international and regional harmonization is under way, including by the EU and the Asia-Pacific Telecommunity. Box 3.1 provides an overview of the DTT and digital dividend activities around the world.

IP-Based Interconnection

Interconnection of different networks is critical to ensure a competitive communications market. It is fundamental for service providers to ensure that their users have the ability to connect with users of any other network or service provider. As the Internet expands and becomes more geographically ubiquitous and as traffic increases, more efficient IP-based Internet interconnection will be required. This is especially relevant for developing countries, where lack of interconnection facilities means that Internet traffic originating there is mostly subject to "tromboning" (that is, using international transit facilities to deliver local traffic).[13] Policies to facilitate national or regional Internet exchange points (IXPs), the physical infrastructure where Internet service providers (ISPs) exchange Internet traffic between their networks, will play a crucial role in ensuring more efficient and cost-effective Internet interconnection in these countries.

Similarly, as the transition toward IP-based NGNs proceeds, questions will arise regarding the manner and terms under which IP-based interconnection will take place between different types of networks and at different functional levels of the network. Especially relevant are issues relating to future wholesale charging mechanisms that may apply to converged broadband networks. The following sections address current trends and expected regulatory developments relating to these issues.

Internet Interconnection and IXPs in Developing Countries

Historically, the exchange of Internet traffic has been focused in developed countries. In the early years of the Internet, traffic was routed and exchanged

Box 3.1: Summary of the Digital Television Transition and Digital Dividend Activities around the World

Digital television transition timelines vary. Developed and developing countries alike have been focusing on the digital TV transition and most have adopted ASO dates or have at least set a goal for completing the transition by a certain year. While countries such as Germany, Finland, Luxembourg, Sweden, the Netherlands, and the United States have already completed the ASO, other countries are focusing on 2015–20 to complete their transitions.

Consideration of the digital dividend is slow. Less progress has generally been made toward developing rules and timeframes for the award of digital dividend spectrum. While several consultations are expected to begin over the next two years, including Chile, Colombia, Ireland, Mexico, and the United Kingdom, most countries have not established technical and service rules or award processes for the digital dividend spectrum, particularly in developing countries.

Approaches to assigning digital dividend spectrum vary. Generally countries are waiting to award the digital dividend spectrum until after the ASO is completed and the spectrum is unencumbered by broadcasters. For example, Finland's ASO in the 800 MHz band was in 2007, but licenses still

have not been awarded. Some countries, however, are following the U.S. approach and are awarding 700 MHz spectrum ahead of completion of the digital TV transition. Ireland is likely to auction its digital dividend spectrum in 2011, but licensees will probably not be permitted to use their new frequencies until completion of the ASO in 2013. Colombia, Mexico, and Peru are also considering auctioning 700 MHz spectrum before the ASO date.

International and regional harmonization is under way. There have been significant international and regional efforts to harmonize the digital dividend spectrum and develop common band plans. The International Telecommunication Union (ITU)'s 2007 World Radio Communication Conference identified spectrum in the 698–960 MHz band for IMT, and the ITU is finalizing a revision to ITU-R Recommendation M.1036-3, which specifies plans for all bands, including the digital dividend, identified for use by IMT. Regionally, the EU and the Asia-Pacific Telecommunity have agreed on common band plans for their member states (the two plans are not the same). To date, there are no formal common band plans for the Americas, Africa, or the Middle East.

Source: Telecommunications Management Group, Inc.

mainly in the United States. As Internet access has expanded and the amount of content available has increased, the exchange of Internet traffic has been distributed to other developed countries in Europe and Asia through the creation of national and regional IXPs (Kende 2011, 25). In the case of developing countries, while IXPs have been progressively implemented and peering is occurring at the national level, the amount of traffic that is exchanged within developing regions is still very small. Most of the traffic is still hauled out of the region for switching and then sent back into the region for delivery.

IXPs in developing countries are important for Internet interconnection for several reasons. By providing an interface for the exchange of local and regional traffic, IXPs facilitate a more efficient and cost-effective management of international bandwidth. Because of their small volume of traffic, ISPs in developing countries mostly have to rely on transit agreements, since the largest providers do not have incentives to enter into shared-cost peering agreements with them. Due to the charging mechanisms for international Internet transit, this means that the developing-country ISP will ultimately bear the costs of outbound and inbound traffic. Local peering through IXPs at the national or regional level helps to resolve this problem and reduces the costs of Internet access for consumers in developing countries.

More local interconnection, in turn, allows for the provision of more reliable services, with lower latency that then can support multiple, innovative, time-sensitive applications. For example, for African ISPs, tromboning adds an estimated 200 to 900 milliseconds to each transmission. This added latency can impede the development of new services, such as Internet telephony, streaming audio and video, video conferencing, and telemedicine. By interconnecting at a local IXP, two ISPs (located near to each other) can overcome this problem and route traffic to each other's networks in 5 to 20 milliseconds.[14]

As noted, IXPs are now being implemented in some developing countries. Before 2002, there were only two IXPs in Africa, with this number increasing to 10 by 2003. By December 2010, there were 20 IXPs distributed among African countries. While this represents significant progress, the great majority of Internet traffic from Africa, around 85 percent, still relies on connections to Europe; just 1 percent of the traffic being exchanged stays within the region.

From a regulatory perspective, a series of barriers can hinder Internet interconnection and the establishment of IXPs in developing countries (McLaughlin 2002). As discussed later in this chapter, Internet interconnection has developed under market-based mechanisms and without the need for regulatory intervention. However, regulators' attempts to extend their mandates to encompass Internet interconnection may result in unwarranted regulation and create disincentives for the deployment of IXPs. These include, for example, legal restrictions that prohibit the deployment of nonregulated ICT facilities, such as IXPs. Unduly restrictive or burdensome licensing regimes may also limit the deployment of IXPs. Similarly, exclusive rights for the provision of international connectivity, which some countries maintain, can also impede efficient Internet interconnection.

In some cases, lack of appropriate regulation of the inputs required to implement effective IXPs, such as national backbone connectivity, may result in above-cost rates for wholesale services. For example, high costs of leased lines can significantly affect an IXP's viability. In addition, deficiencies in regional broadband connectivity play a role in the continued low levels of intraregional Internet traffic exchanged in developing regions, like Africa (Stucke 2006). Lack of relevant local content also affects the extent to which traffic is peered within national or regional IXPs.

Box 3.2 presents a case study of the implementation of the first IXP in Kenya, which illustrates some of the legal and regulatory difficulties outlined above and how they were overcome.

Box 3.2: Challenges and Successes of Implementing an Internet Exchange Point in Kenya

Prior to the Kenya IXP (KIXP), all Internet traffic in Kenya was exchanged internationally, and about 30 percent of upstream traffic was to a domestic destination. In early 2000, the Telecommunications Service Providers Association (TESPOK), a nonprofit ISP group, undertook an initiative to implement and operate a neutral, nonprofit IXP for its six members, launching the KIXP in Nairobi in November 2000. Almost immediately, Telkom Kenya filed a complaint with the Communications Commission of Kenya (CCK) arguing that the KIXP violated its monopoly on the carriage of international traffic. Within two weeks, the CCK concluded that the KIXP required a license and ordered it to be shut down as an illegal telecommunications facility.

After intensive efforts, CCK granted TESPOK an IXP license in November 2001. In February 2002, the KIXP went live again and was relaunched that April, with five ISPs actively exchanging traffic. Within the first two weeks, latency was reduced from an average of 1,200–2,000 milliseconds (via satellite) to 60–80 milliseconds (via KIXP). Monthly bandwidth costs dropped from US$3,375 to US$200 for a 64 kilobits per second (kbit/s) circuit and from US$9,546 to US$650 for a 512 kbit/s circuit.

Currently, the KIXC has 31 members peering traffic, some of which are not ISPs, such as UNON, National Bank, and the Kenya Revenue Authority. TESPOK launched a second IXP in Mombasa in August 2010 to facilitate local peering further. While the throughput of traffic exchanged at the KIXP is low relative to major IXPs (at around 100 megabits per second [Mbit/s]), KIXP ranks among the top 15 IXPs in terms of growth (around 150 percent year-on-year increase in recent years).

Sources: KIXP at http://www.kixp.or.ke; Jensen n.d.; Kende 2011.

IP-Based Interconnection: Wholesale Charging Arrangements

Despite the increasing physical and logical integration between legacy public switched telephone networks (PSTNs), public land mobile networks (PLMNs), and all-IP networks, two separate models are still typically used for exchanging traffic in these networks. Internet traffic is exchanged using IP-based interconnection and relies on privately negotiated peering and transit agreements. PSTN and PLMN traffic, however, may be exchanged using a combination of switched and IP-based interconnection, but it is normally subject to regulation and typically falls within two main wholesale charging arrangements: calling party network pays (CPNP) and bill and keep (BAK).

As convergence toward NGNs advances, these differences create potential arbitrage opportunities between regulated and unregulated services and lead to potential competitive distortions (BEREC 2010, 8). Regulatory authorities are therefore considering what reforms in wholesale charging mechanisms, if any, should be implemented at the national level for termination services to enable IP-based services and broadband further. While it is not clear which wholesale charging arrangements will prevail, some authorities are expecting that a uniform wholesale charging mechanism for IP-based interconnection may emerge in the future.

Current Wholesale Charging Arrangements

The majority of countries around the world use CPNP for PSTN-PLMN interconnection at the wholesale level. Under this system, the originating network is required to pay a charge, generally per minute or per second, to the terminating network for the traffic exchanged. An alternative approach is BAK, which is used for PLMN in countries such as the United States, Singapore, and Canada and is a system where interconnecting operators generally do not charge each other for terminating calls. These terms are equivalent to negotiating termination rates equal to zero and typically include reciprocity obligations, meaning that the same terms are applicable to both parties to the agreement. Under BAK, the costs associated with call termination may in some cases be recovered from the service provider's own subscribers as that provider sees fit—for instance, by levying a charge for calls received.

Efficient IP-based interconnection in the Internet has been achieved for the most part without the need for regulatory intervention. Since no single entity has the ability to connect to all of the networks that form the worldwide Internet, a series of indirect interconnection (transit) and direct interconnection (peering) arrangements have developed to ensure that traffic will reach its intended destination. In Internet interconnection, the combined framework of transit and peering, together with the IP packet routing protocols, removes the a priori case for regulation based on the termination monopoly present in PSTN-PLMN interconnection under CPNP systems. For example, if an ISP denies direct interconnection (peering) to another ISP, the latter ISP is generally capable of accessing customers of the former, although at different costs, as long as it has an indirect (transit) agreement with a third party.[15] This same result is not generally possible in the circuit switched environment. If the PSTN-PLMN provider refuses interconnection, competitors generally cannot terminate calls to its subscribers.

Future Charging Mechanisms

In the long run, the differences in interconnection charging arrangements will not likely be sustainable or efficient in a converged NGN environment, where more traffic will be IP based. Price differences between regulated and unregulated interconnection services result in arbitrage opportunities and potential market distortions. Therefore, a uniform wholesale charging system may be needed for future NGN interconnection. This could be based on the Internet economic model (Marcus and Elixmann 2008, 114), the PSTN-PLMN model, or some third option resulting from a combination of both (European Commission 2009a, 32). Others emphasize that, although NGNs and the Internet use IP as a common technology and are converging in the marketplace by offering similar or substitute services, they are organized differently and so remain separate and distinct, even though they share the same transmission infrastructure, such as fiber networks (Tera Consultants and Lovells 2010, 79–92). Consequently, it is argued that the two types of networks will not converge since the Internet is a collection of "open networks" and NGNs are a collection of "closed" networks (that is, packets cannot be allowed across the interconnection point unless they are authorized), and hence there is no convergence-based argument in favor of a uniform charging system for NGNs based on BAK.

Despite this, there are some early indications that future wholesale price mechanisms may resemble IP network pricing, that is, PSTN-PLMN per minute or per second pricing may migrate to pricing based on barter arrangements (for example, BAK) or on capacity-based interconnection (CBI). A recent attempt by the Polish regulatory authority to lead regulation in the other direction (that is, regulating the terms, conditions, and prices for Internet peering and transit services using tools similar to those applied to PSTN-PLMN) met with significant opposition from the European Commission (EC) and was eventually discarded in March 2010 (European Commission 2010a).

Similarly, the Body of European Regulators of Electronic Communications (BEREC) has recently put forth proposals for a single terminating charging mechanism, specifically a shift toward BAK, which it believes will benefit networks in a converged, multiservice, NGN IP-based environment. If implemented in the future, this approach would result in wholesale arrangements similar to those used under Internet peering agreements. In the United States, the National Broadband Plan provides for the Federal Communications Commission (FCC) to adopt a framework for long-term interconnection reform that creates a glide path to eliminate per minute interconnection charges, while providing carriers an opportunity for adequate cost recovery and establishing interim solutions to address arbitrage. Pursuant to this mandate, in 2011 the FCC began consulting on a major overhaul of the interconnection regime in the United States, noting the need to move away from per minute charges, which "are inconsistent with peering and transport arrangements for IP networks, where traffic is not measured in minutes" (United States, FCC 2011, para. 40).

As policy makers consider ways to reform the interconnection regime to enable broadband development, one of the issues to consider is that termination rates have traditionally been a significant revenue source for PSTN-PLMN operators in many countries. This is especially relevant for developing countries in the case of international voice traffic, where incoming calls significantly exceed outgoing calls. Where termination is a major source of revenue, providers may have the incentive and ability to advocate for maintaining wholesale termination arrangements subject to the current switched model (or some variation similar to the current model), notwithstanding the fact that the underlying technical and market drivers will likely have changed. If call termination rates remain high, many PLMN and some PSTN operators may have incentives to choose not to evolve their networks to IP-based interconnection (European Commission 2009a, 32). This could have a detrimental impact on the development of converged broadband networks.

However, two factors may favor the transition toward NGNs and IP-based interconnection. First, as networks converge toward NGNs and data services become increasingly dominant, per minute costs for voice services are expected to fall. Second, the ongoing worldwide trend toward regulating termination rates to reflect the underlying incremental costs of termination, especially for PLMN operators, has resulted in a significant reduction in termination rates in many countries. For example, recent regulatory proceedings in countries such as Colombia, Kenya, Mexico, and Nigeria have reduced rates to levels comparable to those prevalent in the EU. As BEREC notes, "The lower the costs per minute and the closer they are to zero, the less difference between CPNP and BAK." This may also facilitate a transition to IP-based interconnection in many countries.

Access to Infrastructure

The Regulation versus Investment Debate

In designing policies to foster long-term, facilities-based competition, regulators are tasked with balancing the objective of promoting competition and entry with the need to maintain incentives for investment in new infrastructure and innovation. This entails identifying facilities that are not easily duplicated (that is, bottlenecks) and determining if they are capable of affecting competition in downstream (that is, services) markets. Such a determination would call for the regulation of such bottlenecks to give access to competitors on a nondiscriminatory basis and at cost-based prices, as fostering their duplication would either deter entry or result in a socially wasteful expenditure of resources. The success of such policies ultimately tends to pivot on the regulated prices and terms of access to bottlenecks.

In the absence of functioning market mechanisms, getting access prices just right is a huge challenge for regulators and will affect the incentives of both new entrants and incumbents. If prices are too low, entrants will have no incentive to invest in their own infrastructure, even when it is economically viable and efficient for them to do so. If access prices are too high, competitors either will not enter the market or will choose to deploy their own networks, resulting in inefficient duplication of networks. Conversely, incumbents may refrain from future investment in their networks if their facilities are open to competitors at low rates, as any advantage derived from these investments would be available to rivals, while risks associated with such investment would be borne exclusively by the incumbent.

Regulating Bottlenecks in the Broadband Supply Chain

Supplying broadband services involves a combination of network elements, processing, and business services that can be thought of as the broadband supply chain. More fully described in chapter 5, this supply chain can be divided into four main components: (a) international connectivity, (b) domestic backbone, (c) metropolitan connectivity, and (d) local connectivity. Bottlenecks in any of the links of the chain will stifle competition and the development of broadband. Hence, effective regulatory frameworks must identify and address such instances of market failure in a timely and effective manner.

International Connectivity

As electronic communications traffic—particularly Internet traffic—enters and leaves a country, it is typically routed through one or more international facilities, including submarine cables, cable landing stations, and international gateways.[16] Since international facilities provide the entry and exit point for voice, data, video, and other broadband services, they can become bottlenecks if access and traffic are restricted or prices are set above costs.

As the adoption of broadband services and applications increases, demand for international bandwidth also rises. Between 2002 and 2009, international bandwidth usage increased by 60 percent a year, with the strongest demand growth taking place on links to Africa, Latin America, and Middle Eastern countries, which experienced annual growth rates of over 74 percent during this period.

The most efficient way to lower costs and keep pace with demand is through liberalization and promotion of competition among facilities that provide international connectivity, in particular, international gateways, submarine cables, and landing stations. As such, it is important to ensure that there is more than one international carrier and international gateway and, where possible, that there are redundant international cables and other facilities linking a country to competitive global communication networks. For example, Nigeria supported facilities-based competition in the international connectivity market through the introduction of a unified access service license in 2006, which allowed licensees to "construct, maintain, operate, and use an international gateway" and networks consisting of any type of technology, including wireless or wireline systems (Singh and Raja 2010, 58). While it could be argued that the Nigerian Communications Commission's (NCC) hands-off approach led to a long period of monopoly control by the incumbent provider, NITEL, over the only submarine cable landing in Nigeria, the NCC recently found that a highly competitive market with multiple cable systems is developing (box 3.3; see NCC 2010).

Facilities-based competition in the international connectivity markets may not be feasible in all developing countries, especially those that generate small amounts of traffic. Also, landlocked countries or isolated small island developing states (SIDSs) may not have access to submarine cables and may have to rely on the use of alternative technologies, such as satellites, that often carry a higher price premium.

For countries without a well-functioning international connectivity market, targeted ex ante regulation may be required to address market failure (Hernandez, Leza, and Ballot-Lena 2010). Some countries, such as India, Colombia, and Singapore, have adopted various obligations on international gateways, landing stations, and submarine cable systems (for India, TRAI 2007a; for Colombia and Singapore, IDA 2008). In Colombia, for

Box 3.3: Competition Analysis in the International Internet Connectivity Market in Nigeria

In its 2010 review of competition in the international Internet connectivity market, the NCC found that this market was sufficiently competitive on a forward-looking basis and therefore did not require ex ante regulatory intervention. This determination was based on an expected increase in facilities-based competition by 2012, stemming from the landing of four additional submarine cables, one of which is to be operated on an open-access basis.

In its analysis, the NCC recognized that for the better part of the last decade the market had been dominated by NITEL, which since 2011 was the monopoly operator of Nigeria's only submarine cable, the South Atlantic 3/West Africa Submarine Cable (SAT-3/WASC). During this time, competing providers added only limited extra capacity of their own, mostly via satellite links and limited terrestrial links. At the time of the market analysis, four new submarine cables were scheduled to commence service in Nigeria: two in 2010 (Globacom-1 and Main One) and two more within the next two years (the West Africa Cable System in 2011 and the Africa Coast to Europe in 2012). The NCC noted that the new cables would result in a 33-fold increase in Nigeria's international bandwidth and significantly change the competitive dynamics in the market. As a result, it concluded that any market power NITEL had been able to exercise in the past should be resolved as competitors enter the market.

Source: Telecommunications Management Group, Inc.

example, after conducting a review of wholesale inputs for broadband Internet access, the regulator found that cable landing stations constituted essential facilities and required landing station operators to provide access to their facilities on nondiscriminatory terms and to publish a reference access offer.[17]

Self-regulation can also be a tool for reducing costs and increasing access to facilities required for international connectivity. Consortium agreements for submarine cable systems, for example, are progressively including non-discrimination and open-access clauses, whereby third parties are guaranteed access to facilities and capacity at terms comparable to those offered to the facilities' owners or subsidiaries. For instance, the Eastern African Submarine Cable System (EASSy), which runs from South Africa to Sudan with connections to all countries along its route, includes such safeguards. Launched in 2010, EASSy allows any consortium member to sell capacity in any market in the region to licensed operators on nondiscriminatory terms and conditions (Williams 2008, 42).

Domestic Backbone

Constituting the second level of the network element supply chain, a country's high-capacity domestic backbone network is essential for broadband connectivity since it provides the link from international gateways to local markets as well as domestic connectivity between major cities and towns. However, backbone networks require extensive investments. A major impediment to reducing these costs, particularly in many developing countries, relates to vertical integration in which the backbone network providers are vertically integrated with the local access network operators. This results in a single end-to-end provider that can wield great market power. As such, other service providers may not have access to the backbone or may face high costs for interconnecting, a problem addressed in growing debates on open network access.

From a regulatory perspective, the first step toward facilitating competition in vertically integrated networks is to ensure a liberalized market. In some countries in Sub-Saharan Africa, for example, mobile operators are prohibited from using the incumbent's network for backbone services, resulting in slow growth in broadband infrastructure. The second step toward increasing competition may entail targeted, ex ante regulations requiring the backbone network provider to offer network capacity on a wholesale, open-access, and nondiscriminatory basis to downstream providers. Alternatively, some countries are setting up national backbone operators that only provide wholesale broadband services on an open-access basis in order to prevent any vertical integration. This scheme is being implemented or proposed in countries such as Australia, Brazil, Colombia, Singapore, and South Africa. However, public financing of national backbones should not crowd out private investment or distort competition. Moreover, where a public subsidy is provided to a backbone broadband network, open-access obligations should be imposed.

Cross-sector coordination is also relevant to the efficient deployment of national connectivity. Fiber optic networks are usually built along existing infrastructure networks such as roads, railways, pipelines, or electricity transmission lines. Most of the cost of constructing fiber optic cable networks along these alternative infrastructure networks lies in the civil works. These costs represent a major fixed and sunk investment, increasing the risks faced by network operators. By lowering the cost of access to these infrastructure networks and reducing the risk associated with it, governments can significantly increase incentives for private investment in backbone networks. One way to reduce costs is to make rights-of-way readily available to network developers by simplifying the legal process and limiting the fees that can be charged by local authorities. Additionally, governments can provide direct access to existing infrastructure that they own or control. For example, a railway company could partner with one or more operators to build a fiber optic cable network along the railway lines. In January 2011, for example, Serbian Railways and PTT Srbija agreed to construct telecommunications infrastructure jointly along Serbian Railway's corridors, totaling 2,031 kilometers.[18] The United States, for example, has had a policy since 2004 that assists telecommunications providers seeking access to rights-of-way on federal lands (United States, White House, Office of the Press Secretary 2004).

Metropolitan Connectivity

Metropolitan connectivity, also referred to as the "middle-mile" or "backhaul" infrastructure, connects towns to the backbone infrastructure or remote wireless base stations and then to the operators' core network. Competitive and well-functioning wholesale markets for backhaul capacity (for example, leased lines) are a critical component of broadband diffusion and adoption. Developing countries are beginning to focus on core backbone and backhaul networks as a means to increase broadband deployment. For example, South Africa established a state-owned fiber-based infrastructure provider, Broadband Infraco, to provide national backhaul connections on a wholesale basis.[19] Brazil has also begun focusing on backhaul by entering into an agreement with five wireline operators to build out broadband backhaul networks to 3,439 unserved municipalities in exchange for being relieved of existing obligations to install 8,000 dial-up-equipped telecenters.

Particularly for rural and remote areas, wireless technologies may be the most practical solution for high-capacity backhaul for mobile broadband. A study from ABI Research notes that the global revenues from wireless backhaul leasing are expected to increase fivefold between 2009 and 2014 as operators upgrade to Long-Term Evolution (LTE) and traffic demands on mobile networks rise. Recognizing the importance of backhaul for mobile broadband, the Telecommunications Regulatory Authority of India (TRAI) recommended to the Ministry of Communications that license conditions be amended in order to allow service providers to share their backhaul links from base transceiver stations (BTSs) to base station controllers (BSCs), noting that such sharing should be permitted via wireless and optical fiber links (TRAI 2007b, 19–20). TRAI maintained that, particularly where traffic from BTSs to BSCs is low in rural and remote areas, backhaul sharing would boost coverage, reduce maintenance efforts, and lower costs.

Local Connectivity

Local access networks, also called the "last mile," refer to the links between the local switch and the consumer. This last link in the broadband supply chain has garnered much attention in recent years, as countries seek to expand service into unserved or underserved areas and to promote competition between operators at the retail level. Unlike other parts of the supply chain, local access regulation can be divided into two distinct areas of policy based on technology: wireline and wireless. Although the goals of policy makers are the same in each case—expand network availability and promote competition—the approaches must be tailored to the unique opportunities and constraints entailed in each technology.

Wireline networks. The local access segment (the "local loop") of the wireline network has historically been built and controlled by the incumbent provider of the PSTN. For many years, it was assumed that the local loop services were a "natural monopoly" because they tend to be the most difficult and costly part of the network for alternative operators to replicate. However, as cable networks and commercial wireless services began competing with traditional telecommunications operators, policy makers began reexamining the possibility of facilities-based competition or otherwise promoting service-based competition in the local loop. The degree and extent of regulatory intervention in access networks, particularly on the wireline side, depend on the legacy endowment of infrastructure of each country. In more developed markets, regulation has ranged from a light-touch approach to more extensive restrictions and obligations, such as local loop unbundling (LLU; see chapter 5 for a technical description of how LLU works). However, in developing countries without significant wireline (broadband) infrastructure at the local level, such obligations may have limited impact.

LLU obligations require the incumbent to provide access to exchanges and the physical local loop network so that new market entrants can offer services directly to customers without having to reproduce the incumbent's network. LLU may be used as a surrogate for infrastructure competition or as a way of inducing price competition between facilities- and services-based competitors. The main advantage of LLU is that it permits much faster market entry than would be possible if entrants were obliged to construct their own networks. The main disadvantage is that it can be a disincentive to fresh infrastructure investment by the incumbent operator (for instance, in deployment of a fiber optic network), especially in developing countries where the local loop is not yet fully built out.

LLU has been widely implemented in Europe, where it was initially required by a regulation of the European Commission in 2000 (European Union 2000). It has been credited with stimulating intramodal competition in some countries. Many other countries around the world have also adopted LLU obligations (Berkman Center for Internet and Society 2010; see also Cohen and Southwood 2008), including Japan, Korea, Nigeria, Norway, Saudi Arabia, South Africa, and Turkey.[20] LLU has been applied mainly to wireline telephone networks for DSL services, although in theory it could also be applied to other wireline broadband technologies such as cable modem and fiber to the premises (FTTP). Several countries, including the Netherlands, Sweden, and Slovenia, have proposed or implemented fiber unbundling policies.

LLU has not been widely implemented in developing countries. One reason is that the base of installed wireline telephone lines is generally much lower in developing than in developed nations. Considering the limited regulatory resources in some developing nations, efforts might be better spent in encouraging full, open, and technology-neutral infrastructure competition, particularly in wholesale markets, rather than devoting scarce resources to LLU when there are only a limited number of loops to unbundle.

Wireless networks. Commercial wireless networks have been an important local access technology for more than a decade and have become the predominant means of providing local access to voice and now broadband services in many developing countries. Wireless networks can help to overcome the last-mile wireline bottleneck by giving consumers multiple options for broadband access. For governments seeking to promote greater broadband connectivity, wireless offers some notable advantages, such as a lower cost structure in rural areas and faster rollout, since it is easier to deploy a series of cell towers than to connect each household with a physical wire. With the introduction of 3G and 4G technologies, wireless networks are expected to compete directly with, and be substitutes for, wireline broadband within the next decade. In Austria, for example, the telecommunications regulator (Rundfunk & Telekom Regulierungs [RTR]) determined in 2009 that DSL, cable modem, and mobile broadband connections for residential consumers are substitutes at the retail level. The range of policy options and regulatory changes that could be made to improve wireless broadband development is set forth below:

- *Allocate additional spectrum.* To support the expected increase in demand for advanced services requiring faster download speeds and the greater use of such services, regulators are implementing policies that promote the most efficient and effective use of spectrum resources, including freeing up spectrum bands that are either unused or underutilized.

- *Flexible allocations.* Another major tool for promoting wireless broadband development is for governments to allow flexible use of spectrum so as not to constrain technology or service developments. This will help providers to meet the rapidly changing demands of their customers.

- *Technology neutrality.* Technology neutrality refers to the concept that operators should be allowed to use whatever technology or equipment standard they wish in order to meet market demands. Thus, rather than having regulators mandate that a specific technology must be used in a certain band, operators are allowed to choose whatever technology they wish, subject to technical limitations—to prevent interference, for example.

- *Service neutrality.* With the transition to digital technology and better processing capabilities, advanced systems are now capable of transmitting all kinds of services. Wireless operators can now provide voice, high-speed data services, and video over their networks. Government regulators should modify service and licensing terms to allow operators to realize the benefits of this flexibility.

- *Greater use of market mechanisms.* The move to market mechanisms can be seen in two important trends: assigning spectrum to operators using some sort of competitive mechanism (for example, auctions) and charging market-based prices for acquiring or using spectrum. Having a competitive, transparent means of assignment can also give service providers greater access to spectrum. In conjunction with a regime that allows flexible use of spectrum, such competitive assignment can enable new models of service provision.

- *Spectrum trading.* Once spectrum has been assigned, spectrum trading (secondary market license transfers) allows later entrants to a market to access spectrum by paying a market price for it. This improves competition by allowing companies who want (new or additional) spectrum to acquire it from those who may have excess spectrum in specific areas.

- *Mobile virtual network operators (MVNOs).* Another way to introduce additional competition into the market is for governments to permit MVNOs to contract with existing mobile carriers to gain access to capacity and network services that they then use to establish their own services and brand. The MVNO model, however, has not been universally successful, as its impact appears to depend on the specifics of a country's mobile market structure.

- *Coverage obligations.* Governments can promote wireless broadband availability by establishing coverage obligations at the time of initial licensing. License requirements tied to coverage obligations, however, must be carefully considered. Requirements that are too easy to meet run the risk of not significantly expanding broadband coverage. Conversely, overly strict requirements are unlikely to be met and could result in either no interest in a license or lower payments.

Infrastructure Sharing

As governments seek ways to expand broadband networks and promote competition in broadband services, they inevitably encounter difficulties. In some areas, low population densities may make it unlikely that the market will support multiple competing wireline or wireless infrastructures. In addition, for some buildings in urban areas, there may not be sufficient physical space to run multiple sets of fiber or copper cables to each potential user or to place wireless towers and other equipment. In such cases, policy makers and regulators have begun to encourage—or even require—parties to share the physical infrastructure used to deliver broadband services.

Two types of infrastructure sharing are generally being considered today. "Passive" sharing includes common use of support structures, such as towers, masts, ducts, conduits, trenches, manholes, street pedestals, and dark fiber. "Active" sharing involves electronics, switching, power supplies, and air conditioning, among other elements. Infrastructure sharing can take many forms, with the most common being collocation (the sharing of physical space in buildings), tower and radio access network sharing, access to dark fiber for backhaul, and backbone networks and physical infrastructure sharing (ducts and conduits).

Infrastructure sharing is rapidly becoming an important means of promoting universal access to networks and offering affordable broadband services by reducing capital expenditures and ongoing operating expenses associated with the rollout and operation of networks. In recent years, a noticeable trend has been toward voluntary sharing of active and passive network facilities around the world, especially in the mobile sector. A push to upgrade and expand networks for mobile broadband is resulting in service providers searching for ways to cut costs and raise capital. For example, service providers may create joint ventures that manage the combined infrastructure assets either for shared use by its owners or on an open-access basis. This allows for network optimization and for avoidance or decommissioning of redundant sites, leading to significant cost reductions for the parties involved. The joint venture in the United Kingdom between Hutchison 3G and T-Mobile, now joined by Orange after its merger with T-Mobile in the United Kingdom, and the pan-European agreement between O$_2$ and Vodafone to share infrastructure in Germany, Spain, Ireland, and the United Kingdom highlight this trend toward increased voluntary sharing in the sector.

The trend of sharing mobile infrastructure also extends to developing countries. In India, for example, the regulator, TRAI, proposed sharing rules for the mobile sector in 2007, both for active and passive components. Since then, Bharti Group, Vodafone Group, and Aditya Birla Telecom (Idea Cellular) have created Indus Tower, a joint venture that controls over 100,000 towers and provides passive infrastructure service to its shareholders and other third parties. Also in India, the drive to raise capital for 3G auctions and deployment during 2010 led to significant divestiture of mobile towers to independent companies that operate them on an open-access basis. For example, in January 2010, an Indian tower company, GTL Infrastructure, acquired 17,500 towers from Aircel, making GTL one of the largest independent tower companies in the world. American Tower, another independent tower company, has also been acquiring towers in countries such as Chile, Brazil, Ghana, India, Mexico, Peru, and South Africa, with the aim of providing open access to such infrastructure.

Many other regulatory authorities, including those of Bangladesh, Nigeria, and Pakistan, have adopted policies to promote infrastructure sharing, especially in the mobile sector. Carefully crafted policy measures can increase time to market, introduce new forms of competition, and foster take-up for ICT services. Sharing also addresses the environmental impact of ICT infrastructure, reducing duplicative mobile towers that affect a city's skyline, for example. However, close ties and information exchanges between providers that participate in sharing agreements may create concerns with regard to competition, as they could facilitate collusion and reduce competition at the retail level if sufficient control over the network and services is not maintained and the provider's ability to differentiate retail offers and innovate is curtailed. When promoting voluntary sharing, regulatory authorities and policy makers must balance the potential benefits and costs of such measures, in order to achieve the desired objective of promoting more competitive markets and increased rollout of services.

On the wireline side, several governments are promoting a variety of shared infrastructure approaches. In the most interventionist cases, such as Australia, New Zealand, and Singapore, policy makers have directed the establishment of a single, open-access network that will provide infrastructure services on a wholesale basis to a variety of downstream service providers. Rather than establish an entirely separate network, France has taken a more regulatory approach by setting up sharing requirements and obligations for firms building out fiber networks to more rural areas and to apartment buildings.[21] Other countries are also considering regulations that will require incumbent operators (usually those that hold significant market power or are former monopoly providers) to make their infrastructure available to alternative carriers. This concept might also be extended to other, often government-owned, entities, such as power companies that maintain towers for electricity distribution.

Opening Vertically Integrated Markets

Benefits and Costs of Vertical Integration

Vertical integration, in which a single firm controls multiple levels of the supply chain, is commonly found in ICT markets around the world and often involves the same firm owning and operating network infrastructure as well using this infrastructure to offer retail services to end users. Two main advantages for a vertically integrated firm is the ability to achieve higher economies of scale and lower costs of production by reducing the costs of coordinating upstream and downstream activities. In a competitive market, these efficiencies can benefit consumers through lower retail prices. However, vertical integration may create barriers to entry for new competitors, particularly in the telecommunications sector, where a dominant operator may control essential infrastructure (Crandall, Eisenbach, and Litan

2010, 494–95). In such cases, a dominant, vertically integrated operator may strategically discriminate against competitors and stifle competition.

Remedies to Anticompetitive Conduct by a Vertically Integrated Operator

To address competitive concerns associated with vertical integration, some regulators have required dominant operators to separate vertically to some degree through accounting separation, functional separation, or, in extreme cases, structural separation.

Accounting Separation

The least intrusive and most prevalent remedy, accounting separation, makes transparent the vertically integrated operator's wholesale prices and internal transfer prices, enabling regulatory authorities to monitor compliance with nondiscrimination obligations or to ensure that there is no cross-subsidization. Generally, accounting separation requires the vertically integrated operator to maintain separate records for its upstream and downstream costs and revenues in order to allow the regulator to set wholesale prices for the regulated upstream services. These records are typically subject to independent audit and may also be made publicly available. Although the operator must make its costs transparent, under this remedy it is able to continue benefiting from the efficiencies of vertical integration.

In 2004 the Info-communications Development Authority (IDA) of Singapore issued accounting separation guidelines to allow monitoring of the ICT sector for potential anticompetitive behavior (IDA 2004). These guidelines established two levels of accounting separation: detailed segment reporting (applicable to dominant service providers and entities they control) and simplified segment reporting (certain other entities). This two-tiered approach is intended to provide the IDA with the necessary information, without unduly burdening operators, to ensure that no dominant provider is engaging in cross-subsidization or discrimination. Currently, incumbent SingTel is the only operator designated as dominant in any market, and it is subject to detailed accounting separation obligations.

Functional Separation

Obligations under functional separation range from simply requiring the operator to establish separate divisions for upstream and downstream activities to requiring the operator to separate the wholesale and retail divisions physically. This may involve the separation of employees (for example, physical separation of offices and prohibitions on the same employee working for both divisions) and the separation of information (for example, limitations on the type and amount of information that may be shared between divisions). Since there is no actual change in ownership or ultimate control under functional separation,[22] the operator can continue to enjoy many of the benefits of vertical integration (European Regulators Group 2007). More intrusive than accounting separation, regulators may implement functional separation in "exceptional" cases where there has been persistent failure to achieve effective nondiscrimination in relevant markets and where there is little or no prospect of effective competition within a reasonable period after less intrusive remedies have been attempted (European Union 2009b, para. 61).

The 2009 EU Telecoms Reform formally granted national regulatory authorities explicit authority to require network operators holding significant market power to separate functionally their communication networks from their service branches, but only as a last-resort remedy (European Parliament and Council of Ministers 2009). Prior to requiring functional separation, the national regulatory authority must first find that all less intrusive, market-based remedies have failed to achieve effective competition.[23] Next, it must submit a proposal of functional separation to the European Commission, with evidence justifying the regulatory intervention and an analysis of the likely market impacts. Among the provisions that must be included in the proposal are the precise nature and level of separation, the legal status of the separate business entity, identification of the separate business entity's assets and the products or services to be supplied by that entity, governance arrangements to ensure the independence of the staff, rules for ensuring compliance with the obligations, and a monitoring program to ensure compliance, including the publication of an annual report.[24]

To date, no EU member state has mandated functional separation. In some cases, such as that of the United Kingdom, dominant operators have voluntarily implemented functional separation. There, BT (formerly British Telecom) agreed to establish a separate division for access services called Openreach, which provides most of BT's wholesale products. According to the European Commission, BT's functional separation led to a surge in broadband connections, from 100,000 unbundled lines in December 2005 to 5.5 million by 2008 (European Commission 2009b).

Structural Separation

Structural separation involves full disaggregation of the vertically integrated operator's wholesale and retail divisions into separate, individual companies, each with its own ownership and management structure. All benefits associated with vertical integration are eliminated. Regulated structural separation is considered a last-resort measure and is typically used only if other regulatory interventions have failed and a comprehensive cost-benefit analysis has been conducted.[25] Structural separation is extremely difficult to reverse and can dramatically affect the market, such as by increasing regulatory uncertainty and affecting infrastructure investment. Additionally, it is difficult to allocate the separated firms' assets and liabilities in order to ensure the ongoing viability of both entities. As a result, regulatory authorities rarely impose structural separation as a remedy.

In 2010, the Australian Parliament passed the Telecommunications Legislation Amendment (Competition and Consumer Safeguards) Act 2010 (Australian Government 2010). The act and implementing regulations set out the procedures by which the dominant fixed-line operator, Telstra, must structurally separate control over its copper and hybrid fiber coaxial network infrastructure as well as its provision of wholesale access services, from retail fixed voice and broadband services (Australia, Department of Broadband, Communications, and the Digital Economy 2011). In August 2011, Telstra submitted to the Australian Competition and Consumer Commission its structural separation undertaking plan, which commits Telstra to full structural separation by July 1, 2018.[26] Telstra's structural separation is set to occur through the progressive migration of its fixed-line networks to the National Broadband Network (NBN) Company, which is rolling out a national broadband network to be provided on a wholesale-only basis. Additionally, the plan sets out various measures by which Telstra will ensure transparency and equivalence in the supply of regulated services to its wholesale customers during the transition to the NBN. In exchange for structurally separating and providing the NBN Company with access to its fixed-line infrastructure, Telstra will receive compensation in the amount of $A 11 billion.[27]

Network Neutrality

Network neutrality ("net neutrality") generally refers to the notion that an ISP should treat all traffic equally, whether content, application, or service. Based on this principle of nondiscrimination, proponents of net neutrality seek to restrict the ISP's ability to interfere with or inappropriately manage Internet traffic (Atkinson and Weiser 2006). Blocking or slowing down (also referred to as "throttling") the delivery of certain types of content, applications, or services is one of the main concerns of net neutrality advocates. However, such network traffic practices may be considered necessary to ensure that illegal content is not distributed or to manage networks better during congested periods. Another issue relates to prioritization of certain types of traffic. This may occur where ISPs deliver latency-sensitive traffic, such as voice or streaming video, faster than traffic that is not latency sensitive, such as a music download. Prioritization may also occur where an ISP charges application or content providers to be guaranteed better or faster access to subscribers.

Additionally, net neutrality proponents generally seek to improve the transparency of what the ISPs are doing with regard to traffic management and other Internet-regulating actions. This involves whether an ISP discloses to interested parties its network management practices, such as blocking, degrading, or prioritization. Interested parties may include consumers, the government, and applications, content, and service providers.

Goals of Net Neutrality Regulation

Regulatory authorities have tended to focus on several overarching goals when instituting net neutrality consultations and rules over the last several years, including (a) consumer protection, (b) promotion and preservation of access and innovation, and (c) safeguarding of freedom of speech and freedom of information. Consumer protection issues include transparency and disclosure requirements as well as prohibitions or restrictions on blocking or degrading subscribers' use of lawful content, applications, and services.

The second goal addresses the access that content, applications, and service providers have to an ISP's network, particularly if their services compete with an ISP's services. For example, an ISP may block applications for VoIP services if these services compete directly with the ISP's voice telephony service. Another example may involve paid prioritization in which an ISP favors one content provider over another through a peering agreement, which could affect competition among content providers. Finally, there is also a concern that as new applications and services are developed, providers may find their access blocked or limited—either for (anti)competitive reasons or because new entrants do not have the ability to pay for priority access on an ISP's network.

Regulatory Approaches

As policy makers consider whether net neutrality provisions are needed in their country, they may find it useful to view the possible approaches to net neutrality along a spectrum. At one end of the spectrum, a policy would require "pure" net neutrality of no discrimination; the ISP would be prohibited from managing Internet traffic in any way and would simply work on a "best efforts" basis, delivering all content on equal terms. Companies would be prohibited from charging content providers for priority or favored access. At the other end of the spectrum, a policy would permit an ISP to engage in any network management practice, including allowing it to block users from accessing certain types of legal content, applications, or services without the users' knowledge. Although a country may not have specific net neutrality policies or rules in place, issues related to blocking, delaying, or prioritizing traffic may be addressed by competition laws, while transparency and disclosure may be addressed by consumer protection laws or laws protecting freedom of information or speech.

In practice, regulatory authorities are adopting net neutrality policies all along this spectrum. For instance, a regulator may find that it is not necessary to regulate ISPs' network management practices, but that stronger rules on transparency of traffic management policies are required to ensure that consumers are well informed. This is the case, for example, of the EC policy on net neutrality contained in the April 2011 report, "The Open Internet and Net Neutrality in Europe" (European Commission 2011). The report frames traffic management as a quality of service issue for consumers relating to (a) the blocking or throttling of lawful Internet traffic and (b) Internet traffic management practices. The EC does not impose any rules or restrictions on the blocking or throttling of lawful Internet traffic, but it does recognize concerns over possible consumer protection or competition issues. Instead, the EC recommends that national regulatory authorities conduct further inquiries into such practices before adopting any rules or guidelines on the matter. Similarly, the EC recognizes that traffic management is necessary to ensure the smooth flow of Internet traffic, particularly when there is network congestion. As such, the EC does not impose any rules or restrictions on traffic management practices, such as packet differentiation, IP routing, or filtering between "safe" and "harmful" traffic. The only rules imposed by the EC in the open Internet report, aside from the ability to switch providers in one business day, are associated with transparency and disclosure. These rules require Internet providers to ensure that adequate information about their services is available to consumers, including identifying any possible restrictions on access to certain services, actual connection speeds, and possible limits on Internet speeds. Additionally, providers must make certain that consumers are informed about traffic management practices and their effect on service quality (for example, bandwidth caps), prior to signing a contract.

Under another approach, a regulator may decide to institute both new network management and transparency rules, but fall short of requiring "pure" net neutrality by permitting ISPs to discriminate against certain types of traffic for a specific purpose (for example, to manage congestion) and according to a set of standards, such as "reasonable network management." This is the case in France, where ARCEP released a report entitled "Neutrality of the Internet and Networks: Proposals and Guidelines" in September 2010 (France, ARCEP 2010). In the first proposal, ARCEP recommended that ISPs be required to provide end users with (a) the ability to send and receive the content of their choice; (b) the ability to use the services and run the applications of their choice; (c) the ability to connect the hardware and use the programs of their choice, provided they do not harm the network; and (d) a sufficiently high and transparent quality of service. Under the second proposal, ARCEP recommended that ISPs may not discriminate against different types of traffic, whether by type of content, service, application, device, or address of origin or destination. Under the guidelines, exceptions to the first two recommendations may be acceptable if an ISP follows the third proposal by complying with the "general principles of relevance, proportionality, efficiency, nondiscrimination between parties, and transparency." Pursuant to the fourth proposal, ARCEP will permit ISPs to provide managed services along with Internet access services, but will require them to maintain Internet access service quality at or above a minimum, satisfactory level. However, ARCEP did not specify what this minimum quality of service level should be.

In the fifth proposal of the net neutrality guidelines, ARCEP addressed transparency and disclosure requirements. For example, ARCEP requires that, in their marketing materials, service contracts, and customer information through the duration of the contract, ISPs must clearly and concisely disclose to end users all relevant information regarding (a) the services and applications that can be accessed through these data services, (b) the quality of service, (c) the possible limitations of the service, and (d) any traffic management practices that may affect the user. In particular, any restrictions on data transmission that do not conform to the first two recommendations must be disclosed to users.

Like France, Chile has also adopted net neutrality rules limiting discrimination by ISPs against access to and use of legal online services, applications, and content. In addition, Chile was the first country in the world to enact broad net neutrality legislation under the Chilean Net Neutrality Act, which was signed into law on August 18, 2010. The law focuses on the principles of nondiscrimination and transparency and prohibits ISPs from blocking, throttling, or discriminating against the transmission of any legal application, service, or content. However, ISPs are allowed to manage traffic on their network, but not in an anticompetitive fashion. Chile's regulator, the Subsecretaría de Telecomunicaciones (SUBTEL), issued the implementing regulations of the net neutrality law in March 2011.[28]

Table 3.1   Status of Net Neutrality Initiatives in Select Countries

Stage in process	Position along the spectrum (least to most stringent)	Country
No consultation	Considered net neutrality, but found no problems requiring a consultation and subsequent rule; will continue to monitor	Denmark, Germany, Ireland, Portugal
	Nonbinding neutrality guidelines	Norway
In consultation	Information gathering on current practices potentially to establish rules	Italy
	Transparency or disclosure rules proposed, *but no* traffic management	United Kingdom
	Transparency or disclosure rules *and* traffic management or nondiscrimination rules proposed	Brazil, Sweden
Rules or legislation adopted	Transparency or disclosure rules, *but no* traffic management or nondiscrimination rules	European Commission
	Transparency or disclosure rules *and* traffic management or nondiscrimination rules	Canada, Chile, France, the Netherlands,[a] United States

Source: Telecommunications Management Group, Inc.

a. Lower house of Parliament passed in June 2011, upper house to pass by December 2011.

Table 3.1 summarizes the approaches being taken in selected countries as well as each country's progress in the process of developing net neutrality rules.

Distinction between Wireline and Mobile Broadband Services

Existing mobile networks generally present operational constraints that wireline broadband networks do not typically encounter, particularly relating to efficient use of the spectrum. This puts greater pressure on concepts such as "reasonable network management" for mobile broadband providers. As a result, some regulatory authorities have recognized the need to establish differentiated network management rules for wireline and mobile broadband services.

This is the case in the United States, where the FCC's open Internet order applied transparency rules equally to both wireline and mobile broadband network services, but applied different network management rules to the different technologies. However, while the rule for mobile broadband is less stringent than the rule for wireline, it still prohibits operators from blocking certain websites or VoIP applications, as is occurring in several European countries, such as Sweden and the Netherlands.

Similarly, in France net neutrality rules would be applicable to any broadband access technology (that is, to both wireline and mobile networks). However, ARCEP may implement the rules differently, particularly with respect to the means of assessing which traffic management mechanisms are acceptable. ARCEP might allow mobile operators to restrict access to certain sites or applications for objective, nondiscriminatory, and justified reasons on the basis that mobile networks are currently more vulnerable to congestion due to scarcity of available frequencies and the surge in data traffic generated by smartphones. However, ARCEP proposed that traffic management practices of mobile network operators must satisfy technical imperatives and cannot involve banning or blocking an application or a protocol (including VoIP, peer-to-peer, or streaming) and must not use these practices as a substitute for investing in increasing network capacity.

In the EU, the open Internet consultation also briefly addressed whether principles governing traffic management should be the same for both wireline and mobile networks. The consultation notes that wireline broadband providers have not blocked VoIP services, but that some mobile operators have blocked VoIP services from third-party providers or have charged rates to end users in excess of normal rates for equivalent amounts of data. Since traffic management rules were not imposed by the open Internet report, no distinction is made between wireline and mobile services. However, the transparency, disclosure, and switching rules apply equally to both wireline and mobile Internet providers.

Security in Cyberspace

Broadband services and applications are increasingly expanding into every aspect of our lives. Greater numbers of consumers are now using broadband Internet connections for education, entertainment, banking, and shopping as well as to interact socially and with their governments. Businesses are using broadband to increase their internal efficiency and productivity, and online web representation has become more important for many businesses than traditional marketing channels such as printed publicity materials. Furthermore, essential services, such as water and electricity supply, banking, transportation infrastructure, and public safety, now heavily rely on critical information infrastructure (CII; see ITU-D 2009b, 11).

In an increasingly broadband-connected environment, even brief interruption, degradation, or compromise of service may have significant social, economic, and political consequences that negatively affect consumers, businesses, and governments. Given these consequences, the success of broadband requires a significant focus on security.[29] Due to its broad scope, cybersecurity may be seen as enabling both the supply of and demand for broadband. From the demand side, users need to feel safe online if they are to take full advantage of broadband services and applications, and businesses need to have confidence in their ability to leverage broadband to increase productivity and engage in online activities. From the supply side, to guarantee stable and dependable services, CII must be protected from attacks. In addition, governments must possess the capabilities to enforce cyber laws, which in many cases requires cross-border cooperation. As such, cybersecurity and cybercrime policies, laws, regulations, and enforcement will play a critical role in development of the broadband ecosystem. This legal framework may include criminal codes, laws on privacy, commercial transactions, and electronic communications, and laws relating to criminal procedure and enforcement, among others. Overall, these policies and laws must balance the many inherent trade-offs between, for example, the desire to access information conveniently and easily, on the one hand, and the need to protect data privacy and security, on the other hand.

Data Protection

The ability to protect digital data is essential to promoting a safe and secure broadband ecosystem, which increases consumer confidence and thereby enhances demand. Data protection generally includes the protection of users' personal identifying information, such as banking, medical, credit card, and other private data, as well as the protection of intellectual property and other sensitive, proprietary information of businesses and governments, such as employee data or client information. Effective cybersecurity policies and regulations are needed to combat the many costly violations of data privacy occurring each year, including computer hacking to steal a person's identity or remotely deleting information through viruses.

Security of Critical Information Infrastructure

Securing critical national infrastructure, and specifically CII, is also a key component of facilitating the success and stability of broadband networks. Infrastructure is considered to be critical if its destruction would have a debilitating impact on the defense or economic stability of the country. Thus electricity grids, telecommunications systems, transportation, water supply systems, banking and finance, and emergency services are all deemed to be critical infrastructure (Brunner and Suter 2008–09, 35).

International cooperation is a significant aspect in securing CII. It is important to consider the role of standards and the role of government in developing those standards. Generally, a global standard is developing around critical infrastructure protection aimed at ensuring that any disruptions to CII are brief, infrequent, manageable, isolated, and minimally detrimental (Commission of the European Communities 2005, 1). However, national frameworks vary widely as to which cybersecurity issues are addressed and how CII is protected.

At both an international and a national level, private, governmental, and nongovernmental sectors need to take steps to increase the security of their networks, services, and products. The effectiveness of any critical infrastructure protection program is directly proportional to the extent of cooperation among these actors. For this purpose, computer emergency response teams (CERTs) are being implemented in countries around the world as a means of identifying cyber vulnerabilities and defending against cyber attacks (box 3.4).

Box 3.4: Computer Emergency Response Teams

CERTs are cooperative endeavors among governments, academic institutions, and commercial entities aimed at identifying cyber vulnerabilities and defending against cyber attacks. Generally, CERTs focus on technical issues and information sharing, thereby providing early warning functions of cybersecurity breaches (Satola and Judy forthcoming, 8). They are designed to promote information sharing and strengthen coordination among both the private sector and government agencies. For example, in March 2011, the Sri Lankan computer emergency response team (SLCERT) identified several fraudulent websites located in India and China that were selling fake tickets online to the Cricket World Cup 2011 in order to steal users' credit card information (Dissanayake 2011). SLCERT was able to inform the Indian computer emergency response team about these fake websites and is seeking legal action against those responsible.

Greater international cooperation among CERTs, such as the Sri Lankan and Indian CERTs, is facilitated through the Forum of Incident Response and Security Teams (FIRST), which brings together 238 CERTs across 48 countries, including the national coordination centers for India, Singapore, Brazil, Argentina, Colombia, Qatar, and Saudi Arabia. FIRST aims to foster cooperation and coordination in incident prevention, to stimulate rapid reaction to incidents, and to promote information sharing among members and the community at large. Thus, both FIRST and other CERTs are a positive step toward coordinating international responses to cybersecurity problems.

Source: Telecommunications Management Group, Inc.

Cybercrime

Cybercrime can be broadly described as criminal offenses committed within or against computer networks or by means of computer networks. Cybercrime policies and laws focus on the investigation and criminalization of certain offenses, as well as their prevention and deterrence. As such, cybercrime covers a wide range of conduct, which can generally be divided into four broad categories, as defined by the Council of Europe's (2001) Convention on Cybercrime:

1. Offenses against the confidentiality, integrity, and availability of computer data and systems

2. Computer-related offenses

3. Offenses related to infringements of copyright and related rights

4. Content-related offenses.

All of the offenses in the first category are directed against one of the three legal principles of confidentiality, integrity, and availability. As opposed to crimes that have been covered by criminal law for centuries, the computerization of crime is relatively recent. In order to prosecute these acts, existing criminal law provisions need not only to protect tangible items and physical documents but also to safeguard the above-mentioned legal principles. Some of the most commonly occurring offenses in this category include illegal access (hacking and cracking), data espionage, illegal interception, data interference, and system interference.

Computer-related offenses cover cybercrimes that require computer access to commit. These offenses tend to have more effective and stringent legal repercussions than offenses in the other categories mentioned above. The most common computer-related offenses include computer-related fraud, computer-related forgery (phishing and identity theft), and the misuse of devices. The main difference between computer-related and traditional fraud is the target of the fraud; if a person is targeted, then it is traditional fraud, but if a computer or computer system is targeted, it becomes computer-related fraud. Although some criminal law systems do not yet cover the manipulation of computer systems for fraudulent purposes, offenders often still can be prosecuted. Nonetheless, many governments may need to include computer-related offenses in their definitions of various crimes in order to prosecute. Because of the broad scope of these offenses, some may fall within the ICT regulator's jurisdiction, such as those relating to consumer protection. Computer-related offenses, such as fraud and forgery, generally fall within the purview of criminal law enforcement authorities, while others like the protection of privacy or unsolicited communications (spam) may be the responsibility of data protection authorities or a consumer protection agency. However, if mandates overlap, it is critical for all relevant authorities to coordinate the exercise of their respective functions (Gercke et al. 2010, 6).

Offenses related to infringements of copyright and related intellectual property rights (IPRs) are another category of cybercrime. These violations relate to the unauthorized or prohibited use of protected works, trademarks, or patents, facilitated by using the Internet's inherent ability to disseminate information. With regard to content-related offenses, the development of legal instruments to deal with these offenses is heavily influenced by national approaches. The classification of content-related activity as a criminal offense or as protected free speech is dependent on each country's cultural and legal frameworks. The following sections address these issues.

Cybersecurity and the Need for International Coordination

Cybersecurity is highly globalized because cybercrimes and other attacks can be committed against Internet users, businesses, or governments from anywhere in the world. As such, international coordination is pivotal to the success of cybersecurity. Cybercrime and cyberwar have very clear and direct negative effects on economic activity, but cyber defense can have similar negative effects, due to its high cost and information inefficiencies caused by the deliberate isolation of networks and databases from one another.

Several barriers exist to a successful international cybersecurity framework. One is that different countries take different approaches to cybersecurity, which can lead to a lack of multistakeholder participation in both policy making and legislation. Another problem is that upstream policies promoting an e-agenda conflict with the downstream protections of rights and property. In addition, legal concepts may be outdated in the burgeoning world of cyberspace. The core issues of jurisdiction and sovereignty make it difficult to cross borders to address international cybersecurity events. A fourth issue is simple human error when using the Internet or writing software code. A final barrier to international cybersecurity coordination is that existing cybersecurity tools are often not fully applied. For example, liability in some countries is often imposed on a case-by-case basis rather than pursuant to statutory and regulatory requirements aimed at the particular issue (see, for example, United States, FTC 2010a, 2010b, 2010d; Martin, Judy, and Pryor 2010). These issues, however, are not insurmountable. Rather, concerted, effective national legislation and international coordination frameworks prepared to prevent, identify, and prosecute cybercrimes are needed to ensure the safety of the Internet and ICTs.

Privacy and Data Protection

Threats to privacy and data protection must be addressed to foster demand and promote broadband take-up. Legal and regulatory tools to address these issues can help to build consumer trust and confidence, which are indispensable for a full broadband experience. While consumer privacy and data protection are not novel subjects, broadband diffusion and technology innovation compound the potential risks associated with the collection, use, protection, retention, and disposal of a wide range of personal information. Increased data processing and storage capabilities, advances in online profiling, and the aggregation of online and offline information are allowing a diverse set of entities to gather, maintain, and share a wide array of consumer information and data.

Consumers care about their privacy online. For example, when the social networking service Facebook released new privacy controls in December 2009, 35 percent of its 350 million users worldwide at the time chose to revise and customize their account settings (United States, FTC 2010c, 28). Governments are also concerned with protecting their citizens from practices that may violate their privacy. The worldwide controversy regarding Google's data and image collection practices for its Street View, Maps, and Latitude services and the implications for data privacy highlights this point. Over 20 countries around the world have launched investigations into Google's practice of collecting photos and information to map Wi-Fi networks, reaching different findings and leading to multiple remedies, including fines.[30]

The unprecedented ability to collect data, often without the knowledge of the individual whose data are at issue ("the data subject"), poses new, broadband-specific challenges and opportunities linked to ensuring online privacy and data protection. Issues such as cloud computing, online behavioral advertising, web tracking, and location-based services may create additional privacy risks, but may also provide tremendous benefits for consumers in the form of new products and services. However, increased collection of personal data is not limited to businesses and the private sector. Governments also increasingly collect such data from their citizens as they engage in e-government and other initiatives. Thus, to promote broadband, countries must set up frameworks that strike the appropriate balance between the benefits to citizens and consumers of new and innovative technologies and the risks such technologies may create to their privacy and personal data. Also, due to the cross-border nature of Internet data traffic flows, international cooperation and coordination will be critical to enforce online privacy frameworks.

Scope of Privacy and Data Protection in a Broadband Environment

Privacy and data protection in the broadband environment must continue to focus on assessing risks to consumer information throughout its life cycle—from collection to use to storage to transmission to disposal—and then on adopting safeguards that are reasonable and appropriate to mitigate the identified risks. To date, two broad approaches toward personal data protection have been adopted around the world. Many countries, such as EU member states and many Latin American countries, have opted for a rights-based approach to personal data protection. Under this system, personal data protection is regulated as a fundamental right that applies to all personal data, irrespective of the type of data.[31] By contrast, countries such as the United States have to date relied on "broad self-regulation and targeted sectoral legislation to provide consumers with data privacy protection" (United States, FTC 1998).

More recent developments seem to be bridging this divide, with the European Commission and the U.S. Federal Trade Commission proposing many common changes and upgrades to privacy protection in the wake of rapid technological developments associated with broadband services and the Internet (European Commission 2010b). This includes emphasizing informed consent, requiring increased transparency of data collection, raising awareness, and increasing responsibility of data controllers (that is, privacy by design).[32]

Informed Consent

Informed consent refers to the "freely given specific and informed indication" of an individual's agreement to data collecting and processing activities and allows the consumer to make informed and meaningful choices. Broadband-enabled activities, such as online behavioral advertising, raise new questions regarding informed consent and the extent to which, for example, Internet browser settings may be considered to deliver such consent or whether a more uniform, comprehensive mechanism should be adopted for online behavioral advertising, sometimes referred to as "do not track." While there may be no clear international trend at this time, development of informed consent mechanisms likely will continue to be a key factor for online privacy protection in a broadband world.

Privacy by Design

"Privacy by design" advances the view that privacy cannot be assured solely by compliance with regulatory frameworks, but instead requires privacy considerations to become engrained in everyday business practices (Cavoukian 2011). Both the European Commission and the U.S. Federal Trade Commission are proposing to follow this approach as a means to enhance a data controller's (whether it be a business or a government) responsibility in handling personal data. Under this approach, companies should incorporate substantive privacy protections into their practices, including data security, reasonable collection limits, sound retention practices, and data accuracy. They should also maintain comprehensive data management procedures throughout the life cycle of their products and services. Privacy and data protection authorities are currently looking at ways to encourage compliance with such policies and enforcing possible instances where data controllers have exercised an insufficient level of care.

Broadband and the Scope of Personal Data

Broadband-enabled data profiling is blurring the line of what constitutes personally identifiable information (PII) subject to protection. Certain categories of PII, such as an individual's name, address, or personal identification number, used to be clearly defined and protected. As noted by the U.S. Federal Trade Commission, however, the comprehensive scope of data collection that comes with broadband applications and services allows disparate bits and pieces of "anonymous" information from online and offline sources to be aggregated to create profiles that can be linked back to a specific person, thus making old definitions of PII less relevant. This view is in line with that of Europe, where all information relating to an "identified or identifiable person" should be protected, including "all means likely reasonably to be used either by the controller or by any other person to identify said person."[33] Mexico has recently implemented a broad definition as part of its 2010 data protection legislation, defining personal data to include "any information concerning an identified or identifiable individual."[34] Expansion of the scope of protected personal data will continue to pose challenges as innovation increases the type of data that may be aggregated in innovative ways and then used to trace information back to a specific identifiable individual.

Increased Transparency in Data Collection

In the broadband world, consumers must be given sufficient information to make informed choices regarding the collection and use of their personal data. The proliferation of actors and technical complexity involved in activities such as behavioral advertising make it increasingly difficult for individuals to know when their data are being collected, by whom, and for what purpose. This requires more transparency by data controllers about how and by whom data are collected and processed. In addition, this information needs to be presented in a way that consumers will understand. Even when privacy policies are provided, they often are long and incomprehensible for many consumers. As such, when consumers are faced with the burden of trying to read and understand these policies, they often simply scroll through them and accept the terms provided without really knowing what they are accepting. Increased transparency in a broadband environment may be addressed by facilitating standard privacy notices drafted using plain language or by educating consumers on privacy matters. Also, transparency is enhanced using policies such as data breach notifications.

Awareness Raising

There is also a need to raise awareness, especially among younger users, regarding the impact of broadband and new technologies on personal privacy. In many cases, consumers may not know or understand enough about the data collection and use practices and their privacy implications. For example, as social networking services, or other similar applications, become increasingly popular ways to interact online, it is critical to educate young people about safe social networking and other online issues. Data protection authorities have a key role to play in educating individual users by holding conferences, workshops, and media campaigns and in encouraging industry to engage in awareness-raising initiatives. Chapter 6 further addresses the issue of raising awareness and educating users on matters relating to privacy.

International Enforcement and Policy Cooperation

Proliferation of complex, cross-border data flows and cloud computing services and applications demand increased international cooperation to enforce privacy and data protection. The Internet makes it easier for entities established in one country to provide services in another and to process data online. However, this often makes it difficult for authorities to determine the location of the personal information and the equipment used to process it. As the European Commission notes, however, this fact should not deprive the data subject of protection.

Thus international cooperation and coordination are key elements for enforcement actions. To this end, several international initiatives are under way dealing with cooperation and coordination for the enforcement of privacy laws. One example is the 2007 Organisation for Economic Co-operation and Development (OECD) Council's Recommendation on Cross-Border Co-operation in the Enforcement of Laws Protecting Privacy.[35] On this basis, in 2009, 13 privacy enforcement agencies from around the world created the Global Privacy Enforcement Network to facilitate cross-border cooperation in the enforcement of privacy laws.[36] Similarly, in 2010, the Asia-Pacific Economic Cooperation (APEC) forum established the APEC Cross-Border Privacy Enforcement Arrangement (CPEA), a multilateral cooperation network for APEC privacy enforcement authorities, with the participation of authorities from Australia, Canada, the United States, and Hong Kong SAR, China.[37]

Regulation of Broadband Content

Content is the currency of the Internet: more, better, and timely content means higher visibility, more visitors, and increased revenue. Thus more relevant, more local content is the strongest vehicle to enhance broadband demand. The laws and rules that regulate content in the offline world have been gradually applied to and adapted for online content, even as the pace of innovation online threatens to render them obsolete. Online content can be produced by traditional methods or generated collaboratively by the users themselves—it can be a song played by an Internet radio station, a viral video in an embedded YouTube clip, a blog post, or a news article published by a news website.

Broadband has enabled the easy transfer of all kinds of voice, data, video, and multimedia content. The ability to disseminate and access legal content online is critical to broadband deployment. It affects the development of new services and applications, the launch of innovative online businesses and services, and the active participation of individuals in social and political spheres. Regulation of content over broadband has significant implications both for the supply of broadband services and applications (for example, securing the rights to distribute content) as well as for the demand for broadband (for example, the existence of compelling content to attract users). This section reviews the intersection of supply and demand factors with the legitimate goal of regulating some forms of online content, recognizing the need to establish an appropriate balance between the two. In addition, this section addresses certain regulations of IPRs over broadband and certain content-related business practices that may have anticompetitive effects that hinder broadband development.

Freedom of Opinion and Expression

One of the fundamental rights of persons is the right to freedom of opinion and expression, which includes freedom to hold opinions without interference and to seek, receive, and impart information and ideas through any media and regardless of frontiers.[38] Content regulation, including surveillance and monitoring of Internet use, needs to take into account the standards set by international human rights law and the unique nature of the Internet.

A recent report by the Special Rapporteur on the Promotion and Protection of the Right to Freedom of Opinion and Expression of the United Nations' Human Rights Council notes that any restriction by a state of the right to freedom of expression must meet the strict criteria under international human rights law.[39] The report concludes that the flow of information via the Internet should be as free as possible, except in few, exceptional, and limited circumstances prescribed by international human rights law. It also stresses that the full guarantee of the right to freedom of expression must be the norm, that any limitation should be considered as an exception, and that this principle should never be reversed.

The collaborative web, sometimes called Web 2.0, has revolutionized the way people communicate. Facebook, Twitter, and other social networking websites allow citizens to discuss, debate, and organize. Citizen journalists have democratized the gathering and dissemination of news; postings on personal blogs and user-submitted videos on YouTube are often the first outlets to break a news story. In fact, many have noted that the uprisings in the Arab nations in 2011 were organized in part through the use of social networks such as Facebook and Twitter (see, for example, Giglio 2011).

With faster speeds, and in particular faster upload speeds, broadband can facilitate collaboration as well as access to information. As more and more Internet users employ the web, not just to consume but also to share, the Internet can become a virtual town square for citizen participation. By the same token, restrictions on Internet use, the censorship of certain information, or even restrictions on access posed by "net neutrality" concerns can cut off this vital avenue for citizen engagement. Governments will need to strike a balance between the legitimate need to restrict illegal content and the rights of users to participate freely and lawfully in cyberspace.

Some commentators have proposed that a new economy is emerging where people contribute freely to the production of information goods and services outside of the market (for example, Wikipedia; see Benkler 2006). Such a "networked information economy" has the potential to increase individual autonomy by allowing individuals to do more for themselves and by providing alternative sources of information from both faraway and non-traditional sources such as other individuals.

Regulating Specific Forms of Content

Countries have different social, cultural, and moral traditions. These traditions generally are enforced by legislation that prohibits the display or dissemination of certain types of content. Governments have legitimate reasons to regulate content: protection of minors, prevention of vices, and protection of national security, to name a few. Tensions are inevitable, as countries attempt to strike the right balance between the regulation of content on the Internet and the protection of fundamental rights, such as freedom of expression and information, which are strongly enabled by broadband; broadband-enabled Internet will make such restrictions more difficult to enforce.

When a provider of prohibited content operates within a country's borders, the country's laws should be sufficient to shut it down.[40] However, if the proscribed content comes from overseas, such as from a foreign website, the prohibition can be difficult or impossible to enforce. Nonetheless, more and more countries are implementing Internet controls of ever-increasing sophistication, including monitoring and filtering.[41]

Sometimes content is restricted by a government, possibly in an attempt to protect a domestic industry's interest. Such appears to be the case with online gambling in the United States. In 2006, the federal legislature, in an attempt to impede U.S. residents from gambling online, passed the Unlawful Internet Gambling Enforcement Act. The law prohibits gambling businesses from accepting funds from gamblers wherever it would be unlawful under federal or state law.[42] Passage of the law prompted Antigua and Barbuda to file a complaint with the World Trade Organization, in which it claimed that the United States had violated its commitment under the General Agreement on Trade in Services to free trade in recreational services. The World Trade Organization ultimately ruled in favor of Antigua and Barbuda and awarded it the right to suspend US$21 million annually in IPRs held by U.S. firms (WTO 2007, 55).

Government regulation is not the only option for restricting certain types of content. For example, the movie and video game industries, among others, voluntarily rate their content in order to help consumers to identify content appropriate for themselves and their families. The Family Online Safety Institute, an international nonprofit organization, administers a program whereby websites rate their content in terms of language, violence, and sexual content, in response to a standard questionnaire.[43] In addition, commercial vendors have developed personal computer (PC) applications that employ keyword-based filtering to allow parents to control the kinds of websites their children can visit. Similarly, the development of industry codes of practice relating to online content may be another viable alternative to government regulation. This is the case, for example, in Australia, where the Internet Industry Association has adopted a Code of Industry Co-Regulation Relating to Internet and Mobile Content.[44]

Another relevant issue in the regulation of content over the Internet relates to the issue of ISP liability. If a user posts prohibited content on his or her website, is the ISP that hosts the website liable? In many countries, the answer is no: ISPs and online service providers (OSPs) such as YouTube or Facebook are not liable for the content that users upload to their systems as long as they are not specifically aware of the prohibited content. This applies not only to prohibited content such as child pornography, but also to infringement of IPRs, defamatory statements, and fraudulent activity, among others. If it were not for this "safe harbor," ISPs and OSPs would have to monitor every last bit of user-contributed content and analyze it for possible legal repercussions—likely making many of today's most popular and innovative websites infeasible to operate. However, if an ISP or OSP becomes aware or is made aware of prohibited content on its system, it must act promptly to remove it or risk losing its safe harbor.

Such is the case of copyright in the United States, where the Digital Millennium Copyright Act (DMCA) creates a safe harbor for ISPs and prescribes the procedure that rights holders should follow to request the removal of illegally posted content.[45] The EU policy for ISP liability is very similar to that of the United States.[46] South Africa's Electronic Communications and Transactions Act of 2002 largely follows the example set by the United States and the EU for ISP liability by creating a DMCA-like "notice and takedown" system (the first such system in Sub-Saharan Africa).[47] In May 2010, Chile became the first country in Latin America to amend its legislation in order to regulate ISP liability.[48] Chile's law follows the familiar "notice and takedown" scheme; however, the notice must be issued by a court after the rights holder presents evidence in an expedited hearing.

Intellectual Property Rights

Compared to the limited bandwidth networks of the past, broadband's inherent capacity to transmit large amounts of information has made it easier to share all types of copyrighted works, including songs, books, and videos. And as the software to find and share such works has gotten better and easier to use, the problems associated with the illegal sharing of copyrighted works has become a major issue. IPRs refer mainly to the rights of those persons or entities that hold copyrights, patents, or trademarks. IPRs have long been recognized and protected to encourage investment in and creation of new artistic works, inventions, and businesses. But the very things that make the Internet so powerful—its global reach, low cost, nearly frictionless nature, and potential for anonymity—can enable careless or unscrupulous users to infringe easily on the intellectual property rights of others.

A major concern for copyright holders is illegal file sharing, which is the duplication and dissemination of digital files among Internet users. One of the most powerful aspects of the Internet is how it facilitates the sharing of information between users of all backgrounds, regions, and levels of expertise. But the free sharing of copyrighted works—for example, MP3 files containing copyrighted songs—is likely to be considered a copyright infringement. With digital media, an unlimited number of bit-perfect copies of a work can be made and disseminated. And with faster broadband connections, users can share and download more and larger files—not just songs, but movies, television shows, and PC applications (particularly games) as well.

Copyright holders successfully litigated against the first generation of file-sharing networks, including services such as Napster, that operated based on a centralized index. Victory in court meant taking down the central index, effectively shutting down the network. Users soon started sharing files using new peer-to-peer technologies such as BitTorrent, which because of their decentralized nature, are much harder to shut down than first-generation file-sharing networks (Sisario 2010).

To combat illegal file sharing, some countries have enacted so-called graduated response or "three-strikes" laws. France was perhaps the first country to try this method, introducing such a law in 2009 (France, Ministry of Culture and Communication 2009). Under the law, users who infringed copyrights online would be given a first and second warning. Upon a third infringement, users could be subject to a fine, jail time, and suspension of their Internet access. Monitoring of infringing users was suspended in May 2011 because the software used to collect infringers' IP addresses and send them to the government was found to contain major security flaws; however, there has been no change to the three-strikes law itself (Bright 2011). New Zealand's Copyright (Infringing File Sharing) Amendment Act 2011 also puts in place a three-notice regime to deter illegal file sharing. Other countries, including Malaysia and India, have considered similar laws (Moya 2010). A proposed international agreement known as the Anti-Counterfeiting Trade Agreement has in some drafts included graduated response measures (Kravets 2010).

An alternative solution involves copyright holders working directly with ISPs, forgoing formal legal proceedings. Under this system, copyright holders that detect infringement from a certain IP address contact the ISP in control of that IP address and relay their findings. The ISP then searches its records to correlate the IP address with one of its customers. Finally, the ISP contacts the customer directly, warning the customer that copyright infringement is a violation of the ISP's terms of service and could lead to disconnection. No customer information is revealed to the copyright holder as part of this process. At least one study shows that the majority of Internet users would cease the offending activity after receiving a warning (Hefflinger 2008).

Other issues concerning IPRs in the broadband arena include protection of patents and trademarks. It is easy to infringe patents either intentionally or inadvertently, especially software and business model patents. Trademarks are often involved in cases of cybersquatting (that is, registering a domain name containing someone else's trademark with the intent to deceive or hold it for ransom) and counterfeiting, which is especially common in online auction sites such as eBay. IPRs often run contrary to the concept of a free and open Internet and indeed must be carefully balanced with the rights of users to comment, discuss, and participate freely online.

Notes

1. IPTV Business Act, http://www.glin.gov/view.action?glinID=205548.
2. Telecoms Korea, "IPTV Subscribers Top 1 Million in Korea," October 2009, http://www.telecomskorea.com/market-7674.html.
3. *info*Dev and ITU, "ICT Regulation Toolkit, Module 7: New Technologies and Impacts on Regulation," sec. 3.3.2, Technology Neutrality, http://www.ictregulationtoolkit.org/en/Section.1833.html.
4. Article 3 of the Electronic and Postal Communications Act, 2010, http://www.tcra.go.tz/policy/epoca.pdf.
5. *info*Dev and ITU, "ICT Regulation Toolkit, Module 3: Authorization of Telecommunication/ICT Services," sec. 6.6, Spectrum Authorizations, http://www.ictregulationtoolkit.org/en/Section.1200.html.
6. GSM World, "GSMA Urges Latin American Regulators to Relax Spectrum Caps to Foster Broadband Development," Press Release, January 19, 2009, http://www.gsmworld.com/newsroom/press-releases/2009/2437.htm.
7. In March 2011, following a public consultation opened in 2009, Industry Canada determined that where all conditions for the personal communications services or cellular license had been met, licensees would be eligible to receive a new license for a subsequent term. See Industry Canada (2011).

8. Decree-Law no. 151-A/2000 (July 20, 2000), http://www.anacom.pt/render.jsp?contentId=17094.

9. On March 3, 2010, the minister for broadband, communications, and the digital economy announced that spectrum license reissue would be considered for those existing 15-year spectrum licensees that were already using their spectrum licenses to provide services to significant numbers of Australian consumers or who had in place networks capable of providing services to significant numbers of consumers. As part of an eventual decision, consideration will be given to the five public interest criteria, which were supported by industry, stemming from the 2009 consultation process. The criteria are (a) promoting the highest-value use for spectrum, (b) investment and innovation, (c) competition, (d) consumer convenience, and (e) determining an appropriate rate of return to the community. See Australia, Parliament of the Commonwealth, House of Representatives (2010).

10. Financial Express, "Mobile Operators' Plea for 'Reasonable' Renewal Fees," February 9, 2011, http://www.thefinancialexpress-bd.com/more.php?news_id=125552&date=2011-02-09; TeleGeography, "Bangladesh: Government Agrees to Revise Mobile Licence Renewal Terms," April 13, 2011, http://www.telegeography.com/products/commsupdate/articles/2011/04/13/government-agrees-to-revise-mobile-licence-renewal-terms/.

11. Competitive concerns have been raised regarding municipally owned and operated wireless broadband networks, especially in cases where they may crowd out investment from private parties.

12. ABI Research, "Wi-Fi IC Shipments Forecast to Surpass 770 Million Units in 2010," November 2010, http://www.abiresearch.com/press/1664-Wi-Fi+IC+Shipments+Forecast+to+Surpass+770+Million+Units+in+2010.

13. See *info*Dev and ITU, "ICT Regulation Toolkit, Module 2: Competition and Price Regulation," sec. 4.8.1, The Role of Internet Exchange Points, http://www.ictregulationtoolkit.org/en/Section.2192.html.

14. *info*Dev and ITU, "ICT Regulation Toolkit, Module 2: Competition and Price Regulation," sec. 4.8.3, Internet Exchange Points in Africa, http://www.ictregulationtoolkit.org/en/Section.2195.html.

15. The point that peering and transit arrangements are demand-side substitutes has recently been made by the European Commission in a case involving the Polish regulatory authority's proposal to regulate these services as separate relevant markets. See European Commission (2010a, para. 36).

16. Traffic can also be routed using satellite connectivity, which may be the only alternative in many developing countries, including landlocked countries and SIDSs. However, satellite links have certain drawbacks such as limited capacity, are more expensive, and experience delays in transmission.

17. Colombia, Comisión de Regulación de Comunicaciones, Resolution no. 2065 (February 27, 2009).

18. Serbian Railways, "PE Serbian Railways and PTT Serbia Signed the Contract on Telecommunications Infrastructure Construction along the Lines," January 13, 2011, http://www.serbianrailways.com/system/en/home/newsplus/viewsingle/_params/newsplus_news_id/26885.html.

19. Independent Communications Authority of South Africa, "Infraco ECNS [Electronic Communications Network Services] License," October 2009, http://www.infraco.co.za/Legal/ECNS%20License.pdf.

20. *info*Dev and ITU, "ICT Regulation Toolkit, Module 2: Competition and Price Regulation," sec.7.6.5, Competition and Sharing, http://www.ictregulationtoolkit.org/en/Section.3486.html.

21. ARCEP has promulgated a series of regulations that cover fiber deployments in the country. Different rules apply to installations in rural as opposed to urban areas. In addition, the Law on Modernizing the Economy (August 2008) introduced the idea of fiber "mutualization," whereby the fiber installer must make the fiber available to other companies. ARCEP also contemplated (but ultimately did not adopt) a requirement that multiple strands of fiber be installed initially to accommodate multiple providers.

22. *info*Dev and ITU, "ICT Regulation Toolkit, Practice Note: Structural and Functional Separation of Mobile Network Operators," http://www.ictregulationtoolkit.org/en/PracticeNote.aspx?id=3268.

23. Under the 2009 Telecoms Reform, the market-based remedies are obligations of transparency, nondiscrimination, accounting separation, access to and use of specific network facilities, and price control and cost accounting. See European Union (2009a, art. 13a).

24. Amended EU Access Directive, art. 13a.

25. *info*Dev and ITU, "ICT Regulation Toolkit, Practice Note: Structural Separation Explained and Applied," http://www.ictregulationtoolkit.org/en/PracticeNote.3149.html.

26. Telstra, "Telstra Lodges Structural Separation Undertaking and Migration Plan with ACCC," Press Release, August 1, 2011, http://www.telstra.com.au/abouttelstra/media-centre/announcements/Telstra-lodges-Structural-Separation-Undertaking-with-ACCC.

27. Telstra, "Telstra Signs NBN Definitive Agreements," Press Release, June 23, 2011, http://www.telstra.com.au/abouttelstra/media-centre/announcements/telstra-signs-nbn-definitive-agreements-2.xml.

28. Decree no. 368 of March 18, 2011.

29. The World Summit on Information Society (2003, para. 35), Geneva Declaration of Principles, for example, recognizes that "strengthening the trust framework, including information security and network security, authentication, privacy and consumer protection, is a prerequisite for the development of the Information Society and for building confidence among users of ICTs."

30. Google has to adopt a series of remedial measures spanning from image blurring (required in Canada, Germany, and Switzerland), preannouncing itineraries, and marking its vehicles (required in Italy). In addition to photographs of streets, Google collected information to map Wi-Fi networks and, in doing so, collected (inadvertently, as determined by some authorities to date) personal information, including e-mails, URLs, and passwords. In light of this, Google has agreed to modify its privacy practice and delete personal data collected (in countries such as Austria, Canada, Denmark, and Ireland). Google also has been subject to monetary penalties for these breaches, including a

€100,000 fine imposed in France in March 2011. For a description of Google's practice, see Canada, Office of the Privacy Commissioner (2010).

31. The Treaty of Lisbon, consolidated versions of the Treaty on European Union, and the Treaty on the Functioning of the European Union, Charter of Fundamental Rights of the European Union, art. 16 and 8, respectively, http://europa.eu/lisbon_treaty/full_text/index_en.htm. See also the European Convention on Human Rights, adopted by member states of the Council of Europe, art. 8. The central piece of legislation relating to data protection in the European Union is Directive 95/46/EC, on the protection of individuals with regard to the processing of personal data and on the free movement of such data. In addition, Directive 2002/58/EC, concerning the processing of personal data and the protection of privacy in the electronic communications sector (the e-Privacy Directive), regulates areas that were not sufficiently covered by Directive 95/46/EC, such as confidentiality, billing and traffic data, and rules on spam. This directive was subsequently amended by Directive 2009/136/EC of 25 November 2009 to, among other things, enhance privacy and data protection of Internet users.

32. A data controller is a person (natural or legal) who alone or jointly with others determines the purposes and means of the processing of personal data.

33. Recital 26 of Directive 95/46/EC.

34. See Article 3 § V, of the Law on the Protection of Personal Data Held by Private Parties (July 7, 2010), http://www.dof.gob.mx/nota_detalle.php?codigo=5150631&fecha=05/07/2010.

35. The recommendation provided that OECD member countries should foster the establishment of an informal network of privacy enforcement authorities and should cooperate with each other to address cross-border issues arising from the enforcement of privacy laws. See OECD (n.d.).

36. Current members include authorities from Australia, Bulgaria, Canada, the Czech Republic, the European Union, France, Germany, Guernsey, Ireland, Israel, Italy, the Netherlands, New Zealand, Poland, Slovenia, Spain, Switzerland, the United Kingdom, and the United States. See https://www.privacyenforcement.net/.

37. The CPEA aims to (a) facilitate information sharing among privacy enforcement authorities in APEC economies, (b) provide mechanisms to promote effective cross-border cooperation between authorities in the enforcement of privacy law, and (c) encourage information sharing and cooperation on privacy investigation and enforcement with privacy enforcement authorities outside APEC. See http://www.apec.org/en/Groups/Committee-on-Trade-and-Investment/Electronic-Commerce-Steering-Group/Cross-border-Privacy-Enforcement-Arrangement.aspx.

38. Article 19 of the Universal Declaration of Human Rights and Article 19(3) of the International Covenant on Civil and Political Rights.

39. Any limitation on the right to freedom of expression must pass the following three-part cumulative test: (a) it must be provided by law, which is clear and accessible to everyone (principles of predictability and transparency), (b) it must pursue one of the purposes set out in art. 19, para. 3, of the International

Covenant on Civil and Political Rights, namely to protect the rights or reputations of others or to protect national security or public order or public health or morals (principle of legitimacy), and (c) it must be proven as necessary and the least restrictive means required to achieve the purported aim (principles of necessity and proportionality). See United Nations, Human Rights Council (2011, 8).

40. The case of *Ligue Contre le Racisme et L'Antisémitisme* v. *Yahoo! Inc.*, RG: 00/05308, T.G. (Paris, November 20, 2000), was one of the first national court cases to attempt to restrict content. The case involved the display for sale of Nazi memorabilia via Yahoo.fr. It affected not only cross-border e-commerce, but also ISP liability for content of third parties available on the provider's service as well as jurisdictional issues.

41. See, for example, for example, OpenNet Initiative's research at http://opennet.net/research/regions/asia. ISP filtering is also a key component of the Australian government's cybersafety plan; see http://www.dbcde.gov.au/funding_and_programs/cybersafety_plan/internet_service_provider_isp_filtering.

42. See 31 U.S.C. § 5361 et. seq.

43. Family Online Safety Institute, "ICRA Tools," http://www.fosi.org/icra/.

44. See Code for Industry Co-Regulation in the Areas of Internet and Mobile Content (May 2005), http://www.acma.gov.au/webwr/aba/contentreg/codes/internet/documents/iia_code_2005.pdf.

45. Online Copyright Infringement Liability Limitation Act, 17 U.S.C. § 512.

46. Directive 2000/31/EC.

47. Public Law no. 25 of 2002, http://www.info.gov.za/view/DownloadFileAction?id=68060.

48. Law no. 20.435 (amending Law no. 17.336 on Intellectual Property), art. 85 L-U (May 2010), http://www.leychile.cl/Navegar?idNorma=1012827.

References

Atkinson, Robert D., and Philip J. Weiser. 2006. "A 'Third Way' on Network Neutrality." Information Technology and Innovation Foundation, Washington, DC, May 30. http://www.itif.org/files/netneutrality.pdf.

Australia, Department of Broadband, Communications, and the Digital Economy. 2011. "Telecommunications Regulatory Reform." Department of Broadband, Communications, and the Digital Economy, Canberra. July 8. http://www.dbcde.gov.au/broadband/national_broadband_network/telecommunications_regulatory_reform.

Australian Government. 2010. "Telecommunications Legislation Amendment (Competition and Consumer Safeguards) Act 2010, Act No. 140 of 2010." *ComLaw*, December 15. http://www.comlaw.gov.au/Details/C2010A00140

Australia, Parliament of the Commonwealth, House of Representatives. 2010. "Radio Communications Amendment Bill 2010, Explanatory Memorandum." *ComLaw*. http://www.comlaw.gov.au/Details/C2010B00129/Explanatory%20Memorandum/Text.

Benkler, Yochai. 2006. *The Wealth of Networks: How Social Production Transforms Markets and Freedom*. New Haven, CT: Yale University Press.

BEREC (Body of European Regulators for Electronic Communications). 2010. "Common Statement on Next Generation Networks Future Charging Mechanisms / Long-Term Termination Issues." BoR (10) 24, Rev 1, BEREC, Riga, June. http://berec.europa.eu/doc/berec/bor_10_24_ngn.pdf.

Berkman Center for Internet and Society. 2010. "Next Generation Connectivity: A Review of Broadband Internet Transitions and Policy from around the World." Berkman Center for Internet and Society, Harvard University, Cambridge, MA, February. http://cyber.law.harvard.edu/publications/2010/Next_Generation_Connectivity.

Bright, Peter. 2011. "French 'Three Strikes' Anti-Piracy Software Riddled with Flaws." *Ars Technica*, May 25. http://arstechnica.com/tech-policy/news/2011/05/french-three-strikes-anti-piracy-software-riddled-with-flaws.ars?utm_source=feedburner&utm_medium=feed&utm_campaign=Feed%3A+arstechnica%2Findex +%28Ars+Technica+-+Featured+Content%29.

Brunner, Elgin M., and Manuel Suter. 2008–09. *The Critical Information Infrastructure Handbook*. Zurich: Crisis and Risk Network. http://www.crn.ethz.ch/publications/crn_team/detail.cfm?id=90663.

Canada, Office of the Privacy Commissioner. 2010. "Preliminary Letter of Finding." Office of the Privacy Commissioner of Canada, Ottawa, October. http://www.priv.gc.ca/media/nr-c/2010/let_101019_e.cfm.

Cavoukian, Ann. 2011. "The 7 Foundational Principles: Privacy by Design." Information and Privacy Commissioner of Ontario, Toronto, January. http://www.ipc.on.ca/images/Resources/7foundationalprinciples.pdf.

Cohen, Tracy, and Russell Southwood. 2008. "Extending Open Access to National Fibre Backbones in Developing Countries." GSR08: Six Degrees of Sharing Discussion Paper, International Telecommunication Union, Geneva, November. http://www.itu.int/ITU-D/treg/Events/Seminars/GSR/GSR08/papers.html.

Commission of the European Communities. 2005. "Green Paper on a European Programme for Critical Infrastructure Protection." Commission of the European Communities, Brussels.

Council of Europe. 2001. "Convention on Cybercrime, Budapest, 23.XI.2001." Council of Europe, Budapest, November. http://conventions.coe.int/Treaty/EN/Treaties/html/185.htm.

Crandall, Robert, Jeffrey Eisenbach, and Robert Litan. 2010. "Vertical Separation of Telecommunications Networks: Evidence from Five Countries." *Federal Communications Law Journal* 62 (3, June): 494–539.

Dissanayake, Ridma. 2011. "Beware of Fake Websites Selling WC Tickets." *Daily News* (Sri Lanka), March 30. http://www.dailynews.lk/2011/03/30/news14.asp.

European Commission. 2009a. "Commission Staff Working Document Accompanying the Commission Recommendation on the Regulatory Treatment of Fixed and Mobile Termination Rates in the EU." Explanatory Note SEC (2009) 600, European Commission, Brussels, May. http://ec.europa.eu/governance/impact/ia_carried_out/docs/ia_2009/sec_2009_0600_en.pdf.

———. 2009b. "Main Elements of the Reform." European Commission, Brussels. http://ec.europa.eu/information_society/policy/ecomm/tomorrow/reform/index_en.htm.

———. 2010a. "Commission Decision of 3 March 2010 Pursuant to Article 7(4) of Directive 2002/21/EC (Withdrawal of Notified Draft Measures), Regarding Case PL/2009/1019: The Wholesale National Market for IP Traffic Exchange (IP Transit) and Case PL/2009/1020: The Wholesale Market for IP Traffic Exchange (IP Peering) with the Network of Telekomunikacja Polska S.A." European Commission, Brussels. http://circa.europa.eu/Public/irc/infso/ecctf/library?l=/poland/registered_notifications/pl20091019-1020/act_part1_v4pdf/_EN_1.0_&a=d.

———. 2010b. "Communication from the Commission to the European Parliament, the Council, the Economic and Social Committee, and the Committee of the Regions: A Comprehensive Approach on Personal Data Protection in the European Union." European Commission, Brussels, November. http://ec.europa.eu/justice/news/consulting_public/0006/com_2010_609_en.pdf.

———. 2011. "Communication from the Commission to the European Parliament, the Council, the Economic and Social Committee, and the Committee of the Regions: The Open Internet and Net Neutrality in Europe." European Commission, Brussels, April 19. http://ec.europa.eu/information_society/policy/ecomm/doc/library/communications_reports/netneutrality/comm-19042011.pdf.

European Parliament and Council of Ministers. 2009. "Agreement on EU Telecoms Reform Paves Way for Stronger Consumer Rights, an Open Internet, a Single European Telecoms Market, and High-Speed Internet Connections for All Citizens." MEMO/09/491, European Parliament and Council of Ministers, Brussels, November 5. http://europa.eu/rapid/pressReleasesAction.do?reference=MEMO/09/491.

European Regulators Group. 2007. "ERG Opinion on Functional Separation." ERG (07) 44, European Regulators Group, Brussels. http://erg.eu.int/doc/publications/erg07_44_cp_on_functional_separation.pdf.

European Union. 2000. "Regulation (EC) No. 2887/2000 of the European Parliament and of the Council of 18 December 2000 on Unbundled Access to the Local Loop." *Official Journal of the European Communities*, December 30. http://eur-lex.europa.eu/LexUriServ/LexUriServ.do?uri=OJ:L:2000:336:0004:0004:EN:PDF

———. 2002. "Directive 2002/20/EC of the European Parliament and of the Council on the Authorisation of Electronic Communications Networks and Services (Authorisation Directive)." EU, Brussels, March 7. http://eur-lex.europa.eu/LexUriServ/LexUriServ.do?uri=OJ:L:2002:108:0021:0032:EN:PDF.

———. 2009a. "Directive 2002/19/EC of the European Parliament and of the Council of 7 March 2002 on Access to, and Interconnection of, Electronic Communications Networks and Associated Facilities (Access Directive), as Amended by Directive 2009/140/EC, Art. 13a." European Union, Brussels, November. http://ec.europa.eu/information_society/policy/ecomm/doc/140access.pdf.

———. 2009b. "Directive 2009/140/EC of the European Parliament and of the Council of 25 November 2009 Amending Directives 2002/21/EC on a Common

Regulatory Framework for Electronic Communications Networks and Services, 2002/19/EC on Access to, and Interconnection of, Electronic Communications Networks and Associated Facilities, and 2002/20/EC on the Authorisation of Electronic Communications Networks and Services." *Official Journal of the European Union* 61 (December 18). http://eur-lex.europa.eu/LexUriServ/LexUriServ.do?uri=OJ:L:2009:337:0037:0069:EN:PDF.

———. 2009c. "Directive 2009/114/EC of the European Parliament and of the Council Amending Council Directive 87/372/EEC on the Frequency Bands to Be Reserved for the Coordinated Introduction of Public Pan-European Cellular Digital Land-Based Mobile Communications in the Community." EU, Brussels, September 16. http://eur-lex.europa.eu/LexUriServ/LexUriServ.do?uri=OJ:L:2009:274:0025:0027:EN:PDF.

France, ARCEP (Autorité de Régulation des Communications Électronique et des Postes). 2003. "Consultation publique sur le renouvellement des autorisations des opérateurs GSM." ARCEP, Paris, July 18. http://www.arcep.fr/index.php?id=2231&tx_gspublication_pi1[typo]=8&tx_gspublication_pi1[uidDocument]=72.

———. 2004a. "Le communiqué de presse du Ministère de l'Économie, des Finances et de l'Industrie." ARCEP, Paris, March 19. http://www.arcep.fr/index.php?id=8571&L=1&tx_gsactualite_pi1[uid]=2&tx_gsactualite_pi1[annee]=2004&tx_gsactualite_pi1[theme]=0&tx_gsactualite_pi1[motscle]=gsm&tx_gsactualite_pi1[backID]=2122&cHash=36545c20e5.

———. 2004b. "Décision n° 04-150 de l'Autorité de Régulation des Télécommunications en date du 24 mars 2004 proposant au ministre chargé des télécommunications les conditions de renouvellement des autorisations GSM de la société Orange France et de la SFR." ARCEP, Paris, March. http://www.arcep.fr/uploads/tx_gsavis/04-150.pdf.

———. 2010. "Neutralité de l'Internet et des réseaux: Propositions et orientations." ARCEP, Paris, September 30. http://www.arcep.fr/uploads/tx_gspublication/net-neutralite-orientations-sept2010.pdf.

France, Ministry of Culture and Communication. 2009. "LOI n° 2009-669 favorisant la diffusion et la protection de la création sur Internet." Ministry of Culture and Communication, Paris. http://www.culture.gouv.fr/culture/actualites/conferen/albanel/creainterenglish.pdf.

France, Ministry of the Economy, Finance, and Industry. 2004. "Renouvellement des licences de téléphonie mobile (GSM) des opérateurs Orange et Cégétel." Ministry of Economy, Finance, and Industry, Paris, March 19. http://www.budget.gouv.fr/fonds_documentaire/archives/communiques/2004/c0403191.htm.

Gercke, Marco, Tatiana Tropina, Christine Sund, and Youlia Lozanova. 2010. "The Role of ICT Regulation in Addressing Offenses in Cyberspace." GSR-10, ITU, Geneva, November. http://www.itu.int/ITU-D/treg/Events/Seminars/GSR/GSR10/documents/GSR10-paper6.pdf.

Giglio, Mike. 2011. "The Cyberactivists Who Helped Topple a Dictator." *Newsweek*, January 15. http://www.newsweek.com/2011/01/15/tunisia-protests-the-facebook-revolution.html.

GSA (Global Mobile Suppliers Association). 2010. "Digital Dividend Update." GSA, Sawbridgeworth, U.K., November 5. http://www.gsacom.com/gsm_3g/info_papers.php4.

Guermazi, Boutheina, and Isabel Neto. 2005. "Mobile License Renewal: What Are the Issues? What Is at Stake?" Policy Research Working Paper 3729, World Bank, Washington, DC, October. http://www-wds.worldbank.org/external/default/WDSContentServer/IW3P/IB/2005/09/23/000016406_20050923113019/Rendered/PDF/wps3729.pdf.

Hefflinger, Mark. 2008. "Survey: ISP Warning Would Stop 70% of U.K. File-Swappers." *Digital Media Wire*, March 4. http://www.dmwmedia.com/news/2008/03/04/survey:-isp-warning-would-stop-70%25-u.k.-file-swappers.

Hernandez, Janet, Daniel Leza, and Kari Ballot-Lena. 2010. "ICT Regulation in the Digital Economy." GSR-10 Discussion Paper. International Telecommunication Union, Geneva, November. http://www.itu.int/ITU-D/treg/Events/Seminars/GSR/GSR10/documents/GSR10-paper3.pdf.

Industry Canada. 2011. "Renewal Process for Cellular and Personal Communications Services (PCS) Spectrum Licences." Industry Canada, Ottawa, March. http://www.ic.gc.ca/eic/site/smt-gst.nsf/vwapj/dgso-002-11-pcs-e.pdf/$FILE/dgso-002-11-pcs-e.pdf.

IDA (Info-communications Development Authority of Singapore). 2004. "Accounting Separation Guidelines (Revised with Effect from 24 December 2004)." IDA, Mapletree Business City. http://www.ida.gov.sg/doc/Policies%20and%20Regulation/Policies_and_Regulation_Level1/Revised_ASG.pdf.

———. 2008. "International Sharing: International Gateway Liberalization; Singapore's Experience." Paper prepared for the eighth Global Symposium for Regulators, Pattaya, Thailand, March 11–13. International Telecommunication Union, Geneva, February. http://www.itu.int/ITU-D/treg/Events/Seminars/GSR/GSR08/discussion_papers/IntlSharing_Singapore_web1.pdf.

———. 2011a. "Guidelines for Submission of Applications for Services-Based Operator Licence." IDA, Mapletree Business City, October 13. http://www.ida.gov.sg/doc/Policies%20and%20Regulation/Policies_and_Regulation_Level2/SBOLicence/SBOGuide.pdf.

———. 2011b. "Guidelines on Submission of Application for Facilities-Based Operator Licence." IDA, Mapletree Business City, June 1. http://www.ida.gov.sg/doc/Policies%20and%20Regulation/Policies_and_Regulation_Level3/licensing/FBOGuidelines.pdf.

ITU (International Telecommunication Union). 2004. "Trends in Telecommunications Reform 2004–2005: Licensing in an Era of Convergence." ITU, Geneva, December. http://www.itu.int/ITU-D/treg/publications/Trends05_summary.pdf.

ITU-D (International Telecommunication Union–Digital). 2009a. "Draft Final Report on Question 10-2/1: Regulatory Trends for Adapting Licensing Frameworks to a Converged Environment." Document 1/251-E, ITU-D, Geneva, July 21. http://www.itu.int/ITU-D/treg/Events/Seminars/GSR/GSR09/doc/STudyGroup_draftreportQ10.pdf.

——. 2009b. "Understanding Cybercrime: A Guide for Developing Countries." ICT Applications and Cybersecurity Division, ITU-D, Geneva, April. http://www.itu .int/ITU-D/cyb/cybersecurity/docs/itu-understanding-cybercrime-guide.pdf.

Jensen, Mike. n.d. "Promoting the Use of Internet Exchange Points: A Guide to Policy, Management, and Technical Issues." Internet Society, Reston, VA. http://www.isoc.org/internet/issues/docs/promote-ixp-guide.pdf.

Kende, Michael. 2011. "Overview of Recent Changes in the IP Interconnection Ecosystem." Analysys Mason, Washington, DC, January 23. http://www .analysysmason.com/PageFiles/17527/Analysys_Mason_International_IP_ interconnection_23_Feb_2011.pdf.

Kim, Yongsoo, Tim Kelly, and Siddhartha Raja. 2010. "Building Broadband: Strategies and Policies for the Developing World." Global Information and Communication Technologies Department, World Bank, Washington, DC, January. http://siteresources.worldbank.org/EXTINFORMATIONANDCOMM UNICATIONANDTECHNOLOGIES/Resources/282822-1208273252769/ Building_broadband.pdf.

Kinetz, Erika. 2010. "India's 3G Spectrum Auction Raises $14.6 Billion." *Business Week,* May 20. http://www.businessweek.com/ap/financialnews/D9FQCHS80 .htm.

Kravets, David. 2010. "ACTA Draft: No Internet for Copyright Scofflaws." *Wired,* March 24. http://www.wired.com/threatlevel/2010/03/terminate-copyright- scofflaws/.

Marcus, J. Scott, and Dieter Elixmann. 2008. "The Future of IP Interconnection: Technical, Economic, and Public Policy Aspects." Study for the European Commission, Brussels, January. http://ec.europa.eu/information_society/policy/ ecomm/doc/library/ext_studies/future_ip_intercon/ip_intercon_study_final .pdf.

Martin, Marc S., Henry L. Judy, and Lauren Bergen Pryor. 2010. "FTC Settles with Twitter: More Painful Lessons in Basic Data Security." *TMT Law Watch,* July 1. http://www.tmtlawwatch.com/2010/07/articles/ftc-settles-with-twitter-more- painful-lessons-in-basic-data-security/.

McLaughlin, Andrew. 2002. "Internet Exchange Points: Their Importance to Development of the Internet and Strategies for Their Deployment; the African Example." Global Internet Policy Initiative, June 6. http://www.apdip.net/ documents/policy/strategy/gipi06062002.pdf.

Morocco, ANRT (Agence Nationale de Réglementation des Télécommunications). 2011. "Tableau de Bord Marché Internet au Maroc." Office of the Prime Minister, ANRT, March. http://www.anrt.net.ma/fr/admin/download/upload/ file_fr2164.pdf.

Moya, Jared. 2010. "India Mulling 'Three-Strikes' Plan of Its Own?" *ZeroPaid,* November 16. http://www.zeropaid.com/news/91314/india-mulling-three- strikes-plan-of-its-own/.

NCC (Nigerian Communications Commission). 2010. "Determination on Domi- nance in Selected Communications Markets in Nigeria." NCC, Abuja, March 26.

OECD (Organisation for Economic Co-operation and Development). n.d. "OECD Recommendation on Cross-border Co-operation in the Enforcement of Laws

Protecting Privacy." OECD, Paris. http://www.oecd.org/document/14/0,3343
,en_2649_34255_38771516_1_1_1_1,00.html.

Raja, Siddhartha. 2010. "The Impact of Convergence: Top 10 Broadband Providers." World Bank Blog, July 28. http://blogs.worldbank.org/ic4d/the-impact-of-
convergence-top-10-broadband-providers.

Satola, David, and Henry L. Judy. Forthcoming. "Towards a Dynamic Approach to
Enhancing International Cooperation and Collaboration in Cyber-Security
Legal Frameworks: Reflections on the Proceedings of the Workshop on
Cyber-Security Legal Issues at the 2010 United Nations Internet Governance
Forum." Vilnius, Lithuania, September 15.

Singh, Rajendra, and Siddhartha Raja. 2010. "Convergence in Information and
Communication Technology: Strategic and Regulatory Considerations." Global
Information and Communication Technologies Department, World Bank,
Washington, DC. http://siteresources.worldbank.org/EXTINFORMATION
ANDCOMMUNICATIONANDTECHNOLOGIES/Resources/Convergence_
in_ICT.pdf.

Sisario, Ben. 2010. "U.S. Shuts down Web Sites in Piracy Crackdown." *New York
Times*, November 26. http://www.nytimes.com/2010/11/27/technology/
27torrent.html.

Stucke, William. 2006. "Regional IXPs: The Need for Regional Interconnection in
Africa." In *Commonwealth Ministers Reference Book 2006*. London: Henley
Media Group for the Commonwealth Secretariat.

Tera Consultants and Hogan Lovells. 2010. "Study on the Future of Interconnection
Charging Methods." Report for the European Commission. Tera Consultants,
Paris, November 23. http://ec.europa.eu/information_society/policy/ecomm/
doc/library/ext_studies/2009_70_mr_final_study_report_F_101123.pdf.

TRAI (Telecommunications Regulatory Authority of India). 2003. "Preliminary
Consultation Paper on Unified Licensing Regime." TRAI, New Delhi, November
15. http://www.trai.gov.in/WriteReadData/trai/upload/ConsultationPapers/32/
final%20preconsultation%20paper%205%2015th%20nov%202003.pdf.

———. 2007a. "International Telecommunication Access to Essential Facilities at
Cable Landing Stations Regulations." TRAI, New Delhi, June 7.

———. 2007b. "Recommendations on Infrastructure Sharing." TRAI, New Delhi,
April. http://www.ictregulationtoolkit.org/en/Publication.3632.html.

United Nations, Human Rights Council. 2011. "Report of the Special Rapporteur on
the Promotion and Protection of the Right to Freedom of Opinion and Expression." United Nations, Human Rights Council, Geneva, May. http://www2
.ohchr.org/english/bodies/hrcouncil/docs/17session/A.HRC.17.27_en.pdf.

United States, FCC (Federal Communications Commission). 2011. "In the Matter of
Connect America Fund; A National Broadband Plan for Our Future; Establish-
ing Just and Reasonable Rates for Local; Exchange Carriers; High-Cost
Universal Service Support; Developing an Unified Intercarrier Compensation
Regime; Federal-State Joint Board on Universal Service; Lifeline and Link-Up,
Notice of Proposed Rulemaking and Further Notice of Proposed Rulemaking."
FCC, Washington, DC, February 9. http://www.fcc.gov/Daily_Releases/Daily_
Business/2011/db0209/FCC-11-13A1.pdf.

United States, FTC (Federal Trade Commission). 1998. "Solutions for Data Protection and Global Trade: Remarks of FTC Commissioner Mozelle W. Thompson before the EU Committee of AMCHAM." FTC, Washington, DC, December. http://www.ftc.gov/speeches/thompson/speech123.shtm.

———. 2010a. "Federal Trade Commission (FTC) Decision and Order in the Matter of Dave and Busters." Docket no. C-4291, FTC, Washington, DC, May 20. http://www.ftc.gov/os/caselist/0823153/100608davebustersdo.pdf.

———. 2010b. "FTC Agreement Containing Consent Order in the Matter of Twitter, Inc." File no. 0923093. FTC, Washington, DC, June. http://www.ftc.gov/os/caselist/0923093/100624twitteragree.pdf.

———. 2010c. "Protecting Consumer Privacy in an Era of Rapid Change." Preliminary FTC Staff Report, FTC, Washington, DC, December. http://www.ftc.gov/os/2010/12/101201privacyreport.pdf.

———. 2010d. "Twitter Settles Charges That It Failed to Protect Consumers' Personal Information; Company Will Establish Independently Audited Information Security Program." FTC, Washington, DC, June 24. http://www.ftc.gov/opa/2010/06/twitter.shtm.

United States, White House, Office of the Press Secretary. 2004. "Memorandum for the Heads of Executive Departments and Agencies: Improving Rights-of-Way Management across Federal Lands to Spur Greater Broadband Deployment." Office of the Press Secretary, Washington, DC, April 26.

Wellenius, Björn, and Isabel Neto. 2008. "Managing the Radio Spectrum: Framework for Reform in Developing Countries." Policy Research Working Paper, World Bank, Washington, DC, March. http://www-wds.worldbank.org/external/default/WDSContentServer/IW3P/IB/2008/03/11/000158349_20080311084628/Rendered/PDF/wps4549.pdf.

Williams, Mark. 2008. "Broadband for Africa: Policy for Promoting the Development of Backbone Networks." Global Information and Communications Technology Department, World Bank, Washington, DC, August.

World Summit on Information Society. 2003. "Declaration of Principles: Building the Information Society: A Global Challenge in the New Millennium." International Telecommunication Union, Geneva, December. http://www.itu.int/dms_pub/itu-s/md/03/wsis/doc/S03-WSIS-DOC-0004!!PDF-E.pdf.

WTO (World Trade Organization). 2007. "United States: Measures Affecting the Cross-Border Supply of Gambling and Betting Services: Recourse to Arbitration by the United States under Article 22.6 of the DSU." WT/DS/285/ARB, WTO, Geneva, December 21. http://www.antiguawto.com/wto/84_22_6_ArbitrationReport_21Dec07.pdf.

CHAPTER 4

Extending Universal Broadband Access and Use

Examples from around the world demonstrate that letting the market work can go a long way toward achieving widespread broadband access. Most countries rely primarily on private sector initiatives and investment to achieve broadband access and use throughout their territories. Yet even in well-working market environments, some gaps typically remain between and within countries. Despite declining costs, some locations will not be commercially viable in the foreseeable future. In places with broadband service, some users will not be able to afford it. Persons with disabilities may have difficulty using standard equipment. These are some of the situations where market forces alone are not likely to ensure access to broadband. When government steps in to fill these gaps, it goes ahead of or beyond the market. Achieving broadband access ahead of or beyond the market can be understood as achieving "universal broadband access."

An important question for the achievement of universal broadband access relates to the role of governments when market mechanisms alone do not meet the goals set for broadband access and use. Some degree of government intervention may be required to complement the market and overcome impediments to universal broadband. However, a distinction should be made between enabling, facilitating, and complementing market developments versus substituting government decisions for market forces and public sector investment for private investment. There is also the question of whether and to what extent scarce public sector resources, for which there are many competing demands from other sectors, should be used to extend broadband ahead of or beyond the market.

Governments have a range of instruments at their disposal to narrow gaps or accelerate rollout of broadband (see chapters 1 and 3). The choice of instrument depends on the specific obstacles that the government is trying to overcome. These obstacles may involve the following:

- A proposed investment may not be commercially viable.

- The cost of doing business in the country may be too high.

- Laws and regulations may not be well suited to facilitate the adoption of new technologies and business models.

- There may be a lack or shortage of long-term financing (especially in local currency) commensurate with long economic lives and payback times of the investments.

- Regulatory and political risk may undermine what otherwise would be an attractive business proposition.

- Uncertain prospects for market development may pose excessive commercial risk.

This chapter seeks to provide an overview of what policy makers can do to address perceived shortfalls, to define a broadband development strategy capable of addressing market failures, and to work toward achieving universal broadband service. It discusses the different levels of access that a government strategy may pursue, the role of private-led competitive markets in achieving these objectives, the role of the government in narrowing or eliminating any gaps between markets and the country's development needs, and the design of effective government strategies to meet this challenge. This chapter then examines the use of fiscal resources to support private supply of broadband, including choice of instruments, use of subsidies, and use of mechanisms to collect and disburse funds for subsidy.

Universal Access Strategy and Broadband Development

The history of information and communication technology (ICT) development in low- and middle-income countries shows that private-led, increasingly competitive markets are highly effective at extending new networks and services throughout the population.[1] In numerous countries, chronic, acute telephone service shortages gave way to rapid growth once market-oriented sector reforms were adopted. Mobile phones, initially a premium voice service, now provide a platform for a wide range of information and communication services and applications, covering more than 80 percent of the world's population. The Internet has been growing even faster.

Each new generation of communication services has diffused throughout the population faster than the previous one. Following sector reforms, the number of wireline telephones per capita took over 30 years to multiply tenfold, but it is now in decline worldwide. In contrast, the number of mobile phones took about 12 years to multiply tenfold, and the number of Internet users took only about eight years (table 4.1).

However, differences in access to and adoption of wireline and mobile telephony have remained; the same is also occurring with broadband. Moreover, the private sector often has insufficient interest in investing in broadband in rural and remote areas, even with government incentives. Where market-oriented sector reform falls short of meeting all development needs, public sector support for the deployment, ownership, and operation of a broadband network may be deemed necessary. This may be particularly true where broadband infrastructure is viewed as an essential public utility in which the public sector, such as the local government, is responsible for deployment (ITU 2003, 45).

Thus in each country, policy makers should determine whether private sector–led broadband development in the context of market-oriented reform will achieve economic and social goals or whether more direct, targeted government intervention is necessary. Such an approach is reflected

Table 4.1   Information and Communication Services in Low- and Middle-Income Countries, 1980–2010

Indicator	1980	1990	2000	2010
Population (billions)	3.6	4.4	5.1	5.8
Gross national income per capita (constant 2000 US$)	811	912	1,147	1,811
Wireline phone lines per 100 inhabitants	1.4	2.7	8.3	12.0
Mobile phones per 100 inhabitants	—	0.9	4.6	70.0
Internet users per 100 inhabitants	—	—	1.5	21.0

Sources: ITU, World Telecommunications/ICT Indicators database; World Bank, ICT At-a-Glance database.

Note: — = Not available.

in the European Commission's 2010 Communication on Broadband, which recognizes that, due to the critical role of broadband Internet access, broadband's overall benefits to society appear to be much greater than the private incentives to invest in high-speed networks (European Commission 2010). As a result, stimulating investment beyond the current market-driven levels (while taking into account the recent economic downturn) is seen as key to achieving broadband goals. However, the communication also specifically recognizes that, where intervention is deemed necessary, it is important to limit the government's role as much as possible so as not to distort well-functioning market mechanisms or discourage private investment.

Levels of Access

Countries have adopted various strategies to enable and facilitate universal access to broadband services. Some countries, particularly developed countries with extensive existing wireline penetration, have focused their broadband strategies on providing access to individual users, while other countries with less well-developed network infrastructure have looked more toward providing access to communities and key institutions.

Individual Users and Households

Many countries, some of which are discussed in this chapter, have focused on providing broadband access for individual users and households, including through the extension of universal service definitions and universal service obligations (USOs). In certain instances, these are developed countries that implemented policies to facilitate the deployment of extensive infrastructure and ease the path to high broadband penetration. Therefore, it is more feasible for them to achieve universal broadband by focusing on individual users and households. Finland, for example, was the first country in Europe to include broadband Internet access in its definition of "universal service" and to make broadband a legal right for every citizen.[2] Based on an amendment to the Finnish Communications Market Act,[3] FICORA, the Finnish regulator, designated 26 telecommunications operators as universal service providers. This designation requires such operators to provide, within their operating area, broadband connectivity for consumers and business customers at their permanent place of residence or business, with guaranteed connection speeds of at least 1 megabit per second (Mbit/s).[4] Similarly, the Icelandic government required the country's incumbent operator to guarantee broadband access to the 1,800 remaining unserved consumers in order to achieve universal broadband access (BEREC 2010).

In Denmark, Canada, and Ireland, governments have focused on expanding universal access to households. In June 2010, Denmark announced an ambitious broadband goal of providing access of at least 100 Mbit/s to all households and businesses by 2020 (Denmark, National IT and Telecom Agency 2010). To achieve this goal, the Danish government stated that it would continue to pursue its market-based and technology-neutral approach, focusing on the deployment of broadband infrastructure in Denmark. Measures to promote broadband have included promoting competition in the access network and rolling out wireless broadband to cover hard-to-reach areas (Petersen n.d.). By the middle of 2009, out of 2.8 million households, fewer than 9,000 did not have access to a broadband connection, and by the end of 2010, all households had access to a broadband connection of at least a 512 kilobits per second (kbit/s).

Canada's 2009 Economic Action Plan provided Industry Canada with Can $225 million over three years to extend broadband coverage, with the biggest component of this strategy being the Broadband Canada: Connecting Rural Canadians Program. The program sought to extend broadband service to as many unserved and underserved Canadian households as possible, recognizing that, since communities vary greatly in size, the fact that a community has broadband access does not always mean that service is available to individual households.[5]

Ireland has attempted to ensure nationwide provision of broadband through its National Broadband Scheme (NBS).[6] The NBS was a government project funded under the National Development Plan to provide broadband coverage to areas in Ireland in which broadband services were deemed to be insufficient. Under the scheme, users' connections must be "always on" and capable of 1 Mbit/s downloads and 128 kilobits per second (kbit/s) uploads. The lowest possible cap on downloads was defined as 10 gigabytes (GB) per month, and the connections had to support virtual private networks and voice over Internet Protocol (VoIP) applications.

Communal and Institutional Access

Providing universal access at the individual user and household levels may not always be possible, particularly in developing countries or even in developed countries with significant rural or hard-to-reach areas. As a result, some countries have opted to give greater attention to communal or institutional solutions for providing broadband to end users, especially service to unserved or underserved areas. These projects are often funded, at least in part, by resources from universal service funds (USFs). While traditionally these funds were used primarily or exclusively to support the deployment of telephony services, they have been expanded to support broadband deployment. In certain instances, these USFs are aimed at facilitating the supply of broadband services and are often coupled with initiatives focused on generating demand for such services.

For example, in India, the Universal Service Obligation Fund (USOF) is used to support communal access by providing wireline broadband connectivity to rural and remote areas of the country from the existing rural wireline exchanges of Bharat Sanchar Nigam Limited (BSNL).[7] BSNL provides one kiosk connected to each designated rural exchange, and the connectivity is subsidized by the USOF. The kiosk maintains a workstation with facilities to provide Internet browsing and support other broadband applications such as video chat, video conferencing, telemedicine, and online learning.

Similarly, in Jamaica, the Universal Access Fund (UAF) Company was established in 2005 to accelerate the deployment of broadband through public access in high schools, public libraries, post offices, and other government agencies or institutions. In April 2011, the UAF Company funded a J$543 million (US$6.37 million) project with telecommunications companies LIME (Cable and Wireless) and FLOW (Columbus Communications) to build out a high-speed, islandwide broadband network, again focusing on all secondary schools, post offices, and public libraries in Jamaica.[8]

Some countries take a hybrid approach by establishing a general universal access plan focused on connecting individuals or households while also targeting access at the community level. For example, the U.S. National Broadband Plan (NBP) generally seeks to ensure affordable access to at least 100 million U.S. households with 100 Mbit/s or more download speed and 50 Mbit/s or more upload speed by 2020 (United States, FCC 2010a). The NBP additionally seeks to ensure that every U.S. community has affordable broadband Internet access at speeds of at least 1 gigabits per second (Gbit/s), highlighting the importance of having institutions (such as schools, libraries, and health clinics) both serve as anchors for these local communities and deliver digital literacy, job training, continuing education, and entrepreneurship programs with support from government funds.

Universal Broadband Targets within the Broadband Strategy

Achieving universal broadband access is a challenge for all countries. In the case of developing economies, broadband is also seen as a key component of fostering growth and supporting the provision of a range of services to rural regions. To this end, countries are defining more comprehensive universal access and service (UAS) strategies and aiming to set universal broadband targets in the context of the country's UAS and overall development strategies. As detailed in box 4.1, the Dominican Republic's e-Dominicana Strategy

Box 4.1: Rural Broadband Connectivity in the Dominican Republic

Since 2004, one of Indotel's primary efforts in promoting broadband and the use of computers has been directed toward installing local community informatics training centers (*centros de capacitación en informática,* CCIs) and supplying them with computers. Indotel also provides the entire technical infrastructure including hardware, software, and a backup electric supply system. By January 2009, more than 867 CCIs were in operation, and 462 were in the process of being created.

In 2007, Indotel also launched the Rural Broadband Connectivity project as part of the e-Dominicana Strategy. The project's objective is to provide 508 unserved municipalities with residential and public telephones as well as broadband Internet access through Internet cafés. This is in line with the overall goals of the e-Dominicana Strategy, which include (a) providing Internet access within 5 kilometers of all households at speeds of at least 128 kbit/s; (b) reaching an Internet penetration rate of 40 percent of the population, with at least 30 percent with Internet access speeds of 128 kbit/s or more; and (c) ensuring that at least 50 percent of the population has access to a personal computer. Indotel has funded the cost of these projects with resources from its USF.

Source: Adapted from San Román 2009.

focuses on the long-term promotion of universal access to ICTs, with the objective of ensuring that the country's population develops the necessary skills to use ICTs through the creation of conditions, such as the availability of ICT resources and infrastructure at a reasonable distance from the place of residence and at affordable price levels.[9] The strategy also puts particular emphasis on the link between the development of ICTs, long-term economic growth, and the development of human capital (ITU 2008). In the United States, the NBP advocates an expanded funding commitment to the Community Connect Program, which provides free Internet access to residents with the goal of facilitating economic development and enhancing educational and health care opportunities in rural communities. The European Commission, in line with the European Union's common interests of territorial, social, and economic cohesion, has indicated its support for state financial resources to assist in providing broadband services to those areas currently unserved and where private investors do not have plans to deploy broadband networks in the near future (European Commission 2009).

As these examples show, it is becoming increasingly important for UAS policies and broadband policies to influence each other.[10] UAS policies can promote the spread of broadband services and stimulate demand. Broadband policies can use a range of regulatory and fiscal options to reduce costs (for example, international gateway liberalization) and facilitate broadband network investment, which, in turn, leads to better access at lower prices.

Although several countries have separate broadband and UAS policies (India, Jordan, Malaysia, Pakistan, and South Africa), the boundaries between UAS and broadband policy are not as clear in other countries. Recent trends, however, show that policy makers are increasingly merging the two topics to accommodate universal broadband challenges. UAS and universal broadband availability have become fundamentally linked in nearly all countries' universal access strategies, except in least developed countries (Dymond 2010). As shown in box 4.2, Chile has a new Information

Box 4.2: Chile's Digital Connectivity Plan

In 2010, the Chilean government launched a program to provide digital connectivity to 1,474 localities with about 3 million people in rural areas that lack access to the Internet. Households, businesses, schools, health centers, and government offices will be able to connect to the Internet at 1 Mbit/s download and 512 kbit/s upload speeds, with service quality and prices similar to those prevailing in larger towns. The objective is to enable rural communities with productive potential to participate more effectively in the economy, through innovation and increased competitiveness. The program also seeks to increase the reach of the Internet among low-income rural population groups. The program will invest about US$100 million, including a US$43 million subsidy financed equally by the central government's Fondo de Desarrollo de las Telecomunicaciones and the regional governments. Locations were selected based on demand expressed by local and regional authorities and civil society organizations and also reflect development priorities in agriculture, small and medium enterprises, and tourism. Costs were estimated with an engineering model using combinations of fiber and wireless technologies, including investment and operation and maintenance costs. Benefits reflected forecast revenues, business productivity gains, and benefits from e-government. This model underestimated total benefits, as the impact on education, employment, and other externalities, while recognized, could not be quantified in monetary terms. Nonetheless, the program was estimated to yield a small, but positive, economic net present value (NPV). The maximum subsidy was set at the equivalent of US$63 million, which would make the financial NPV = 0, rendering the program commercially viable. The actual subsidy needed was determined through open competitive bidding, with the eventual winning bid at US$43 million. Implementation is under way and due for completion in 2012.

Sources: Chile, SUBTEL 2008; see also Acta de Apertura de los Proyectos Financieros (Sobres S4) para la Asignación del Proyecto Infraestructura Digital para la Competitividad e Innovación and Proyecto Bicentenario, Red Internet Rural: Todo Chile Comunicado, both at http://www.subtel.cl; *info*Dev and ITU, "ICT Regulation Toolkit, Module 4: Universal Access and Service," sec. 4.1.3, Relationship to Broadband Policy, http://www.ictregulation toolkit.org/en/Section.3258.html.

Society Universal Access Policy. The new policy brings together Chile's broadband policy and the Universal Access and Service Fund (UASF) and seeks to enable rural communities with productive potential to participate more effectively in the economy through innovation and increased competitiveness.

A universal broadband policy should be a central part of the ICT framework and not construed as simply a result of corporate social responsibility or acts of "goodwill" by investors in the ICT sector. Policies and measures should be formulated carefully, and universal broadband policies should be given a proper space in the national policy and legislative frameworks for development as well as in the institutional framework for telecommunications regulation. Thus it is important not only to set universal broadband targets in the context of the country's UAS policy, but also to take account of the country's overall development strategy. Universal broadband targets should be developed based on the country's short- and long-term goals for economic growth and broadband deployment. Development policies (for example, e-education) should also consider telecommunications-specific regulations and policy goals, such as competitive parity between players (Atkinson, Correa, and Hedlund 2008, 19).

As illustrated in a World Bank study on the Republic of Korea, a key factor in achieving widespread broadband access was the country's holistic approach to defining and implementing numerous policy developments and initiatives, including policies to promote universal access to broadband. The Korean government has sought to promote ICTs, particularly broadband networks and services, by implementing a series of "master plans" that extend over several years and provide strategic, long-term development frameworks. Each framework has set out overarching policy objectives as well as the supporting policies to achieve these goals. Among the key elements taken into consideration were the policies promoting universal access to broadband (Kim, Kelly, and Raja 2010, 20–21).

Mechanisms to Drive Universal Broadband Access

Government Intervention

In many cases, broadband infrastructure projects are being led by the private sector, with the government's role focused on developing policies to encourage and facilitate these private sector initiatives. Within this context, countries have recently adopted more integrated strategies for developing and financing telecommunications services. This is particularly true in the case of universal access and the financing of large infrastructure projects, including projects to fund broadband, since implementation of such projects has generally been seen to require the involvement of both private sector financing and public authorities.

Policy makers have moved toward creating a multipronged approach to promoting universal broadband access. Such complementary strategies have been defined in addition to market liberalization and regulatory initiatives aimed at promoting broadband in general as well as focusing on universal access obligations or special conditions that favor projects in high-cost or low-income areas. Table 4.2 shows some of the mechanisms defined as part of the Philippine's UAS strategy.

Other countries also have introduced a mix of measures to promote universal broadband access:

- *Canada.* While the Telecommunications Act of 1993 clearly recognizes the role of the private sector, it also calls for "reliable and affordable telecommunications services of high quality accessible to Canadians in both urban and rural areas in all regions of Canada." In reality, policy makers recognize that the high cost of rural and remote broadband access requires public sector funding and initiatives to supplement market forces.[11]

Table 4.2   The Multipronged Universal Access and Service Strategy in the Philippines

Strategy	Details
1. Implement improved regulations	• Key is spectrum and tower sharing • Will improve market efficiency • Increased broadband rollout and extended coverage into rural areas due to reduced costs
2. Stimulate demand (households and government)	• Key is household personal computer loan program and government demand aggregation • Will stimulate market • Increased broadband rollout and extended coverage into rural areas due to more demand
3. Universal Access and Service Fund and competitive subsidy bids (for residual)	• Key is public Internet access and connectivity • Will address the access gap • Increased broadband rollout and extended coverage into rural areas due to subsidies

Source: Taken from Beschorner 2010.

- *Finland.* The government's approach has relied on market competition to drive growth. Limited public financial intervention, however, has been implemented since 2004, when the first national broadband strategy to achieve universal broadband access was articulated. The strategy provided that broadband access in sparsely populated and rural areas should be supported by structural funds from the European Union (EU) and the national government. Since then, it has been adapted and allows for public sector intervention when necessary. The main objective of the December 2008 national broadband scheme for 2009–15 was to ensure that more than 99 percent of the population in permanent places of residence, as well as businesses and public administration offices, are no farther than 2 kilometers from a 100 Mbit/s fiber optic or cable network. The government expects telecommunications operators to increase the rate of coverage to 94 percent by 2015, depending on market conditions, while public finances are being used to extend services to sparsely populated areas where commercial projects may not be viable, bringing coverage to the target of 99 percent.[12]

- *Peru.* Following privatization of the state telecommunications enterprises in 1994 and opening of the market to new entrants and competition in 1999, Peru's telecommunications sector progressed dramatically from being the second least developed in Latin America to about average for the region. The government also played an important role in extending telecommunications services to places where they are not commercially viable. The Fondo de Inversión en Telecomunicaciones (FITEL, Telecommunications Investment Fund) finances the provision of telecommunications services in rural and other priority development areas that do not have service. As shown in box 4.3, the experience of FITEL in Peru suggests several lessons that are widely applicable to broadband development.

Government intervention is not just limited to the national level. Some countries have recognized that municipal and local governments often possess the technical expertise to deploy networks as well as the ability to engage in long-term financing strategies for building out infrastructure (OECD 2008). In the United Kingdom, the government is taking a new approach to delivering connectivity in rural and hard-to-reach areas where the market is unlikely to provide service. Where local authorities have designated superfast broadband as a development priority, Broadband Delivery U.K. will work with the local government to coordinate projects and financing. Such collaboration will be the foundation for the government's US$859 million commitment until 2015 (United Kingdom, Department for

Box 4.3: Broadband Development in Remote and Underserved Locations: Lessons from Peru

In Peru, the government has played an important role in extending telecommunications services to places where they are not commercially viable on their own. Several lessons from the experience of Peru are widely applicable to broadband development.

With a pro-competitive regulatory framework, the market on its own goes a long way toward rolling out broadband, but gaps are likely to develop between the more profitable areas and the rest of the country and among income groups. In this situation, a telecommunications development fund can accelerate rollout and reach by leveraging private investment, focusing subsidies on clearly defined target population groups, and using the market to determine and allocate subsidies.

The private sector and the communities themselves respond vigorously to the opportunities presented by new technologies and demands. In Peru's rural areas, some of the most creative initiatives to use wireless technologies came from agricultural associations, local communities, and small entrepreneurs, rather than from established telecommunications companies. In urban areas, intense competition among numerous small informal shops offering public Internet access resulted in the number of Internet users rising quickly—well above what could have been expected with regard to the number of servers and connections. This contributed to widespread dissemination of ICT throughout the population.

For new infrastructures and services to contribute to development, it is also necessary to support the demand side. The national broadband project, financed by FITEL, includes developing content and building capacity among users and local entrepreneurs. Operators are expected to establish a portal in each community, with information on economic activities, tourist attractions, and other material of local significance. The content will be updated by the community itself, with support from the operators. Funding for these activities, however, remains very limited.

A national broadband policy is needed to guide the scope and direction of efforts to extend new services ahead of or beyond the market. For many years, FITEL lacked such guidance. The choice of services to be supported, target population, and technical requirements were reasonable and generally responsive to changing technologies and demands, but appeared to lack consistency. For example, the required Internet speed varied among projects between 9.6 kbit/s and over 600 kbit/s. A commission to propose a national broadband policy aimed at achieving affordable access countrywide was established in 2010.

A funding mechanism that allows collecting mandatory contributions and only later decides how to use them tends to disburse funds slowly and inefficiently. Between 1995 and 2008, FITEL spent slightly over 40 percent of the US$279 million it had collected. At some point, it had about US$100 million in cash. For five years, no new investments were approved. Partly this reflected

(continued)

Business, Innovation, and Skills and Department for Culture, Media, and Sport 2010). The French government has also given local authorities a greater role in developing broadband infrastructure. The Caisse des Depots et Consignations (a government-owned bank) provides concessional loans to municipalities for broadband development.

Improve the Legal, Regulatory, and Business Environments

Address Universal Access and Service Challenges through Policy and Regulatory Solutions

Before making public investments in rolling out broadband networks, governments should first look at regulatory tools that might be able to increase entry and competition and hence maximize what the market can deliver on its own, including in unserved and underserved areas. Whereas a more "traditional" framework would likely impose universal service obligations on designated operators as a first-step measure, governments should consider focusing on policies and regulatory tools that incentivize and encourage operators to extend service into underserved and unserved areas, including through increased privatization and liberalization (European Commission and ITU 2011).

In establishing the framework necessary to provide universal broadband access and service to rural populations, policy makers are introducing licensing regimes that will allow them the flexibility to take advantage of technological development and convergence. Operators are also expressing a preference for alternatives, such as accepting reasonable build-out targets in their licenses or negotiating ex ante specific rural UAS targets with the regulator in exchange for relief from UASF levies or taxes. Countries have thus adapted their licensing regimes to achieve such targets. In Brazil, for example, Anatel established licensing provisions allowing operators to obtain additional authorizations after they fulfilled their universal service obligations.[13] Brazil Telecom, which met its USOs by 2004, was granted the right to roll out additional mobile and long-distance call services in southern areas where it previously only was licensed to provide local services. In addition, Anatel is now pursuing broadband UAS targets, planning to connect all of its 5,600 municipalities with minimum broadband capacity, as well as creating and connecting 8,500 telecenters and 50,000 urban schools.

Within the context of defining a universal broadband strategy, it is also important to consider what other key legal, regulatory, and business environment constraints are holding back universal broadband development and what government can do to overcome these constraints. If governments are to provide support to encourage the universal deployment of broadband, they should do so in a manner that guarantees equitable access for all.

In Sweden, for example, government policies require recipients of public funds to operate open-access networks. Promoting such nondiscriminatory access might come more readily from municipal governments, many of which own and operate local networks. The Czech Republic has implemented legislation aimed at enabling the government to capitalize on the privatization of Český Telecom by putting 1 percent of the proceeds from the privatization into a fund that will be used to co-finance infrastructure projects for metropolitan and local networks. Conditions on receiving the funds include participation by the relevant regions and operation of the network under open-access rules.

Revise the Scope of Universal Access and Service to Include Broadband

In many countries, the scope of UAS policies as a whole has traditionally focused on the provision of basic telephony, either to individual households or through communal or institutional access. As stated in a 2010 World Bank study, however, the availability of new, less expensive technologies allows countries to adopt more ambitious UAS policies while avoiding higher costs and the need to engage in continuous subsidies (Muente-Kunigami and Navas-Sabater 2010, 7).

In order to expand broadband connections to rural areas where they are currently unavailable, some countries are considering turning broadband into a USO and reforming their universal service policies. In numerous countries, the scope of UAS has evolved to include broadband. According the International Telecommunication Union (ITU), over 40 countries now include broadband in their universal service or universal access definitions.[14] The following are some of the countries that have revised the scope of their universal service policies and USFs to include broadband:

- *India* was one of the first countries to include broadband in the USOF in 2006. The USOF allows for the support of broadband connectivity and mobile services in rural and remote areas of the country.

- In *Morocco*, the USF's priorities were expanded through a revision of the law in 2004 to include rural public telephony, installation of community Internet centers, and an increase in broadband capacity through various programs.[15]

- In *Switzerland*, the government decided that, beginning in January 1, 2008, universal service providers must provide a broadband connection to the whole population, via digital subscriber line (DSL), satellite, or other technologies. Connections must offer at least 600 kbit/s download speeds and 100 kbit/s upload speeds, and the monthly subscription cannot be more than SW F 69 (US$85).

Support Private Sector Network Build-Out: Supply

Governments may adopt a range of instruments to accelerate the supply of broadband ahead of or beyond the market. These can include subsidies for investment, equity in public-private partnerships (PPPs), facilitated access to rights-of-way, preferential tax treatment, long-term loans for investment in local currency, on-lending loans, credits or grants from international development organizations, and guarantees to offset regulatory or political risk. Implementation of such policies can encourage operators to focus on deploying networks and services in unserved areas since they will be able to earn a higher rate of return on their investments over the long term (World Bank 2005).

The most common practice is for the government to contribute money when needed to ensure that important investments in rural development are commercially viable. This is done primarily by providing one-off subsidies for investment and start-up, focusing on unserved and underserved areas in particular. Alternatively, governments can contribute equity to PPPs with similar objectives. For example, the government can help to build broadband backbone networks that are then made accessible in equal terms to all interested downstream providers. When well designed, these practices can mobilize substantial private sector investment, enable large projects that otherwise would not materialize, contain the cost and risk borne by the government, and jump-start sustainable markets from which the government can exit quickly.

It may be possible to reduce the cost of broadband development by giving investors access to rights-of-way along railways or roads, on rooftops, and on other public property. Absent alternative uses for these rights-of-way, their opportunity cost to the public is negligible, and if they are made equally available to all interested parties, their use will not distort competition. Therefore, granting rights-of-way may help to reduce the total investment cost of broadband development.

Granting exceptional tax treatment is also sometimes considered. Good tax practice in general suggests that a particular economic activity should not be singled out for tax conditions that do not apply to all like activities throughout the economy. This means that taxes or duties that apply only to broadband should be phased out and, conversely, that exemptions from generally applicable obligations should be avoided. The Hungarian government, for example, took efforts to institute tax incentives to further the build-out of broadband. Specifically, Hungary's government grants a tax reduction of 50 percent on profits as a way to support the construction of broadband infrastructure. The concessions are available only to telecommunications companies if their expected profits exceed Ft 50 million (US$250,000) and if they have invested at least Ft 100 million (US$500,000). The tax allowance cannot be applied to Internet service providers (ISPs) if the infrastructure is built in areas where Internet service is already provided or where the investment does not contribute to the growth of infrastructure.

Investment in new open and competitive networks, including broadband networks, can also be supported by the actions of national and local authorities in lowering costs. The European Commission's 2009 Guidelines on the Application of State Aid Rules, for example, lay down the conditions for public financial support on nonmarket terms for broadband deployment in areas where commercial investments are unlikely to take place in the foreseeable future. The main objective of the 2009 guidelines is to assist the actions of national and local authorities. The guidelines are presented as part of the broadband package, together with the two other broadband commitments made by the commission in the Digital Agenda for fast and ultra-fast Internet: the Next-Generation Access (NGA) Recommendation to provide regulatory guidance to national regulators and the Radio Spectrum Policy Program to improve the coordination and management of spectrum and hence facilitate, among other things, the growth of wireless broadband. In the guidelines, the commission recognizes that broadband networks tend to cover only part of the population since they are generally more profitable to roll out where potential demand is higher and concentrated (that is, in densely populated areas) rather than in areas with less population, specifically because of the high fixed costs of investment and high unit costs. The commission distinguishes acceptability of state intervention among (a) areas where no broadband infrastructure exists or is unlikely to be developed in the near term and where support is considered to promote territorial social and economic cohesion and address market failures (so-called white areas); (b) areas where market failure or a lack of cohesion may exist despite the existence of a network operator, thus requiring a more detailed analysis and careful compatibility assessment prior to allowing state intervention; and (c) so-called black zones, which are defined as a given geographic zone where at least two broadband network providers are present and broadband services are provided under competitive conditions (facilities-based competition). In these black zones, the commission does not consider that there is a market failure, so there is little scope for state intervention. In the absence of a clearly demonstrated market failure, state funding for the rollout of an additional broadband infrastructure is not available.

Instruments of Fiscal Support for Universal Broadband Access

Subsidies as an Instrument of Fiscal Support

Subsidies are the most commonly used instrument to support universal broadband development ahead of or beyond the market.[16] Subsidies are used extensively in the telecommunications, electricity, transportation, water supply, and sanitation sectors. If well designed, subsidies can be accurate and transparent and can effectively target the desired beneficiaries. Subsidies may be financed by government budgets, user surcharges, international grants, and other sources. A central agency or financial institution, a specialized fund, or some other mechanism may be used to collect and distribute the subsidies.

The Rationale for Subsidies

Generally speaking, a subsidy exists when the costs incurred in supplying a service are not fully recovered from the revenues raised by selling this service. The economic rationale for a subsidy is based on the existence of consumption and production externalities, network externalities, and scale economies. Also, access to service at affordable prices may be considered essential for enabling the population to participate equitably and effectively in the modern economy.

In the context of market-oriented economic policies, subsidies are aimed at developing sustainable markets for the private provision of services. They are designed to turn investments that are desirable from the viewpoint of the economy at-large, but not profitable by themselves, into commercially viable undertakings. Projects that are not demonstrably good for the economy at-large or likely to stand on their own do not justify subsidy and are rarely undertaken.

Good Subsidy Practice

Good subsidy practice in infrastructure projects commits all participants to contribute to financing the provision of services:

- Service providers invest and risk their own resources to set up the facilities and provide the services during a given time under specific conditions.

- The government helps service providers to meet some of their investment and start-up costs.[17]

- Customers pay for the use of services at least as much as is needed to meet operating and maintenance costs. Where domestic installations are involved, customers are also required to pay part of the investment cost, as a confirmation of demand for service and commitment.

The design of subsidies is closely tied to the available service delivery mechanisms. Subsidies are channeled through the service supply chain in ways that aim at being neutral with respect to competition, service providers, service options, and technologies.

Competition for Subsidies

Subsidies for broadband development are increasingly being determined and allocated among firms participating in a competitive public tender that is awarded to the firm that bids the least subsidy. This modality is sometimes referred to as "least-cost subsidies." Compared with traditional public sector funding of investments, least-cost subsidies result in lower cost to the government, mobilization of substantial private investment, and enhanced transparency. Other forms of competition for subsidies include competition among projects proposed by communities or firms, competition among regional governments for central funds, and competition among sectors for a share of these funds. Implicit in all modalities is competition among technologies and business models for delivering these services.

Competition among firms for least-cost subsidies to provide infrastructure services was pioneered in Chile in the mid-1990s for the provision of rural pay phones. Since then, it has become a recognized good practice in telecommunications and has also spread among upper-middle-income countries and services that appeal most to private investors (telecommunications, electricity) and to lower-income countries and less attractive services (water and sanitation, transportation).

Competition among firms for subsidies is increasingly being used to support broadband development programs and comprises the following main steps:

- The government defines the broad objectives, target population, and levels of funding of the subsidy program. It also establishes key service conditions such as service quality, maximum prices, and duration of service commitments (see the experience of Mongolia in box 4.4).

- Specific service needs and choices are identified primarily by prospective beneficiaries and communities. Economic, financial, and technical analysis is used to select and prioritize projects that are likely to be desirable from the viewpoint of the economy at-large, but not to be commercially

Box 4.4: Universal Service Subsidies in Mongolia

With the lowest population density in the world, Mongolia launched two pilot programs in 2006: one to provide public access telephones for nomadic herders in 27 communities, the other to extend wireless Internet and voice service to one *soum* (rural administration center). For these pilots to be commercially viable, estimated subsidies of US$5,100 to US$6,200 would be needed for the herder community and US$63,000 to US$73,400 for the *soum*. Mongolia's regulatory authority conducted separate competitive bidding processes for each pilot. Each request for proposals specified the maximum allowable subsidy and included a draft service agreement, which specified how the subsidies would be paid out—linked to progress on construction and initial operation. Bidders were required to submit evidence of corporate and financial qualifications and experience in Mongolia. The bids were evaluated first on technical and operating compliance with the specifications. Those bids that passed were then evaluated on their requested subsidies. The bidder requesting the lowest subsidy was awarded the subsidy. Each tender attracted two bids from operators already active in the target markets. Three of the four were in substantial compliance. The networks in both pilots were implemented in September–November 2006 and were fully operational before the onset of the winter. Both operators met targets for service availability and technical quality ahead of schedule.

Source: Dymond, Oestmann, and McConnell 2008.

viable on their own, and to determine the maximum subsidy justified for each project.

- Private firms submit competitive bids for these projects. Subject to meeting service conditions and complying with rules that apply to all bidders, bidders are free to develop their business strategies, including choice of technology.

- Subsidies are awarded to the bidders that require the lowest one-time subsidies. Alternatively, bids are invited for fixed amounts of subsidies and awarded against other quantifiable measures, such as the lowest price to end users or the fastest rollout of service.

- Subsidies are paid in full or in installments, linked to implementation of investments and start of service.

- Service providers own the facilities and bear all construction and commercial risks. No additional subsidies are available downstream for the same services.

- The government monitors and enforces service quality and pricing standards, protects users against arbitrary changes of service, and provides investors with stable rules of the game.

Success Factors of Competition for Subsidies

Competition among firms for subsidies can be used to extend broadband development ahead of or beyond the market. Compared with traditional public sector infrastructure funding, competition among firms for subsidies can mobilize private investment, reduce government outlays to meet given policy objectives, promote cost-effective solutions and the emergence of new entrepreneurs, and enhance transparency.

But competition among firms for subsidies is likely to succeed only when certain critical factors are present. These factors relate to demand, supply, and the enabling environment. The experience from telecommunications and other infrastructure service sectors is likely to be relevant to broadband as well (table 4.3). Individual situations can be examined initially by reference to these factors. Whether the factors of success are in place or not can only be assessed case by case. A simple classification of countries and broadband projects cannot do justice to the complexity of the issues.

The demand side. Competition for subsidies to extend broadband ahead of or beyond the market is likely to work well only if users are willing to pay at least as much as is needed to keep the service running after initial

Table 4.3 Competition among Firms for Subsidies: Factors Critical to Success

Demand factors	Supply factors	Enabling environment
• Users are able and willing to pay for services.	• Several firms are qualified to bid for subsidies.	• Elements of a market-oriented legal and regulatory framework are in place.
• Service features are tailored to user needs and preferences.	• Business opportunities are aligned with operators' strategies.	• Government has access to stable and reliable sources of subsidy finance.
• Services have considerable growth potential.	• Project components are cost-effectively packaged.	• Private investors have access to long-term financing.
		• Donors and different tiers of government are able to coordinate financing policies.
		• Institutional capacity is in place to implement and manage a competitive subsidy mechanism.

Source: Adapted from Wellenius, Foster, and Calvo 2004.

investment and start-up. There is ample evidence that even low-income users, given the opportunity, spend a significant part of their income on communication services. In developing countries, about 10 to 20 percent of rural household income is spent on infrastructure services (for example, communications, electricity, water, and transportation).[18] In some countries in Africa, rural households spend over 5 percent of their income on telecommunications. A survey in Nigeria found that about 7 percent of household income was spent on mobile telephone service.[19] These levels of expenditure on communication may suffice for the provision of communal broadband facilities (which aggregate the local population's purchasing capacity) even in very poor localities, but the income threshold will be higher for individual household connections. For competition for subsidies to achieve its intended purposes, broadband service targets must be consistent with realistic estimates of the users' willingness to pay.

Besides income, other factors influence the demand for rural infrastructure services. These factors include location, information on options, ease of use and payment, and reliability of communal services as well as hassle-free connection, low fixed periodic charges, easy control of expenditures, accurate billing, and prompt repair of household connections. Demand growth potential is a major determinant of sustainability. For example, some companies that provided subsidized rural pay phones in Chile also offered individual telephone lines and Internet access to homes and small businesses on commercial terms using the subsidized infrastructure at marginal cost.

The supply side. The primary concern on the supply side is having enough qualified providers competing for the subsidies. Competition for subsidies works best when several firms compete for each subsidized project. In such situations, the lowest-subsidy bid is typically between one-third and half the maximum available, with occasional zero-subsidy bids. When there is only one bidder for a project, bids tend to be close to the maximum subsidy available. In that situation, the subsidy awarded is determined by the calculus of costs and benefits used to design the bidding process more than by the market, and errors of calculus become errors of investment.

The number of prospective bidders depends partly on how well the market for communication infrastructure services is already developed in the country. Whether eligible firms actually bid for the subsidies depends on the extent to which the projects offer attractive business opportunities that fit the firms' overall business strategies. For example, ENTEL-Chile, one of the largest Chilean telecommunications companies, did not regard rural telephone service as a strategic business interest and never bid for the subsidies being offered from 1995 to 2000, despite the competitive advantage of its countrywide network and substantial presence in rural areas. Yet the same firm in 2008 bid for subsidies to roll out broadband, came in second, and eventually agreed to take on the project after the winner (which had requested zero subsidy) failed to firm up financing. An insufficient number of bidders may also result from lack of confidence in the regulatory regime, entry limitations still in place from earlier times (for example, only one company authorized per region), or a process that is competitive in its initial rounds but ends up with providers consolidating their markets on a regional basis.

There is still debate over the merits of offering exclusive operating rights to enhance the value of a business opportunity offered in an otherwise pro-competitive market environment. Exclusivity is generally no longer granted for the provision of telecommunications services, but the practice is mixed in other sectors. Exclusivity, besides running against market-oriented reform principles, is unlikely to add value to concessions or licenses in markets that operators are not prepared to serve on their own. Exclusive rights to subsidy, in contrast, make sense since the objective of the subsidy program is to extend service where none is available rather than to promote competition in the market. Subsidizing demand rather than supply can reconcile both objectives.

How demand is aggregated into projects and the extent to which this is left to individual bidders may affect considerably their ability to compete. Bundling the provision of several infrastructure services may help to spread out costs and attract more bidders.[20] A bidding process that is simple, transparent, expeditious, and not unduly burdensome on the bidders also contributes to attracting bidders.

The environment. Competition for subsidies among firms is designed to be used to narrow gaps between the market and development needs, not to substitute for the market or to compensate for regulatory distortions of the market. It makes practical sense only when the private sector is responsible for service provision, new entry and competition are encouraged, and cross-subsidies within firms have been largely phased out. A clear, stable, and credible legal, regulatory, and general business framework is needed for prospective service providers to make reasonable estimates of costs and revenues and assess the risks they are being asked to assume. Service providers are especially concerned about the rules and practices with regard to competition, pricing, interconnection, and access to scarce resources within the sector as well as with regard to private ownership, foreign exchange, and taxation of businesses in general.

Competition for subsidies also requires that all key players have access to financing. A major aspect of qualifying firms to bid for subsidies is their capacity to mobilize equity and debt financing for the components of investment and start-up that are not subsidized. This is not likely to be a problem when programs are large enough to be attractive to foreign investors, who have access to long-term financing in the international markets. However, smaller-scale schemes targeted primarily at local operators may face difficulties if longer-term financing is not available through the domestic capital markets.

The government must have in place sustainable sources and transparent mechanisms to collect money for and disburse the subsidies it is offering. In the telecommunications sector, the revenues often come from levies on sector revenues or sometimes from the proceeds of spectrum or operating license auctions. Funding within the sector, although second best to funding from the government in terms of economic efficiency, fairness, and fiscal discipline, can improve the reliability of access to subsidy resources. Governance of subsidy resources is critical. Even where sector funds have been established, in some instances the resources remaining have not been disbursed or have been diverted to meeting pressing fiscal needs.

A key challenge in implementing rural infrastructure delivery models based on competition for subsidy is coordinating strategies among donors and among different tiers of government. A concession of a private operator, premised on a partial investment subsidy and a financially self-sustaining operation over the medium term, would be destroyed if, within a couple of years, a local municipality or a nongovernmental organization started to offer a free service in the same geographic area.

Institutional capacity is needed to establish and run a competitive subsidy system for rural infrastructure services. This includes originating and shepherding specific legislation and regulations, setting up and managing the financing mechanisms, designing and implementing the bidding processes, monitoring service development, and enforcing service commitments. In countries with well-established public administration traditions, a well-designed program of competition among firms for rural subsidies can be implemented by a rather small team of professional and support personnel.

Long-term sustainability of the model will depend on how well and realistically the risks have been apportioned among the players and the extent to which commitments can be enforced. Since subsidies are paid early in the project life cycle, should expected revenue streams later fail to materialize, the operator may face a sustained negative cash flow and prefer to close down. Service obligations and penalties for noncompliance may deter such behavior in some cases. Ultimately, even if construction and commercial risks are initially assumed entirely by the private operators, if they fail, the government may have no choice but to step in and take measures to maintain service, since that was its objective in the first place.

Sources of Funds to Support Broadband Development

Government Programs

Because extending broadband ahead of and beyond the market is intended to benefit society at large, for reasons of both economic efficiency and equity, the first choice for financing support is the government budget. This can be done at different levels of government, such as central, state, or municipal governments. It may include direct use of funds from the government budgets or use of funds from government programs aimed at addressing specific economic or regional development objectives.

For example, in 2002 the federal government of Canada launched the Broadband Rural and Northern Development Pilot (BRAND) to extend broadband to about 400 unserved localities, with priority given to disadvantaged First Nation, Inuit, and Métis communities.[21] BRAND was implemented by Industry Canada. The total cost of BRAND was Can $174 million, of which Can $78 million (44.7 percent) was financed from Industry Canada's budget. The rest was financed mainly by the firms that provided the services, community leaders, provincial and municipal governments, and other federal and provincial programs. Table 4.4 summarizes the sources of funds. BRAND brought broadband to 896 localities, more than double the number initially planned, without additional federal funding (Industry Canada 2006). During the same period, other federal, provincial, regional, and municipal programs connected about 1,100 additional localities. The number of rural communities without broadband access was thus roughly halved between 2001 and 2006. In 2009, the federal government allocated a further Can $225 million to Broadband Canada, a new program to extend broadband to about 22 percent of rural households still without service or having only low-speed access (Industry Canada 2009). Additionally, several other federal and provincial development programs supported broadband projects.[22]

Mandatory Contributions

For reasons of economic efficiency and fairness, government financing through the budget is the preferred choice for fiscal support of broadband development. This, however, puts broadband in direct competition with other demands on the budget. It also subjects the broadband strategy to the uncertainties of annual budgetary appropriations.

An alternative is to raise the funds from mandatory contributions by telecommunications operators, which are generally placed in UASFs. Operators are generally willing to contribute reasonable amounts to such

Table 4.4   Investment in the Broadband Rural and Northern Development Pilot in Canada, by Source of Funds, 2002–06

Source of funds	Can $ (millions)	% of total
Federal government	78.0	44.7
Other sources		
Service providers	50.7	29.1
Community leaders	15.0	8.6
Provincial and municipal governments	16.0	9.1
Other federal or provincial programs	14.2	8.2
First Nations	0.2	0.1
Other	0.3	0.2
Total other sources	96.4	55.3
Total	174.4	100.0

Source: Industry Canada 2006.

UASFs if the contributions and the management thereof are transparent, contributions are allocated fairly, and operators are eligible to receive support financed by the proceeds of this contribution. This is essentially a tax, and, although it is generated and used off-budget, it should be subject to the same considerations regarding economic justification of fiscal support discussed earlier. Since operators pass these levies on to their customers, mandatory contributions by operators raise the question of fairness in which users of existing services, including low-income users, are required to help to pay for a new service that will initially benefit mostly higher-income and better-educated groups. There is also some evidence that this practice has an overall negative impact on the economy as a whole (Wallsten 2011).

Where used, levies of typically around 2 percent of gross revenues have sufficed to support programs to extend wireline telephone service to rural areas in developing countries. To spread the burden among as many customers as possible, eventually approximating the effect of a broad-based tax, in principle all operators should pay.[23] Mandatory contributions are generally assessed in proportion to gross revenues, adjusted to avoid double counting (for example, to exclude payments received from other operators along the supply chain). Revenue information is easy to collect and audit, and this method results in customers paying markups that are proportional to their bills, much like a progressive tax.

International Loans, Credits, and Grants

Several international organizations support the development of communication services. The ITU provides technical assistance, training, standards, and forums for policy and regulatory debate. Multilateral development banks, such as the World Bank and regional development banks for Africa, Asia, Europe, and Latin America, have a long history of providing assistance to governments to extend communication services to localities where service provision is not commercially viable. Traditionally, this assistance consisted of financing rural components of national telecommunications development programs carried out by state enterprises. In the wake of sector reforms from the 1990s, emphasis shifted away from financing public sector investment to creating a policy and regulatory environment that enables private investment in increasingly competitive markets. More recently, the scope of support has widened from telecommunications to ICTs, from voice and data to broadband, and from reaching end users to building high-speed backbones connected to international networks. Multilateral development organizations that support private companies rather than governments, such as the International

Finance Corporation, have also found a growing role in broadband development.[24] Bilateral development agencies, such as the U.S. Agency for International Development, the U.K. Department for International Development, and the Japan International Cooperation Agency, have followed roughly similar trends.

Support from international development organizations for broadband development typically involves some combination of technical assistance, grants, loans, and credits. Output-based aid (OBA), which links financial support to results, is increasingly used to accelerate or expand access to a range of basic services (such as infrastructure, health care, and education) for the poor in developing countries.[25] Well-designed OBA schemes sharpen the targeting of development outcomes, improve accountability for the use of public resources, and provide stronger incentives for efficiency and innovation. OBA schemes often use competition among firms for assigning cash subsidies. The Global Partnership on Output-Based Aid (GPOBA) is a group of donors and international organizations that works together to support OBA approaches. GPOBA has funded telecommunications projects in Bolivia, Cambodia, Guatemala, Indonesia, and Mongolia, as well as a study on new tools for universal service in Latin America and a study on ICTs in the Pacific. GPOBA provided a US$5.5 million grant as seed money to establish a USF in Mongolia, which would collect a levy on telecommunications bills and use it to extend service to rural and nomadic areas that are not commercially viable on their own. Both GPOBA and the World Bank's Private-Public Infrastructure Advisory Facility (PPIAF) provided grants for technical assistance to design the universal service program and set up the fund (Dymond, Oestmann, and McConnell 2008). Box 4.5 discusses the role that the International Development Association is playing in Africa.

Universal Access and Service Funds for Broadband Development

The financing of UAS has gone through various stages, ranging from cross-subsidies that finance nonprofitable areas under a monopolistic scenario to the creation of UASFs financed by operator levies that support projects in more competitive markets. A range of other solutions lies between these two points. Historically, first-generation fund projects have been primarily top-down (for example, Colombia and Peru), with the fund defining the locations and requirements. However, in the last few years, bottom-up projects have been tried in Chile and other countries. In Sub-Saharan Africa, the tendency has been toward top-down projects, primarily allocated through competitive processes such as least-cost subsidy bids. Chapter 2 discusses

Box 4.5: Regional Communications Infrastructure Program in East and Southern Africa

Credits from the International Development Association totaling US$164.5 million are financing the first stage of the Regional Communications Infrastructure Program (RCIP) in East and Southern Africa.[26] Several submarine optical fiber cable projects that are under way will provide global broadband connectivity at potentially low marginal costs at several landing points in this region. But developing regional broadband backbone networks to connect the population to these points is, for the time being, not commercially viable on its own. The RCIP is financing investment in PPPs to develop these regional broadband networks so that global connectivity from submarine cables can be extended throughout the region, to ensure open access to this infrastructure by all providers, and to promote utilization of the infrastructure to realize the benefits of broadband connectivity and commercial sustainability. The RCIP is open to 26 countries, subject to readiness and eligibility for financing. The first phase comprises projects in Burundi, Kenya, and Madagascar. Eight other countries have expressed interest in joining, supported by International Development Association credits for about US$260 million.

Source: World Bank, Africa Regional Program, http://worldbank.org/ict.

mechanisms such as public-private partnerships, local efforts, and bottom-up networks, and the following sections discuss the use of UASFs to collect and disburse funds.

Over the last two decades, UASFs were created in many countries to finance network expansion. UASFs are being used in competitive markets to supplement market-based policies and address access gaps and market failures in remote and underserved locations. They are often seen as a competitively neutral solution for open-market environments, where all operators in the market are obliged to share the responsibility for (and the benefits of) providing universal access. They are also relevant to extending broadband ahead of and beyond the market.

Advocates of UASF state that, where properly designed and implemented and with sufficient internal resources and expert capacity, the UASF model can act as a country's centralized "clearinghouse" through which funding from a range of sources flows in while development projects are assigned and awarded in order to improve the efficiency and coordination of various ICT development and financing initiatives (Task Force on Financial Mechanisms for ICT for Development 2004). During the 1990s, several Latin American countries (such as Chile, Peru, and Colombia) used the UASF to support widespread deployment of public telephones to rural and remote areas. These countries are now also using the UASF to extend broadband services to rural areas; for example, Chile is using UASFs to provide Internet access and multipurpose telecenters to unserved areas.

*Info*Dev's ICT Regulation Toolkit illustrates how so-called "second-generation" UASFs are applying their resources to the financing of Internet points of presence (POPs) in rural districts, telecenters, and cybercafés, school connectivity, and other ICT initiatives. Uganda is one of the first countries to establish a more comprehensive USF, and many of its recent initiatives use technology-neutral competitions, which are increasingly being won by mobile operators.

However, many countries have raised legitimate concerns regarding UASFs due largely to a few instances of mismanagement and lack of transparency in fund collection and disbursement. Countries have also been concerned with the complexity involved in implementing and managing a UASF. It may be a daunting task for governments to get all operators to accept the conditions of the fund, particularly who will contribute to the UASF and how much those contributions will be (Maddens 2009).

Brazil, for example, has struggled with its fund, the Fundo de Universalização dos Serviços de Telecomunicações (FUST). FUST was established with the purpose of creating a financial resource that could complement the deployment of universal obligations of the wireline operators, but in reality the cost of expanding services is being borne directly by the operators. FUST's most critical challenge is that it is not technologically neutral. It favors wireline service operators over other telecommunications providers, as the funds can only be applied toward wireline service projects. However, all telecommunications service providers are required to contribute, which favors one service over another.

Acknowledging that this is not the best funding mechanism, the Ministry of Communications carried out a public consultation in April 2008 with the intention of reforming the Brazilian telecommunications framework. In the consultation, the ministry proposed that the FUST should be, at the very least, technology neutral in its distribution mechanism. The Brazilian Congress is presently considering a variety of other ways to distribute funds and to determine appropriate projects. Currently, various draft laws are under consideration that, if passed, will amend the FUST regulations to allow the use of FUST moneys for projects that seek to increase access to broadband services in Brazil.

Policy makers have also found that mechanisms need to be put in place to make UASFs accessible to a wider range of telecommunications service providers. Limiting access to funds to a specific category of licensee or to licensed operators, for example, can create barriers that continue to support existing conditions (that is, the expansion of wireline networks to provide universal service or access) and discourage the implementation of new technologies to provide service in unserved or underserved areas. In Peru, telecommunications service providers with concession contracts for final public services (wireline, including pay phones, and mobile) and value added services (data services, including broadband Internet access) can access FITEL funds. If the entity requesting the funds does not have a concession contract for the area for which it is requesting the funds, it must request the appropriate expansion of the concession contract from the Ministry of Transportation and Communications (Peru, OSIPTEL 2005). Letting a variety of entities have access to UASFs allows countries to benefit from a greater number of possible resources to help them to achieve their universal service goals. In addition, these resources can sometimes provide innovative solutions for small-scale projects that would not normally be considered profitable.

In addition, the development and presentation of project proposals for UASF consideration should not be restricted to the fund authority or to telecommunications providers, but instead should be open to all entities with an interest in contributing to the fulfillment of universal service or access. In Chile, project proposals can be presented by telecommunications service providers, regional, provincial, or municipal authorities, universities, nongovernmental organizations, neighborhood communities, and others. Chile's Subsecretaría de Telecomunicaciones (SUBTEL), the entity responsible for administering and managing the country's UASF, uses these project proposals to design and develop the fund's annual project agenda (Chile, SUBTEL 2001). A system where multiple parties can submit project proposals allows all interested parties to contribute to achieving the country's USO objectives. Having multiple sources for project proposals can provide a more realistic vision of the needs and conditions of the market, such as what type of service is required by localities and which technology is best suited and more likely to result in creative and resourceful projects. This has become even more relevant in a broadband context.

The UASF should not only support a country's present universal service objectives, but also be able to adapt to the demands and trends of a converging telecommunications sector by fostering the use of new and innovative technologies to achieve future USO goals. A 2006 study undertaken for the Forum of Latin American Telecommunications Regulators (Regulatel) concluded that Latin American UASFs had played an important role in network development and identified some of the challenges (Bossio and Bonifaz 2006). The study made specific recommendations for improving, streamlining, or realigning the activities of Latin American UAS policies and UASF programs. Since universal access to telephony is, in the opinion of the study's authors, close to being achieved in Latin America, the study recommended reorienting UASFs toward supporting "ubiquitous deployment of advanced technologies and services (Bossio and Bonifaz 2006, 45)." The study advised that, as the communications technology revolution continues, the new generation of UASFs could become leaders, not followers, in ensuring that populations have access to the most modern and effective networks, services, and applications available. This would include broadband, wireless, multiservice platforms permitting full access to all functions and features of telephony, Internet, data transmission, e-commerce, e-government, multimedia entertainment, and interactive communications. The new USAFs' role in promoting broadband would be through support to intermediary facilities, such as backbones (including POPs), towers, and other passive infrastructure.

The study recognized, however, that a new generation of funds managed by public sector administrators is still unlikely to have the capacity to lead developments in the field of advanced technologies and services; instead, the private sector is likely to continue to be the leader in technology and service innovation and in service expansion, which is in line with market-driven developments. Therefore, new UASFs may not lead, but, by putting emphasis on broadband, can at least mirror in rural and underserviced areas what the market is achieving on its own in urban areas. Once government has agreed on an aggressive broadband promotion policy, new UASFs will not wait until a large portion of the population has access to broadband to start filling in the gaps, but rather will act in parallel to the market, while taking care not to subsidize areas that the market would serve on its own.[27] As illustrated in boxes 4.6 and 4.7, other countries from around the world are also shifting the focus of UASFs.

Best Practices for Effective Management of Flow of Funds

Whether funds flow through a UASF or other public financing body such as a PPP or municipal-led project, several key principles are applicable to ensure that funds go to projects aimed at achieving universal broadband access:

- *Transparency*. Relates to the effective and transparent management of the flow of funds in accordance with the mandate of the entity managing the funds. Transparency of procedures can be enhanced through

Box 4.6: Reform of the USF in the United States

In the United States, the Federal Communications Commission released a Notice of Proposed Rulemaking on its efforts to transform the high-cost portion of the USF to support broadband as well as to reform the intercarrier compensation system in a new Connect America Fund.[28] The proposed reforms are based on four pillars:

- Modernizing USF and intercarrier compensation to support broadband networks

- Ensuring fiscal responsibility by controlling costs and constraining the size of the fund

- Demanding accountability from both USF recipients and the government itself

- Enacting market-driven and incentive-based policies to maximize the impact of scarce program resources and the benefits to all consumers.

Source: United States, FCC 2010b.

Box 4.7: Reform of the RCDF in Uganda

In Uganda, the Rural Communications Development Fund (RCDF) was established in 2003 to support the development of a commercially viable communications infrastructure in rural Uganda, thereby promoting social, economic, and regional equity in the deployment of telephone, Internet, and postal services. Subsidies are awarded through a competitive process and are only available to geographic areas where service provision is not feasible or is unlikely to be provided by operators within the next one to two years without subsidy. The RCDF mandates the provision of Internet POPs and wireless access systems at district centers as well as national Internet exchange points (IXPs) to facilitate inter-ISP traffic. The approach taken in the RCDF Internet POP Program focused on the delivery of broadband services to districts with existing demand. The RCDF took the approach that, if such services are to be sustainable and viable, they should be deployed first where private and public clients are ready to support them. Once the service is established in the more densely populated district centers, further deployment beyond their boundaries should be reviewed as demand and capacity become evident.

*Source: info*Dev and ITU, "ICT Regulation Toolkit, Practice. Note on Uganda." http://www.ictregulationtoolkit.org.

Note: Uganda's Rural Communications Development Fund, http://www.ictregulationtoolkit.org/en/PracticeNote.3144.html.

a manual or handbook for recipients of public financing, whether funding is through a UASF, a PPP, or another financing mechanism. Such manuals generally set out the specific rules with respect to critical issues such as procurement, accounting standards, project selection criteria, technical partner selection criteria, tendering processes and procedures, and disbursement procedures or participation rules in the case of a PPP.

- *Accountability.* Relates to the level and detail of reporting on activities and is aimed at ensuring transparency of operations. In general, accountability requires periodic reports to be provided to the relevant stakeholders, including the communities, with respect to the monitoring, evaluation, and impact of the projects being undertaken. This helps to ensure both accountability and stakeholder and community awareness. In addition, there should also be a requirement for annual auditing—the funds and accounts of the USAF should be audited independently on an annual basis, and the audit results should be made public. Similarly, in the case of a specific project, the recipient of public financing (whether from a fund or other source) should provide regular, audited reports on its progress and performance. And finally, there is generally the requirement for an annual report on the flow of funds.

- *Efficiency.* Requires several elements in order for funds to flow efficiently and to promote broadband ahead of and beyond the market. Required elements include an understanding of the environment and responsiveness to market realities; management autonomy allowing flexibility to adapt to market realities; sufficient financial resources to allow efficient selection of projects; adequate human resources and capacity to enable effective project implementation; quality of service targets and measures; monitoring, dispute resolution, and sanctioning powers; and evaluation and review mechanisms. An inefficient structure can be too slow in implementing projects, in which case the steps taken may be inappropriate, too late, or too expensive, among other problems.

Reviewing the Flow of Funds

National UAS programs should be reviewed regularly in terms of strategy and management. Such reviews should also be applied to the flow of funds to achieve broadband ahead of and beyond the market, whether through the use of UASFs, PPPs, or other funding mechanisms. Best practice indicates that such reviews should be carried out by an independent entity (with relevant expertise in the fields of UAS, project finance, and operational management). Where public funds are applied to move broadband access ahead of or beyond the market, the evaluation should consider the following elements:

- The achievement of specific targets, as indicated in a UAS or NBP, to move broadband ahead of and beyond the market and, if applicable, the achievements of the UASF against its objectives

- The role of the commercial sector and of development or financing partners in contributing to universal broadband implementation, including through PPPs

- If applicable, the collection and disbursement of the UASF against projections and the costs and effectiveness of the UASF's management and management structure

- The impact and contribution of universal broadband projects and services on social diffusion and use of ICT services

- The impact and contribution of universal broadband projects and services on the development of the country, including the impact on the country's macroeconomic situation, social development, and entrepreneurship and innovation

- The impact and contribution of universal broadband projects and services on the development of infrastructure supply in the telecommunications sector

- The strategic options for future development of the UAS program to meet the objective of achieving universal broadband access

- The financial requirements to meet these objectives and recommendations with respect to future levies, if applicable, fund raising, and partnerships

- Other strategic recommendations regarding the direction of the program to move broadband ahead of and beyond the market and management of the fund, if applicable.

The government can use the results of the review and its recommendations to guide future UAS and broadband policy, renew and revise its objectives, or, where applicable, change the mandate of the UASF.

Notes

1. An alternative interpretation is that only those innovations that ultimately pass the market test are sustainable and considered successful.

2. Finland, Ministry of Transport and Communications (2009); "Finland Makes Broadband a Legal Right," *BBC News*, July 1, 2010, http://www.bbc.co.uk/news/10461048.

3. Amendment 331/2009, as integrated into section 60c of the consolidated Finnish Communications Market Act, http://www.finlex.fi/en/laki/kaannokset/2003/en20030393.pdf. begin_of_the_skype_highlighting

4. FICORA, "1 Mbit/s Broadband for Everyone on 1 July 2010: Telecom Operators' New Universal Service Obligations Enter into Force," June 29, 2010, http://www.ficora.fi/en/index/viestintavirasto/lehdistotiedotteet/2010/P_27.html.

5. Government of Canada, "Canada's 2009 Economic Action Plan," http://www.actionplan.gc.ca/eng/index.asp.

6. Ireland, Department of Communications, Energy, and Natural Resources, "National Broadband Scheme," http://www.dcenr.gov.ie/Communications/Communications+Development/National+Broadband+Scheme.htm.

7. "Universal Service Obligation Fund Supported Scheme for Wire Line Broadband Connectivity in Rural and Remote Areas," http://ittripura.nic.in/USOF_broadband_scheme.pdf.

8. "Deal Reached to Build Broadband Network across Jamaica," *Caribbean 360*, April 7, 2011, http://www.caribbean360.com/index.php/business/329864.html.

9. E-Dominicana is a national strategy that seeks "to promote the use and appropriation of information and communication technologies in the Dominican Republic by means of initiatives that create synergies between the governmental sector, the civil society, and the productive sector, to offer all its inhabitants better opportunities which will contribute to their development, by bringing them welfare and progress in the exercising of their capacities." Its vision is "to place the country in a position that will allow it to compete in the new scenario of a globalized world, by achieving sustainable development in the economic, political, cultural, and social scope, and to assume the challenge of converting inequality and social exclusion from the digital divide into a digital opportunity" (San Román 2009, 20).

10. *Info*Dev and ICT, "ICT Regulation Toolkit, Module 4, Universal Access and Service," sec. 4.1.3, Relationship to Broadband Policy, http://www.ictregulationtoolkit.org/en/Section.3258.html.

11. Industry Canada, "Minister Clement Provides Update on Government of Canada's Rural Broadband Plan," May 2010, http://www.marketwire.com/press-release/Minister-Clement-Provides-Update-on-Government-of-Canadas-Rural-Broadband-Plan-1158936.htm.

12. "Finland on Track with National Broadband Plan," *Business Monitor International*, April 2010, http://store.businessmonitor.com/article/343046.

13. *infoDev* and ITC, "ICT Regulation Toolkit, Module 4: Universal Access and Service," sec. 2.4.2, Revising the Licensing Regime or Issuing New Licenses, http://www.ictregulationtoolkit.org/en/Section.3214.html.

14. ITU, "ITU Statshot," January 2011, http://www.itu.int/net/pressoffice/stats/2011/01/index.aspx.

15. Law no. 55-01, adopted in November 2004, made important modifications in the setup of universal service in Morocco. The universal service definition was extended to include the supply of value added services, including Internet. A new approach relating to the operator's contribution to the mission of universal service was also introduced, including regional development obligations and the introduction of the "pay or play" mechanism; see the note on telecommunications and the ICT sector in Morocco, available at http://www.apebi.org.ma/IMG/pdf/E-Morocco.pdf.

16. The discussion of subsidies, including verbatim without quotes, draws from Wellenius, Foster, and Calvo (2004). The examples are taken from the individual references noted.

17. Subsidies can be designed to reduce access barriers to which target groups (for example, low-income families) are especially sensitive, such as initial connection, equipment, or installation charges. This is common in electricity and water supply.

18. About 2–3 percent of community income is often used for initial discussion of rural telecommunications programs. Surveys of rural communities in countries as different as Argentina, India, Nicaragua, and the Philippines show that households spend about 5 percent of monetary income on energy, fairly consistently across countries and with even higher proportions for the lower-income households. The World Health Organization's target is that water supply should not cost more than 5 percent of household income. See references in Wellenius, Foster, and Calvo (2004).

19. See *info*Dev and ITU, "ICT Regulation Toolkit, Module 4: Universal Access and Service," sec. 6.1.2, Per Capita and Household Expenditure on Communications, http://www.ictregulationtoolkit.org. In high-income countries the proportion is smaller, although the amount is higher in absolute terms. In Finland in 2006, households on average spent 3 percent of income on telecommunications.

20. Combining rural infrastructure projects of different sectors (for example, telecommunications and electricity) can reduce total government costs (for example, demand surveys, road shows, supervision) and supply costs (especially operation and maintenance).

21. First Nation comprises aboriginal groups (except Inuit and Métis) organized in over 600 governments, mainly in the provinces of Ontario and British Columbia. Inuit are aborigines who live in the Arctic of Canada, Greenland, and Alaska, formerly referred to as Eskimos. Canadian Inuit live mainly in northern Québec, coastal Labrador, and parts of the Northwest Territories, especially on the Arctic Ocean coast. Métis are the descendants of First Nation and Europeans, mainly French, during colonial times. They are found in British Columbia, Alberta, Saskatchewan, Manitoba, Québec, New Brunswick, Nova Scotia, and Ontario, as well in the Northwest Territories.

22. Industry Canada, "Broadband Canada Connecting Rural Canadians: Other Federal and Provincial Programs," http://www.ic.gc.ca/eic/site/719.nsf/eng/h_00032.html.

23. Some categories of providers might be exempted for competition policy and other reasons. For example, new entrants should not be expected to help to finance the incumbent with which they are trying to compete.

24. The International Finance Corporation is the World Bank Group's arm for investing and lending to the private sector.

25. OBA is also known as performance-based aid or results-based financing (in the health sector). It is part of a broader effort to ensure that aid is well spent and that the benefits go to the poor.

26. The International Development Association is the arm of the World Bank Group that provides long-term financing for development at low interest rates to low-income countries.

27. Regulatel and the World Bank, "New Models for Universal Access in Latin America, Summary of Main Report," http://www.ictregulationtoolkit.org/en/Section.3286.html.

28. United States, FCC, "Universal Service," http://www.fcc.gov/wcb/tapd/universal_service/.

References

Atkinson, Robert D., Daniel K. Correa, and Julie A. Hedlund. 2008. "Explaining International Broadband Leadership." Information Technology and Innovation Foundation, Washington, DC, May. http://www.itif.org/files/Explaining BBLeadership.pdf.

BEREC (Body of European Regulators for Electronic Communications). 2010. "Report on Universal Service: Reflections for the Future." BEREC, Riga, June. http://www.erg.eu.int/doc/berec/bor_10_35_US.pdf.

Beschorner, Natasha. 2010. "Universal Access and Service: New Direction, Examples from the East Asia and Pacific Region." Paper prepared for the World Bank and International Telecommunication Union, "Seminar on Broadband and USO," Bangkok, November. http://www.tridi.ntc.or.th/library/components/com_booklibrary/ebooks/3C994EDFd01.pdf.

Bossio, Jorge, and Luis Bonifaz. 2006. "Perú." In "Nuevos Modelos para el Acceso Universal de los Servicios de Telecomunicaciones en América Latina – Informe de Países, REGULATEL, Bogotá: FITEL portal." http://fitel.gob.pe.

Chile, SUBTEL (Subsecretaría de Telecomunicaciones). 2001. "Decree Approving the Guidelines for the Telecommunications Development Fund [Fondo de Desarrollo de las Telecomunicaciones]." SUBTEL, Santiago, December 28.

———. 2008. "Infraestructura digital para competividad e innovación, informe nacional." SUBTEL, Santiago. http://www.subtel.cl.

Denmark, National IT and Telecom Agency. 2010. "Ambitious New Broadband Goal for Denmark." National IT and Telecom Agency, Copenhagen, June. http://extranet.broadband-europe.eu/Lists/StrategiesData/Attachments/41/Danish%20broadband%20goal%20June%202010_EN.pdf.

Dymond, Andrew. 2010. "Universal Service: The Trends, Opportunities, and Best Practices for Universal Access to Broadband Services." Intelecon Research and Consultancy, Vancouver, November. http://www.inteleconresearch.com/pages/documents/OOCUR_Paper_Dymond_UAStoBroadband.pdf.

Dymond, Andrew, Sonja Oestmann, and Scott McConnell. 2008. "Output-Based Aid in Mongolia: Expanding Telecommunications Services to Rural Areas." OBApproaches Note 18, World Bank and Global Partnership on Output-Based Aid, Washington, DC, February.

European Commission. 2009. "Community Guidelines for the Application of State Aid Rules in Relation to Rapid Deployment of Broadband Networks." European Commission, Brussels, September. http://eur-lex.europa.eu/LexUriServ/LexUriServ.do?uri=OJ:C:2009:235:0007:0025:EN:PDF

———. 2010. "European Broadband: Investing in Digitally Driven Growth." European Commission, Brussels. http://ec.europa.eu/information_society/activities/broadband/docs/bb_communication.pdf.

European Commission and ITU (International Telecommunication Union). 2011. "Update of SADC Guidelines on Universal Access and Service and Assessment Report." ITU-Digital, Brussels, March. http://www.itu.int/ITU-D/projects/ITU_EC_ACP/hipssa/events/2011/SA2.2.html.

Finland, Ministry of Transport and Communications. 2009. "Memo on the Communications Decree on the Minimum Rate of a Functional Internet Access as a Universal Service." Ministry of Transport and Communications, Helsinki, October 7. http://www.lvm.fi/c/document_library/get_file?folderId=913424&name=DLFE-10508.pdf&title=Background.%20Ministry%20of%20Transport%20and%20Communications%20decree%20on%20the%20minimum%20rate%20of%20a%20functional%20Internet%20access%20as%20a%20universal%20service%20(7.10.2010).

Industry Canada. 2006. "Formative Evaluation of the Broadband for Rural and Northern Development Pilot." Industry Canada, Ottawa. http://www.ic.gc.ca/eic/site/ae-ve.nsf/eng/01425.html.

———. 2009. "Canada's Economic Action Plan: Broadband Canada, Connecting Rural Canadians Application Guide, September 1, 2009." Audit and Evaluation Branch, Ottawa. http://www.ic.gc.ca/eic/site/719.nsf/eng/h_00015.html.

ITU (International Telecommunication Union). 2003. *Birth of Broadband: ITU Internet Reports*. Geneva: ITU. http://www.itu.int/osg/spu/publications/sales/birthofbroadband/index.html.

———. 2008. "The Experience of the Dominican Republic in the Development of Innovative Strategies for the Promotion of Information and Communication Technologies" "Eighth Global Symposium for Regulators," Pattaya, Thailand, March 11–13. ITU, Geneva.

Kim, Yongsoo, Tim Kelly, and Siddhartha Raja. 2010. "Building Broadband: Strategies and Policies for the Developing World." Global Information and Communication Technologies Department, World Bank, Washington, DC, January. http://www.thinkinnovation.org/file/research/5/en/Building_broadband.pdf.

León, Laura. 2009. "Informe de acción de incidencia regional Perú: Fondo de Inversión en Telecomunicaciones" Centro Peruano de Estudios Sociales, Lima. http://www.apc.org/en/system/files/CILACIncidenciaRegional Peru_20090707.pdf;

Maddens, Sofie. 2009. "Trends in Universal Access and Service Policies." GSR-09 Background Paper, ITU, Beirut, November 10. http://www.itu.int/ITU-D/treg/ Events/Seminars/GSR/GSR09/papers.html.

Muente-Kunigami, Arturo, and Juan Navas-Sabater. 2010. "Options to Increase Access to Telecommunications Services in Rural and Low-Income Areas." *Info*Dev and World Bank, Washington, DC. http://siteresources.worldbank.org/ EXTINFORMATIONANDCOMMUNICATIONANDTECHNOLOGIES/ Resources/282822-1208273252769/Options_to_Increase_Access_to _Telecommunications_Services_in_rural_and_Low-Income_Areas.pdf.

OECD (Organisation for Economic Co-operation and Development). 2008. "Broadband Growth and Policies in OECD Countries." OECD, Paris. http://www.oecd.org/dataoecd/32/57/40629067.pdf.

Peru, OSIPTEL (Organismo Supervisor de Inversión Privada en Telecomunicacio- nes). 2005. "Resolution # 025-2005-CD/OSIPTEL." OSIPTEL, Board of Directors, San Borja, Peru, May.

Petersen, Finn. n.d. "Broadband Policies: The Danish Approach." National IT and Telecom Agency, Helsinki, http://www.ictaustria.at/web/MDB/media_folder/ 79_ICT-austria.ppt.

San Román, Edwin. 2009. "Bringing Broadband Access to Rural Areas: A Step by Step Approach for Regulators, Policymakers, and Universal Access Program Administrators; The Experience of the Dominican Republic." Paper prepared for the "Ninth Global Symposium for Regulators," Beirut, November 10–12.

Task Force on Financial Mechanisms for ICT for Development. 2004. "Financing ICT4D: A Review of Trends and an Analysis of Gaps and Promising Practices." ITU and World Summit on Information Society, Geneva. http://www.itu.int/ wsis/tffm/final-report.pdf.

United Kingdom, Department for Business, Innovation, and Skills and Department for Culture, Media, and Sport. "Britain's Superfast Broadband Future." Department for Business, Innovation, and Skills, London, December. http://www.culture.gov.uk/images/publications/10-1320-britains-superfast- broadband-future.pdf.

United States, FCC (Federal Communications Commission). 2010a. "Connecting America." In *The National Broadband Plan*, ch. 2. Washington, DC: FCC. http://www.broadband.gov/plan/2-goals-for-a-high-performance-america/.

———. 2010b. "Report and Order and Further Notice of Proposed Rulemaking." FCC, Washington, DC. http://transition.fcc.gov/Daily_Releases/Daily _Business/2011/db1122/FCC-11-161A1.pdf.

Wallsten, Scott. 2011. "The Universal Service Fund: What Do High-Cost Subsidies Subsidize?" Technology Policy Institute, Washington, DC, February. http://www.techpolicyinstitute.org.

Wellenius, Björn, Vivien Foster, and Christina Malmberg Calvo. 2004. "Private Provision of Rural Infrastructure Services: Competing for Subsidies." Policy Research Working Paper 3365, World Bank, Washington, DC.

World Bank. 2005. "Financing Information and Communication Infrastructure Needs in the Developing World: Public and Private Roles." Working Paper 65, Global Information and Communication Technologies Department, World Bank, Washington, DC.

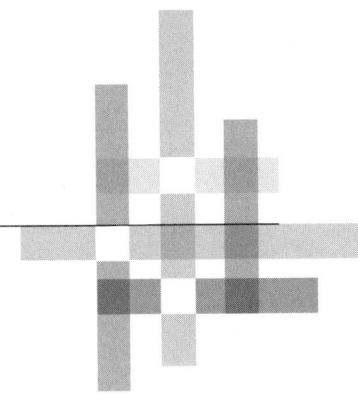

CHAPTER 5

Technologies to Support Deployment of Broadband Infrastructure

This chapter examines the building blocks for constructing broadband networks. It looks at high-speed connectivity from a hierarchical perspective, moving from international, to national, to metropolitan, and finally to local access deployment solutions. The chapter describes the various wireline and wireless technologies for deploying broadband infrastructure, including examples of various deployments throughout the world, and discusses some of the issues associated with implementing these technologies. The focus is on the physical networks and associated protocols for routing traffic rather than the end user services and applications that are accessed over the networks, which are discussed in chapter 6.

Overview of Broadband Networks

As policy makers consider plans and strategies for developing broadband networks, it is important to recognize that such networks have many components. All of these parts must work together for the network to function effectively and efficiently. This handbook categorizes these component parts into four hierarchical levels, which together constitute the broadband supply chain: international connectivity, the national backbone network, metropolitan access links, and the local access network (figure 5.1). Besides physical connectivity, networks require traffic routing intelligence to ensure that information is correctly sent and received. This section describes the physical components of broadband networks and discusses the evolution of network intelligence based on Internet Protocol (IP) routing.

Figure 5.1   Broadband Supply Chain

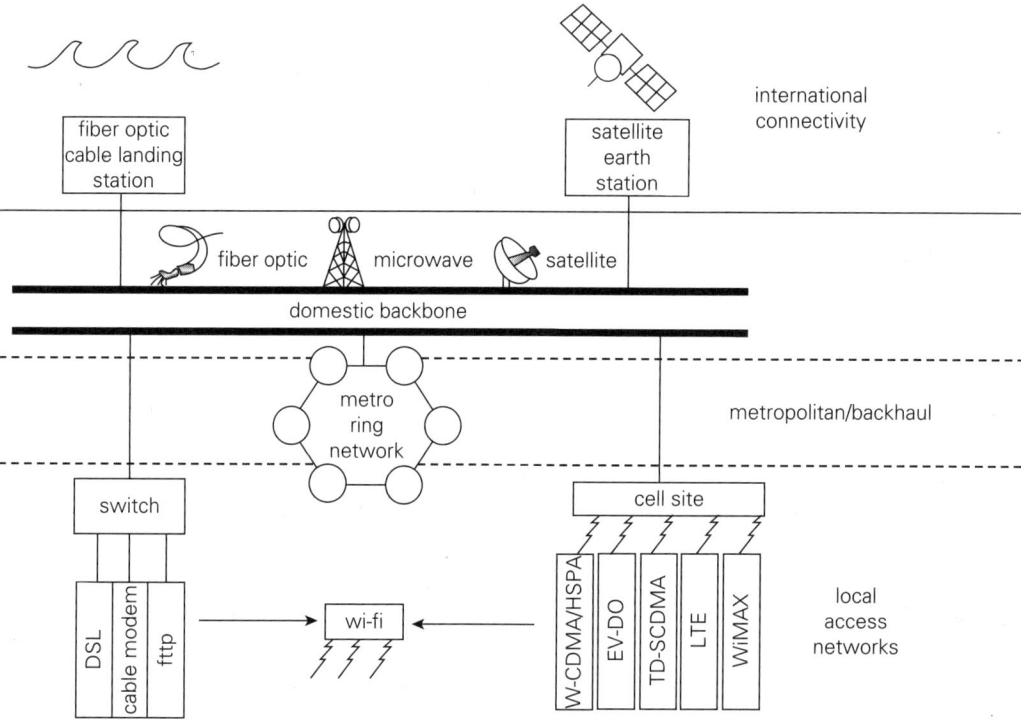

Source: Telecommunications Management Group, Inc.

Note: Generalized typical infrastructure implementation and topology, excluding technologies not widely used. DSL = digital subscriber line; EV-DO = CDMA2000 Evolution Data Optimized (mobile communication standard); fttp = fiber to the premises; LTE = Long-Term Evolution (mobile communication standard); TD-SCDMA = Time Division–Synchronous Code Division Multiple Access (mobile communication standard); W-CDMA/HSPA = Wideband Code Division Multiple Access (family of mobile communication standards)/High-Speed Packet Access; Wi-Fi = Wireless Fidelity, a wireless local area network standard based on the IEEE 802.11 standards; WiMAX = Worldwide Interoperability for Microwave Access (fixed and mobile communications standard).

The Broadband Supply Chain

The broadband supply chain has four main infrastructure components (figure 5.1):

- *International connectivity* provides links to broadband networks in other countries usually via satellite and fiber optic cable. This requires network intelligence to exchange and route international Internet traffic.

- *National backbone network* provides pathways for transmitting Internet data across a country, typically via microwave, satellite, and fiber optic links. This also includes traffic management, exchange, and routing as well as issues related to enhancing efficiency and quality over IP networks such as Internet exchanges, metropolitan rings, and next-generation networks (NGNs).

- *Metropolitan or backhaul links* provide the connections between local areas and the national backbone network, usually via fiber optic and microwave and, to a lesser extent, satellite. In a wireless network, these links are used to bring traffic from cell sites back to a switching center (known as backhaul).

- *Local access networks* provide the wireline and wireless infrastructure that end users utilize to connect to the broadband network.

The boundaries between these network components are sometimes blurred. For example, Internet traffic exchanges route domestic traffic. However, they are also related to international traffic in that the exchange may be a peering point for an overseas network. Internet exchanges also reduce reliance on international connectivity by ensuring that domestic traffic is kept within the country. Metropolitan ring networks provide a bridge between the domestic backbone network and the various local access networks. There are also regional implications in that one country's national backbone could provide an international connectivity link for a neighboring landlocked country.

Two additional points should be noted. First, the different levels of the overall broadband network should ideally be in sync. High speeds in the local access network segment can only be accomplished if the speed and capacity in the national and international network segments are adequate to support them. Second, technology deployment is dependent on a country's existing level of infrastructure. Countries without significant wireline infrastructure in the local access network may find it financially impractical to deploy ubiquitous wired networks, but they may be able to upgrade existing wireless networks. Similarly, countries often find it more financially attractive to leverage existing networks through upgrade or evolution than to deploy the latest state-of-the-art technology by building completely new networks.

Government involvement in the deployment of broadband networks has important repercussions, as addressed in this chapter. Most local access networks around the world use copper typically installed by formerly state-owned enterprises. While many countries leave the construction of broadband networks to the private sector, governments in other countries either guarantee bilateral or multilateral loans for the construction of backbone networks or are full or partial owners of wholesale or retail service providers. Governments may also play a pivotal role as a promoter for large projects such as international connectivity or national backbones where the private sector has been hesitant to invest. Even where the private sector has assumed the main role for investment in broadband networks, governments remain influential through their decisions with regard to spectrum allocation, rights-of-way, and infrastructure sharing. In addition, governments themselves are important users of broadband.

The Transition to All-IP Networks

An important trend affecting broadband network development is the convergence of broadcasting, telecommunications, and information technology networks and services. Convergence has proceeded mainly through operators making incremental changes to upgrade their networks, while minimizing large investment outlays. In some cases, this can lead to operators having to support an array of technologies. For example, wireless providers may simultaneously support data solutions that include General Packet Radio Service (GPRS), Enhanced Data Rates for Global System for Mobile Communications (GSM) Evolution (EDGE), Wideband Code Division Multiple Access (W-CDMA), High-Speed Packet Access (HSPA), and Long-Term Evolution (LTE). In a very few cases, governments have sought to jump-start broadband network development by sponsoring the development of a completely new network to replace legacy networks (for example, Australia).

NGNs exploit the advantages of IP to packetize all information and route it to its destination. NGNs simplify network maintenance and operation by standardizing protocols with IP or Multiprotocol Label Switching (MPLS) at the core.[1] In this architecture, common applications and services can be provided independently of the underlying physical transport network, making it easier for multiple providers to compete effectively in different parts of the broadband supply chain (figure 5.2).[2] Many

Figure 5.2   Design of IP Next-Generation Networks

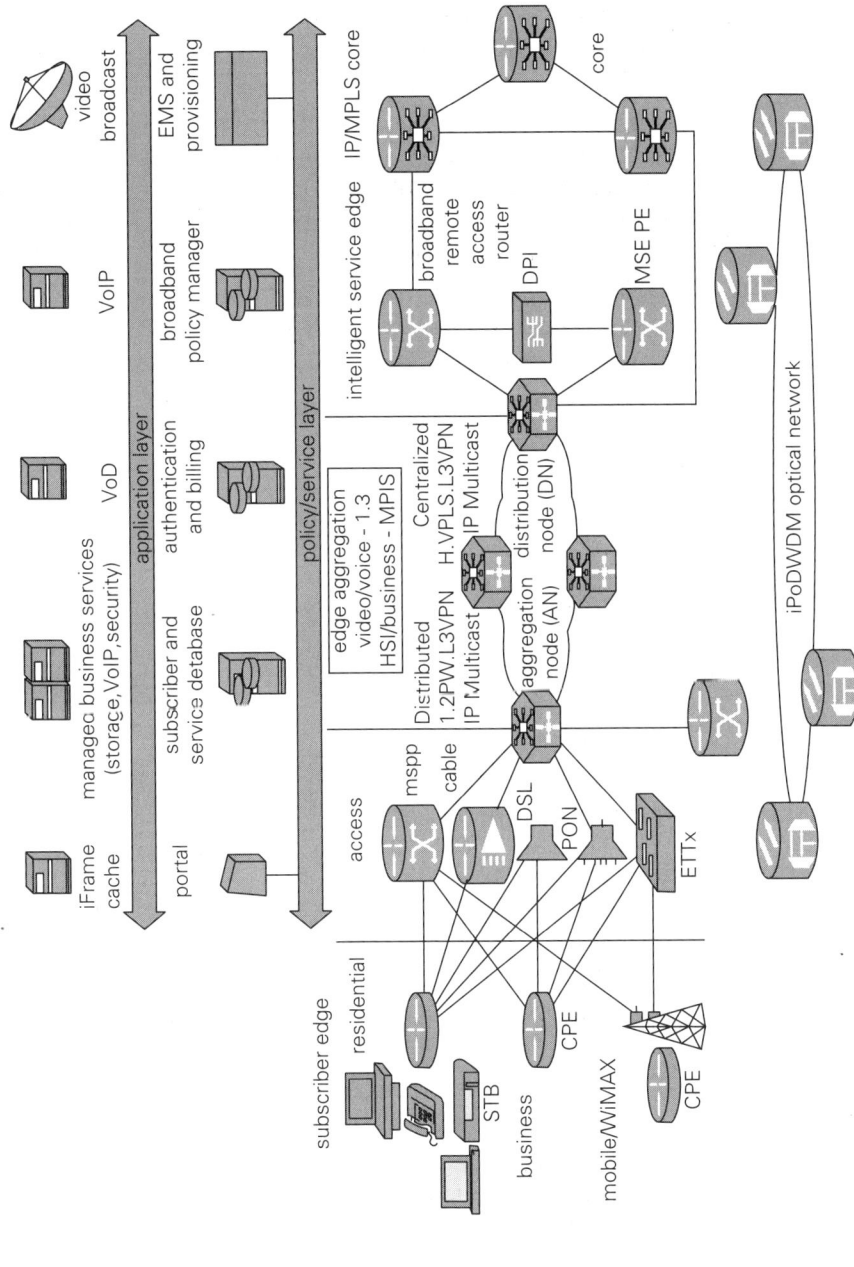

Source: Cisco 2009.

Note: DSL = digital subscriber line; IP/ MPLS = Internet Protocol/Multiprotocol Label Switching; WiMAX = Worldwide Interoperability for Microwave Access (fixed and mobile communications standard).

incumbent operators around the world are now converting their legacy networks to NGN.

For example, KPN (Royal Dutch Telecom), the incumbent operator in the Netherlands, was one of the first operators in the world to transition to an all-IP network in 2005 (figure 5.3). As part of the project, KPN's backbone infrastructure was upgraded to fiber. Despite the ambitious plans, KPN has found the conversion to all-IP to be slower than expected. One reason is the high cost to install fiber in the local access network. Nevertheless, by the end of 2010, over half of KPN's access customers were IP based, including almost half a million using triple-play offers of voice over Internet Protocol (VoIP), broadband, and Internet Protocol television (IPTV).[3] An interesting aspect of KPN's upgrade plan was that the transition to a much smaller number of IP exchanges allowed for the disposal of land and buildings made obsolete by the transition, which largely paid for the required investment with substantially lower future operating costs.

Figure 5.3 Transition of KPN Netherlands to an All-IP Network

Network efficiency through All-IP
Focus on cash flow generation on while to IP

strategy as announced in 2005	focus going forward

• Ramping up new services like VoIP, broadband and TV	• Legacy infrastructure operated longer than initially planned
	  – high cash flow generation
	  – optimizing customer migration to IP

• Maintaining market shares in traditional services	• Focus on cash flow generation
	  – increasing life time of copper assets - 'sweating the asests'
	  – significant cost savings realized before complete legacy switch-off

• Structurally lower cost by migrating to all-IP network	
• Switching off legacy IT and platforms	• Managed phase-out legacy networks – e.g., ATM in business market
• Network upgrade financed by real estate disposals of ~€1 bn	• Real estate disposals ongoing – executed on ad hoc basis in line with economic conditions

Source: KPN, "Update on KPN's Fiber Rollout," Press Release, December 15, 2009, http://www.kpn.com/v2/upload/4140a0cd-d7b7-4104-b7b1-76ba7c3419fc_Presentation_Fiber_update.pdf.
Note: €1 billion = US$1.2 billion.

Basic Technologies for Broadband Connectivity

This section examines the key technologies that are being used to construct today's broadband networks. Although each of these technologies can be used throughout the supply chain, they tend to be used most heavily in the international, domestic backbone, and metropolitan link segments. Each of these levels is discussed in more detail later in this chapter, as are the technologies that support wireline and wireless local access networks.

Fiber Optic

Much of the Internet's content travels via fiber optic cables, particularly for long-haul transmissions. Fiber optic cable provides closed circuit transmissions with very large bandwidth and at very high transmission speeds. These two complementary features occur because these cables, made of thin strands of coated glass, can transmit signals modulated over laser-generated beams of light. Rather than transmit using lower-frequency radio waves, fiber optic cables operate at the frequencies of light, where the spectrum is larger than in the radio frequencies (the visible spectrum contains more than 100,000 gigahertz, GHz), making it possible to carry large volumes of traffic at a rate of up to several hundred gigabits per second (Gbit/s) or even terabits per second (Tbit/s). Additionally, carriers can transmit traffic at several different frequencies using a technology called Dense Wave Division Multiplexing (DWDM). Multiplexing makes it possible for carriers to aggregate traffic onto a shared channel. Demultiplexing unpacks and separates the aggregated traffic back into separate transmission streams for delivery to the intended recipients.

Because of the high expense incurred when installing cable across an ocean floor or buried underground, carriers deploying fiber optic cables typically install dozens of glass strands into one cable. Initially, not all of these individual fibers will be used; carriers can activate ("light") individual strands as demand grows. Installed but unused "dark fiber" can be activated later, as required. In addition to installation costs, the comparative disadvantages of using fiber optic cables over copper lie primarily in the cost of the equipment and labor. While this technology can interconnect with existing copper networks, additional cross-connect switching equipment must be installed. Carriers with a large installed copper wire network may undertake a cost-benefit analysis and conclude that simply retrofitting and upgrading the existing network may help to conserve capital in the short run. Carriers opting to upgrade will install replacement fiber optic cable first on backbone routes with high volumes of traffic. As demand for bandwidth grows and investments can be justified, fiber progressively replaces copper cables throughout the network, reaching closer to the end users.

At present, most backbone networks are fiber based, even in developing economies, and the use of fiber in metropolitan and "middle-mile" links is rapidly increasing as well, particularly in developed countries. As the demand for wireless broadband grows, there is also increasing use of fiber to provide backhaul from cell sites to mobile carrier switching facilities. Fiber penetration in the local access network is still very limited, even in developed countries. But the emerging trend, especially for building out new housing and commercial developments, is to install fiber from the outset. Several deployment scenarios are possible for fiber optic cable:

- International connectivity—international undersea networks and international terrestrial networks

- National backbones—national undersea networks and national overland backbone networks

- Metropolitan rings and cellular backhaul

- Subscriber access—fiber to the premises.

Satellite

In the broadband supply chain, satellites are used primarily for international connectivity and some domestic backbones; they are used less frequently for metropolitan and local access networks. Geostationary communication satellites receive and transmit information from orbital slots located 35,786 kilometers (22,282 miles) above Earth. At this height, the satellite appears in a fixed location when viewed from Earth; this stable location is an advantage since subscriber satellite dishes do not need to move or track the satellite.

A satellite's communication capabilities can be analogized to an invisible "boomerang" or "bent pipe," with signals transmitted (uplinked) to the satellite, which then relays (downlinks) them back to Earth. Data are transmitted via the communication satellite's transponders. Satellites usually have between 24 and 72 transponders, with a single transponder capable of handling up to 155 Mbit/s (megabits per second).[4] Next-generation satellites will offer speeds in excess of 100 Gbit/s.[5]

From a geosynchronous vantage point, satellites can transmit signals covering as much as one-third of Earth's surface. This stable "footprint" coverage makes satellites an ideal medium for distributing television (TV) and Internet content on both a single point-to-point basis and a point-to-multipoint basis. Today's advanced satellites also make use of "spot beams" (principally in the Ka band) that allow higher power to be concentrated in specific regions to improve bandwidth and signal quality. These beams can also be steered or reconfigured to match bandwidth to specific areas of demand.

A satellite network can be configured in various ways, ranging from a simple one-direction link to a more complex mesh network. Communications with the satellite take place via an earth station or individual antenna. The size of the antenna depends partially on the frequency being used and also affects the volume of information that can be exchanged with the satellite. Large antennas are typically installed at earth stations for high-bandwidth applications, while smaller antennas, such as very small aperture terminals (VSAT) or direct to home (DTH) dishes are used for applications such as lower-bandwidth Internet access in rural areas or satellite TV reception. An estimated 3 million commercial VSATs are used for commercial and consumer purposes around the world, with the majority supporting broadband Internet or high-data-rate services.[6]

Each communication satellite requires several hundred million dollars in investment to cover its construction, insurance, launch, and tracking. These satellites have a limited usable life (usually around 20 years) because operators cannot make repairs or add fuel to the propulsion motors to keep them in proper orbit and pointed at the correct angle toward Earth. Additionally, satellites have comparatively less transmission capacity than terrestrial options, such as fiber optic cables. The large distance between the satellite and users on Earth also results in delays, known as latency, due to the time it takes for instructions to reach a satellite and content to arrive on Earth. Despite these limitations, satellites excel in their ability to distribute broadband content, such as television, to many locations and are advantageous for different developing-country characteristics such as archipelagos or difficult terrain as well as for emergency and disaster situations.

Microwave

"Microwave" systems are named for the wavelengths they use to communicate and are generally implemented using frequencies between 6 GHz and 38 GHz (Hansryd and Eriksson 2009). Microwave systems provide a point-to-point or point-to-multipoint broadband transmission option using very high frequencies that transmit a highly directional, pencil-thin beam of energy. Unlike satellite beams that cover thousands of square miles, microwave is usually used to transport broadband data signals from one specific location to another over relatively short distances (generally 40–70 kilometers, depending on the frequency used). The installation of several microwave receiving and transmission facilities arranged in a chain is needed for longer links, with each transmission link known as a "hop."

Microwave radio transmissions use antennas that concentrate radio energy to generate a naturally amplified signal. To achieve this signal gain, the very high frequencies of microwave—in the single or multiple GHz range—are concentrated using antennas shaped in a parabola. With advanced modulation, typical microwave networks can support up to 500 Mbit/s. In 2010, Ericsson demonstrated a microwave radio connection with a capacity of 2.5 Gbit/s (Ericsson 2010). WiMAX (Worldwide Interoperability for Microwave Access) is a specific type of microwave standard that is designed for connecting end users, but it can also be used for backbone connectivity at high costs. Ranges up to 120 kilometers (75 miles) have been advertised, with speeds up to 100 Mbit/s.[7]

Before the advent of fiber optic cables, microwave systems were a leading provider of backbone and metropolitan (long-distance) connectivity. As fiber technology improved and costs fell, however, operators began to replace their microwave systems with fiber cables. This trend started on the highest-volume traffic routes and continues to push into more local parts of the network. Today, microwave technology is used almost extensively for point-to-point backhaul and last-mile line-of-sight communications, especially when available capital expenditures are limited. The main advantages of a microwave system are its relative immunity to interference, its straightforward deployment, and easy reconfiguration. Thus, it can be a practical alternative in some cases compared to the cost and logistics of laying cable.[8] The main drawbacks are that it requires line-of-sight and transmission capacity that may be too limited for heavy broadband uses.

Copper

Another terrestrial technology still in use for long-haul transmission is copper wire. While fiber optic cable will eventually replace legacy copper, replacement costs create financial incentives to use and upgrade existing networking technology. Copper wire offers significantly less channel capacity and commensurately slower bit transmission speeds than other media, but it can often suffice for low-traffic routes. Even in developed nations, backbone fiber optic routes may exist only for links between major cities, with copper wire links still serving smaller towns and rural areas. A recent issue with copper cabling is theft, due to the high price of copper (Gallagher 2010).

International Connectivity

The Internet is an international "network of networks." In order to provide the physical connections between widely separated broadband resources and consumers, countries must establish international links (gateways) to connect to the world's Internet and telephone networks. The technologies providing long-haul transmission, such as fiber optic cable and satellites, typically have very high investment costs. While initial "sunk" costs are high, they have very low incremental costs to accommodate additional users. These technologies also enable carriers to activate additional capacity on an incremental, graduated basis as demand grows.

International Links

The vast majority of international telecommunications traffic is carried by undersea cable systems—more than 95 percent according to some estimates (Bressie 2010). This reflects the advantages of fiber optic cable in terms of bandwidth and latency compared to satellite. Undersea fiber optic cables can transmit data at speeds measured in Tbit/s, while even the newest communication satellites offer speeds below 1 Gbit/s as well as higher latency. As of early 2011, there were more than 120 major submarine cable systems, with another 25 planned to enter service by 2015.[9] Submarine cables are quite expensive to deploy, with costs that routinely reach into hundreds of millions of U.S. dollars. As such, many are financed by consortiums of operators rather than a single investor. For example, the Eastern Africa Submarine Cable System (EASSy) has landing points in nine countries and connects to several additional landlocked countries; it is funded by 16 African and international shareholders, all of whom are telecommunications operators and service providers.[10]

While undersea fiber optic cables may be the preferred option for international connectivity, it is not a viable option for some countries and operators. Landlocked countries, for example, do not have direct access to the sea and thus are constrained in their ability to exploit submarine technology fully. Transit costs to tap into an undersea cable can be significant (national and regional fiber backbones may not be available to tap into the undersea cable), but this is becoming less of an issue over time, as landlocked countries complete some type of fiber connection to international cables through neighboring countries. Landlocked countries may be able to negotiate a virtual coastline so that they own and operate a cable landing station in a neighboring country's territory but otherwise depend on the neighboring country to provide reliable and reasonable prices for transit. Many small island developing states (SIDSs), mainly in the Pacific Ocean, are distant from undersea fiber routes, and the economics of connecting to undersea cable are problematic. Regulatory restrictions or high costs may restrict service providers from accessing undersea cables. These factors tend to encourage the use of satellite connectivity. Another issue is that, even where countries have access to undersea cable, they still may want to deploy satellite as a backup to ensure redundancy.

Service providers need to contract physical international connections in order to support their end user broadband requirements. They do so either by participating in ownership consortiums of the physical facilities or by leasing connectivity through wholesale operators. A relatively small number of Internet service providers (ISPs)[11] have the financial resources needed to invest directly in capacity in international backbone broadband networks, so most lease capacity from larger international operators. This can present several business and regulatory challenges including the following:

- *Monopoly or dominant control of international backbone routes.* Physical backbone networks are generally owned by a few operators or consortiums. The restricted ownership can be a barrier for nonaffiliated ISPs that need international connectivity.

- *Monopoly or dominant control of international landing points.* Landing stations for undersea cable and satellite earth stations are generally controlled by a few entities. Even if an ISP has successfully contracted for capacity on a fiber optic cable or satellite transponder, it may be constrained in its ability to connect that capacity to its domestic backbone network. For example, a service provider may lease capacity on a satellite transponder, but may have to come to a separate agreement with the owner or operator of the international gateway that receives the satellite's transmission.

The potential for international connectivity to be a bottleneck in the development of broadband connectivity cannot be overstated. Submarine cables connect to domestic backhaul networks at a cable termination station, which is—but may not be—the same facility as the cable landing station (that is, where the cable makes landfall). Because all operators in a market, particularly new entrants, may not have the resources to own and operate a cable landing station, the owners of such stations—generally the incumbent operators in newly liberalized markets—may be required to provide access to the station, and therefore to the cable, on reasonable terms to competing service providers. Limited access to landing stations can have a chilling effect on the diffusion and take-up of broadband services. Conversely, limited opportunities or burdensome regulations related to cable landing can discourage interest in that market among cable operators, again creating a connectivity bottleneck. Governments and regulators may need to implement competitive policies with respect to issues such as submarine cable landing stations, open access, and infrastructure sharing to eliminate such bottlenecks (see chapter 3).

In Singapore, for example, the single cable landing station was owned by the incumbent operator at the time of market liberalization. Singapore's regulator undertook two parallel approaches to improve international connectivity (IDA 2008). The Info-communications Development Authority of Singapore (IDA) required the incumbent operator to offer collocation in its submarine cable landing station to alternative operators, later imposing connection at regulated prices and granting alternative operators access to the capacity of submarine cables on behalf of a third party. In addition, the IDA streamlined the administrative procedures for submarine cable companies to obtain landing permits and authorizations in Singapore. As a result of these and related actions, prices of international leased circuits in Singapore decreased 95 percent, total submarine cable bandwidth capacity increased from 53 Gbit/s in 1999 to 28,000 Gbit/s in 2007, and broadband penetration reached 77 percent of households in 2007.

Internet Links

Whether via fiber optic cable or satellite, securing physical international links is only the first step in procuring international Internet connectivity. ISPs also need to arrange for exchanging and routing their traffic. Such arrangements ensure that Internet traffic can be delivered anywhere in the world, eliminating the need to have physical connections to every country. An ISP will typically arrange to hand off its traffic at the points where its contracted physical connectivity terminates. Such arrangements are usually of two kinds:

- A *peering* arrangement is where two ISPs freely exchange Internet traffic. The peering requirements of large ISPs often exceed the capability of smaller ISPs. For example, in order for an ISP operating in the Asia-Pacific region to peer with Sweden's TeliaSonera, it would have to provide traffic equaling at least 500 Mbit/s and the ratio of inbound and outbound traffic exchanged between the ISP and TeliaSonera could not exceed 3:1 (TeliaSonera 2010).

- A *transit* arrangement is where a small ISP pays a large ISP to provide Internet traffic exchange. The fee is generally a function of the traffic or physical connection. Smaller ISPs generally make transit agreements with global IP carriers that can guarantee that their Internet traffic will get routed anywhere in the world. Global IP carriers with worldwide IP networks are often referred to as "Tier 1" carriers, with the distinguishing characteristic that they do not generally pay any transit fees and have the capability to reach all networks connected to the Internet (Van der Berg 2008).

While large global carriers from developed countries operate most of the Tier 1 networks, carriers from developing countries are starting to emerge as significant players. India's Tata Communications, for example, operates a global network that makes it the world's largest, farthest-reaching, wholesale Internet transit provider. It provides Internet connectivity to over 150 countries across six continents, with speeds up to 10 Gbit/s.[12]

Tata's reach and Internet routing can be illustrated by running a trace route from a broadband subscriber in Washington, DC, accessing a website in Gaborone, Botswana. Once the packet reaches the west coast of the United States, it is turned over to Tata for delivery to Botswana over physical connections transiting Singapore, India, and Johannesburg, South Africa (figure 5.4).

Implementation Issues for International Connectivity

The huge costs of deploying undersea fiber optic and satellite networks present a challenge for many developing countries and ISPs. Capacity on these networks tends to be owned by a few carriers, and wholesale arrangements are not always optimum for smaller players. Likewise, a few global IP carriers dominate wholesale access to the Internet, and smaller ISPs are forced to pay one-way interconnection charges. Landlocked countries face special problems since they lack coastal regions that could support a landing station for undersea cable, while SIDSs face a connectivity challenge since they are distant from undersea cables and lack large markets. In countries with just one physical international link, access and pricing can become an issue, particularly if the operator of the gateway is also a provider in other

Figure 5.4   Internet Protocol Packet Route from Washington, D.C., to Gaborone, Botswana

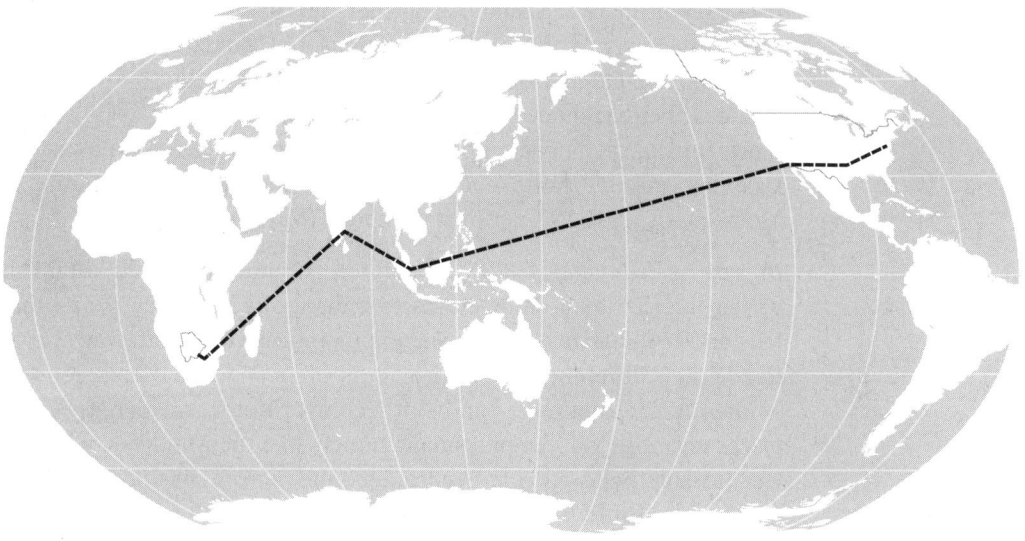

IBRD 39041
JANUARY 2012

This map was produced by the Map Design Unit of The World Bank.
The boundaries, colors, denominations and any other information
shown on this map do not imply, on the part of The World Bank
Group, any judgment on the legal status of any territory, or any
endorsement or acceptance of such boundaries.

Source: Telecommunications Management Group, Inc.

parts of the supply chain and has an incentive to restrict competition or demand high payments.

Countries are exploring various ways to overcome the challenges of international connectivity, including the following:

- *Forming public-private partnerships (PPPs) to establish direct international links.* The high cost of connecting to international networks may be insurmountable for smaller service providers. In Kenya, the government took the lead in procuring an undersea fiber connection through an agreement to construct a cable from Kenya to the United Arab Emirates by enlisting service providers to take a shareholding in The East Africa Marine System (TEAMS) cable (World Bank 2011).

- *Establishing points of presence (POPs) in major Internet hubs.* This can be cheaper than paying transit fees. Sri Lanka Telecom established a subsidiary in Hong Kong SAR, China, and acquired domestic and international voice and data services licenses allowing it to offer undersea fiber optic cable capacity services from Hong Kong SAR, China, to Asia, Europe, and North America.[13]

- *Enhancing cross-border cooperation.* It is critical for landlocked countries to coordinate and establish partnerships in order to ensure end-to-end connectivity to undersea land stations. In Uganda, the ISP Infocom leased fiber capacity from the country's electrical utility, allowing it to create a fiber backbone to the Kenyan border (Kisambira 2008). From there, Infocom arranged with Kenya Data Networks (KDN) to transport Ugandan Internet traffic to a new undersea fiber optic cable landing in Mombasa using KDN's national backbone (Muwanga 2009).

- *Improving redundancy and competition.* Countries should establish varied international connections to enhance redundancy if one link fails and to enhance competition among international gateway operators. When different service providers offer additional connections, wholesale international bandwidth competition also increases and prices generally fall. Even a small country like the Maldives, where it was initially believed that even one connection to an undersea fiber optic cable would be prohibitively expensive, has found that an open telecommunications market with a liberal international gateway license regime can motivate operators to invest in high-quality connectivity. The Maldives now has two links to undersea fiber optic cable systems (box 5.1).

- *Creating Internet exchange points (IXPs) close to data servers and international bandwidth.* By establishing its own IXP, a country can reduce expensive international traffic by keeping local traffic local and by attracting leading global content providers. For example, Google has a liberal peering policy and has established POPs in several locations, including recently in South Africa at the Cape Town Internet Exchange.[14]

Domestic Backbone

Backbone networks are a critical component of the broadband supply chain. They consist of very high-speed, very high-capacity links that connect the major nodes of the network—often the major cities of a country. These links need to have large capacities because their function is to aggregate traffic from the different areas of the country and then carry it on to the next node or city. Although the comparison is not perfect, broadband backbone networks serve an analogous function to a country's highways, allowing fast

Box 5.1: Connecting the Maldives to the International Submarine Cable Network

Like most small countries, the Maldives has been relying on satellite technology to connect to the outside world. The main reason is the cost-effectiveness of satellite as compared to fiber cable for the level of international traffic generated by this small country. Global submarine optical fiber cable networks like SE-ME-WE (South East Asia–Middle East–West Europe) have passed the Maldives, but the high cost of joining these cable consortiums prevented the country from reaping the technical benefits of optical fiber technology.

Although satellite technology was sufficient in the past when voice telephony was the driver of international communications, the bandwidth consumed by data applications has surpassed the bandwidth usage of voice applications. In 2005, the government decided that it was economically feasible to install an optical fiber system, and a consortium was established among three service providers: Wataniya Telecom Maldives, Focus Infocom Maldives, and Reliance Infocom of India. The consortium, WARF Telecom International, brought the first fiber into the country in October 2006, which connects the Maldives to the Falcon Network at a node in Trivandrum, India. In early November, Dhiraagu brought in a cable connecting the Maldives to Colombo, Sri Lanka.

Source: Ibrahim and Ahmed 2008, 204.

connections between source and destination. Figure 5.5 shows how backbone networks have been deployed in Botswana.

In developed countries and liberalized telecommunications markets, there may be more than one backbone network. Competing firms, for example, often lay fiber cables across a country to compete with incumbent long-distance carriers. This is not usually the case in developing countries, where voice and data traffic demands have not historically required such high-capacity links. Recently, however, developing countries have been promoting the development of high-capacity domestic backbones as part of a broader effort to develop regional fiber networks. In Zimbabwe, for example, incumbent TelOne announced in March 2011 that it had connected a fiber link to the EASSy system through Mozambique.[15] The link is the first phase of a planned national backbone rollout that will also include the Harare-Bulawayo-Beitbridge and Harare-Chirundu routes.

It may not be necessary for each operator to have a backbone network that covers the entire country. An operator may have an extensive network covering one part of a country, but not others. Operators can interconnect their respective networks in order to use the backbone network of another by purchasing backbone network services or entering interconnection

Figure 5.5 Backbone Networks in Botswana

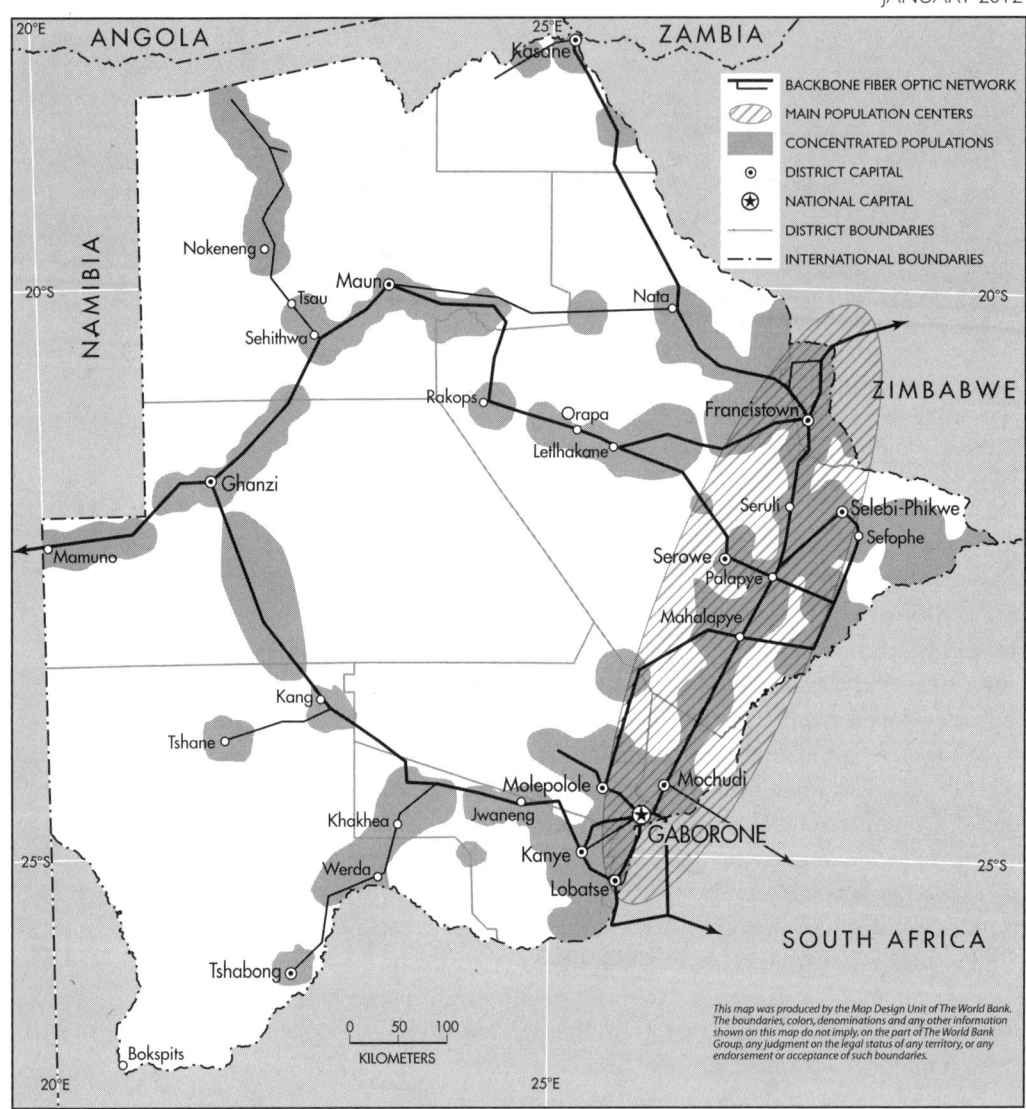

IBRD 39042
JANUARY 2012

Source: Hamilton 2007.

agreements. In many developed countries, the owners of the backbone networks and elements of the market are consolidated into a few large companies with very high-capacity networks, while the downstream components tend to be smaller and more geographically disaggregated. But this is not always the case; in many countries, especially those that have only recently liberalized their markets, one dominant provider may still control both the backbone and downstream (metropolitan and local access) networks.

The economic impact of backbone networks lies in their ability to reduce costs by spreading them over higher volumes of traffic. However, this benefit is highly dependent on the market situation in a given country. In Nigeria, for example, one of the reasons that the incumbent operator historically has been able to maintain high wholesale prices for backbone services is the lack of effective competition in the backbone services market (Williams 2010, 6). This is a pattern seen throughout Sub-Saharan Africa and in other parts of the world, where neither competition nor regulation has effectively controlled wholesale prices. Conversely, all broadband providers benefit where there is competition and the backbone is open and interconnected to multiple downstream providers, particularly for smaller players who can buy network services at reasonable prices rather than build their own end-to-end networks.

Countries face several challenges in deploying national backbone networks. One is the challenge of ensuring high-speed links throughout the country to minimize the broadband divide. Because each country has unique geographic (size, terrain) and demographic features, each will have to pick those technologies that best fit its situation. As a result, different mixes of technologies will be employed, and private investors and policy makers will need to examine the trade-offs between bandwidth needs, capital expenditures, operating expenses, upgradeability, and regulatory impacts, among others.

National Links

The choice of a national backbone strategy is highly dependent on a country's size, topography, regulatory environment, and broadband market size. In reviewing the different technologies, it is important to bear in mind that the selection of the appropriate backbone connectivity option often depends on the distance to be covered and the forecasted capacity requirements (table 5.1).

Fiber optic cable is typically perceived as the optimum solution for national backbone connectivity given its high capacity and upgradeability. Almost every operator in the world is upgrading its network backbone by

Table 5.1 Optimum Choice of Backbone Technology, by Distance and Capacity

Distance (kilometers)	Capacity (Mbit/s)		
	< 8	8–450	> 450
<100	Satellite, microwave	Microwave	Fiber optic
>100	Satellite	Microwave, fiber optic	Fiber optic

Source: Williams 2010, 18.

installing fiber, although the extent and pace vary. Some operators have fully fiber backbones, with other technologies kept for redundancy, while others may have only a few kilometers of fiber for high-traffic routes, supplemented by satellite, microwave, or even copper cables.

Satellite is employed in some countries to provide national as well as international backbone connectivity. Some countries have launched their own satellites to ensure coverage, while in other countries, operators lease capacity from satellite operators. Satellites are a particularly attractive solution for providing connectivity to remote areas where the cost of terrestrial solutions can be high. They are not, however, an ideal solution for short, high-traffic routes.

Microwave remains a legacy solution for domestic backbone in many countries. It is less expensive to deploy than fiber optic or satellite, although operating expenses may be higher compared to other solutions. Although capacity is less than fiber optic, current traffic demands may not be high enough to justify switching from microwave. Rights-of-way and space constraints are also less of an issue with microwave.

Internet Exchanges

In addition to the physical infrastructure needed to transmit broadband traffic, national broadband traffic needs to be routed and exchanged using IP. One key issue here is Internet exchanges. Internet exchanges provide a venue for transferring traffic between different ISPs that can help to reduce the cost of international connectivity. An Internet exchange can also connect to local data centers featuring content and applications that reduce the need for international connectivity.

The economic rationale for establishing an IXP is based on the trade-off between the cost of the physical connection to the IXP and the cost of an international connection to process the local traffic. There are two main models for implementing an IXP (figure 5.6). With a Layer 2 IXP, each ISP provides its own router, and traffic is exchanged via an Ethernet switch.

Figure 5.6 Internet Exchange Point Models

Layer 2 exchange

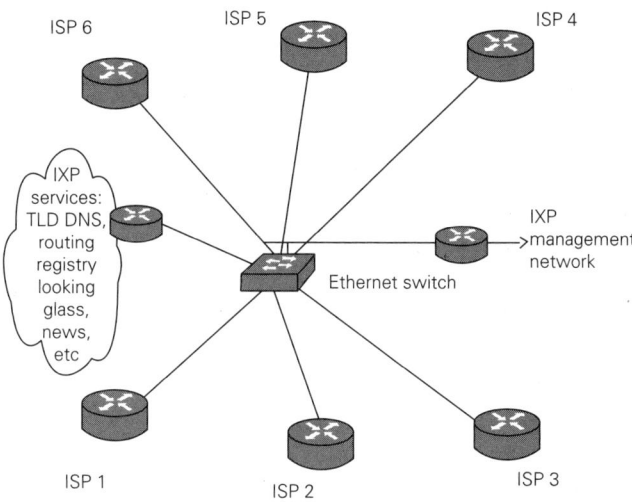

Layer 3 exchange/wholesale transit ISP

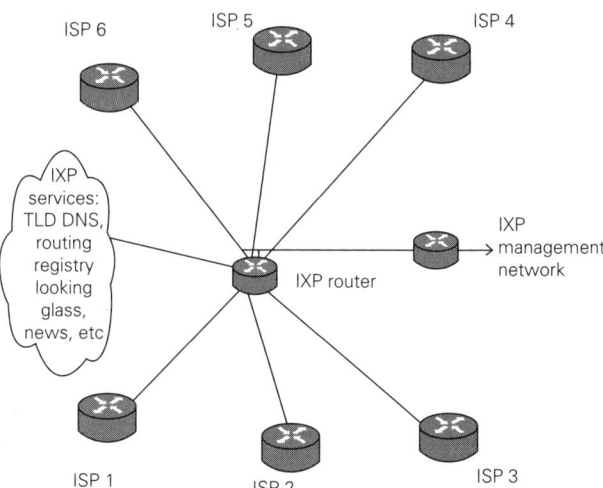

Source: Cisco, "Internet Exchange Point Design," http://www.pacnog.net/pacnog6/IXP/IXP-design.pdf.

Note: TLD DNS = Top-Level domain - domain name system.

With a Layer 3 IXP, traffic is exchanged between members through a single router. ISPs make a physical connection to the IXP typically using a bandwidth of between 100 Mbit/s and 1 Gbit/s. In addition to the cost of the physical connection to the IXP, ISPs usually pay a joining fee plus a monthly fee that is sometimes based on the size of the connection or the volume of traffic.

There are several administrative models for IXPs. One of the most common is where a group of ISPs operates the IXP, typically through some kind of association. Another model is commercial, where an unaffiliated third party (for example, not an ISP) provides IXP services. In some countries, nonprofit organizations operate IXPs for government or educational Internet traffic. In some cases, large ISPs operate Internet traffic exchange points, which often involve the large ISP charging for transit. Large ISPs have also been known to disrupt an IXP's effectiveness either by not participating or by underprovisioning their link to the IXP (Jensen 2009). Countries might consider facilitating the development of IXPs and providing support or appropriate regulation to help to overcome resistance to their establishment and effective operation. In Chile, for example, the government requires all ISPs to interconnect (Cavalli, Crom, and Kijak 2003).

Where competitively priced international fiber optic connectivity is available, an IXP can attract international participants such as foreign ISPs and major content companies. This can lower the cost of international connectivity through peering with foreign ISPs as well as the potential for storing heavily accessed content locally. For example, Google participates in various IXPs through its Global Cache service, which stores its applications and content, such as YouTube, closer to the end user (Guzmán 2008).

Implementation Issues for Domestic Backbone Networks

As countries build out their backbone networks, several issues must be considered. For example, the initial fixed costs are significant. A study by the Organisation for Economic Co-operation and Development (OECD), for example, concluded that around 68 percent of the costs in the first year of rolling out a fiber network to the premises are in the civil works associated with the digging of trenches and the installation of cables (OECD 2008). In countries with large physical distances to cover, this fixed cost may be substantial and difficult to justify when demand is uncertain. The risk associated with the high up-front costs of fiber backbones can be alleviated through various mechanisms such as risk guarantees and demand aggregation.

The average cost of a backbone network (that is, the unit cost per subscriber) also varies enormously, depending on the subscribers' geographic location. In urban areas, where subscribers are concentrated, the average cost of backbone networks is much lower than in smaller towns or rural areas. In practice, the ability of a backbone network to reduce costs is one of the key determinants of the financial viability of providing broadband services. The absence of a backbone network in a particular area of a country to aggregate traffic and thereby reduce costs may mean that broadband services are unlikely to be commercially viable. In such instances, policy makers and network planners will need to agree on which cities and towns should be connected and over what period of time the work can be accomplished.

The high cost of installing backbone networks in developing countries has often resulted in incomplete national coverage, with operators deploying microwave networks on some routes. A single fiber optic network with wider coverage is often a more optimum solution and can be as cost-effective. An analysis for Nigeria found that a single high-capacity backbone had significantly lower costs than traffic carried over multiple low-capacity networks. However, operators in many countries have generally not been very cooperative in sharing backbone networks. Governments are overcoming such resistance through various arrangements for deploying fiber optic backbone networks:

- *Borrowing from multilateral and bilateral agencies.* Developing countries unable to afford the immediate cost of deploying national fiber optic backbone networks have been turning to development agencies and bilateral development partners for funding. For instance, Uganda borrowed from the China Export and Import Bank to finance construction of its national data transmission backbone infrastructure (Uganda, Parliament of 2009).

- *Encouraging the existing operator to build out the network.* Current service providers can be incentivized to extend their backbones and offer cost-based wholesale connectivity. In Sri Lanka, the government decided to work with the incumbent to extend its fiber optic network rather than to build its national backbone network from scratch.[16] In Pakistan, a universal service fund is used to subsidize the cost of fiber optic rollout to rural areas. Awards from the fund will extend some 8,413 kilometers of fiber optic cable to underserved locations.[17]

- *Leveraging electrical utilities and railways.* Companies in other infrastructure sectors such as electricity and railways have large fiber optic networks running along grids or railroad tracks. National connectivity can be enhanced by facilitating telecommunications regulations that allow electricity providers and railways to act as wholesale bandwidth providers. Kenya Power and Lighting Company Limited, an electrical utility, is leasing dark fiber running along its backbone to service providers.[18] In Norway, the fiber backbone of Ventelo spans the entire railway infrastructure, covering 17,000 kilometers. Ventelo is the second largest wholesale provider in Norway, offering dark fiber and collocation services.[19]

• *Ensuring open access.* Despite the general benefits of competition, it may be inefficient in some areas to have competing backbone networks. In such cases, a single network can be built, but protections must be put in place to ensure competition among service providers, including nondiscriminatory access for all downstream providers. Singapore, for example, has imposed structural separation for the deployment of its next-generation national broadband network (NBN) in an effort to minimize infrastructure duplication, increase wholesale transparency, and promote retail competition for the benefit of consumers (IDA 2010).

Metropolitan Connectivity

Beyond network backbones, connectivity is needed to connect smaller towns and villages to the backbone and provide links in and around metropolitan areas. These links are sometimes called the "middle mile." Such links can be provided by satellite, microwave, or fiber optic cable, with the latter becoming increasingly common due to its high capacity. Metropolitan area networks are often established for high-traffic locations such as major cities by routing traffic along high-capacity fiber optic rings. This part of the broadband supply chain also includes links used to transport traffic from distant points, such as a wireless base station, to an aggregation point in the network, such as a mobile telephone switching office or other network node (United States, FCC n.d.). This particular function in wireless networks is often referred to as "backhaul" (that is, hauling traffic back to the network).

Regional and Metropolitan Links

In many cases, as governments develop policies to encourage backbone development or the rollout of local access networks, the metropolitan portion of the broadband supply chain can be forgotten. But building out the two ends of the network—backbone and last mile—will be ineffective unless capacity exists in the middle to tie all the pieces together. Hence policies to address middle-mile and backhaul problems, such as promotion of facilities-based competition or open-access requirements, are just as important as policies in other parts of the network.[20]

Metropolitan ring networks are a special case worth noting. In most countries, the majority of broadband traffic is generated in urban areas. Initial links are typically point-to-point, but over time this architecture can become increasingly complex and inefficient. The topology of a ring network is highly practical for metropolitan areas where a significant amount of traffic is destined for other users in the area. A ring network simplifies network architecture by connecting premises in central business areas together over fiber optic cable. Traffic flows along the ring, with each node examining every data packet (figure 5.7). The standard for metropolitan ring networks is Institute of Electrical and Electronics Engineers (IEEE) 802.17.[21]

One of the dangers with ring networks is that if a node goes down or the fiber optic cable breaks, the whole ring could fail. This can be overcome by transmitting the information in two directions (clockwise and counterclockwise) or by building in other types of redundancy. Rings have tended to use Synchronous Optical Network/Synchronous Digital Hierarchy (SONET/SDH) technology for transport. Wavelength Division Multiplexing (WDM) is emerging as a transport standard because of its efficiency and integration with gateways to national and international backbones.

Implementation Issues for Metropolitan Connectivity

Many of the implementation issues associated with the middle mile are the same as those involved with backbone development, namely cost and competition. However, the choices of where such links should be built (or, perhaps more accurately, upgraded, since lower-capacity links may exist) and how the network should be designed can be more difficult, both politically and technically. Government interventions are usually part of a plan to connect rural areas and are combined with other measures to roll out networks to those areas as well as part of metropolitan government initiatives. Even if broadband networks reach rural areas, most countries have a significant gap in broadband speeds between rural and urban areas. For instance, in Europe most of the lowest broadband download speeds (256–512 kbit/s) are found in rural areas.

In the context of limited funds for network build-out, choices will have to be made that balance the government's desire to spread the benefits of

Figure 5.7 Metro Fiber Ring

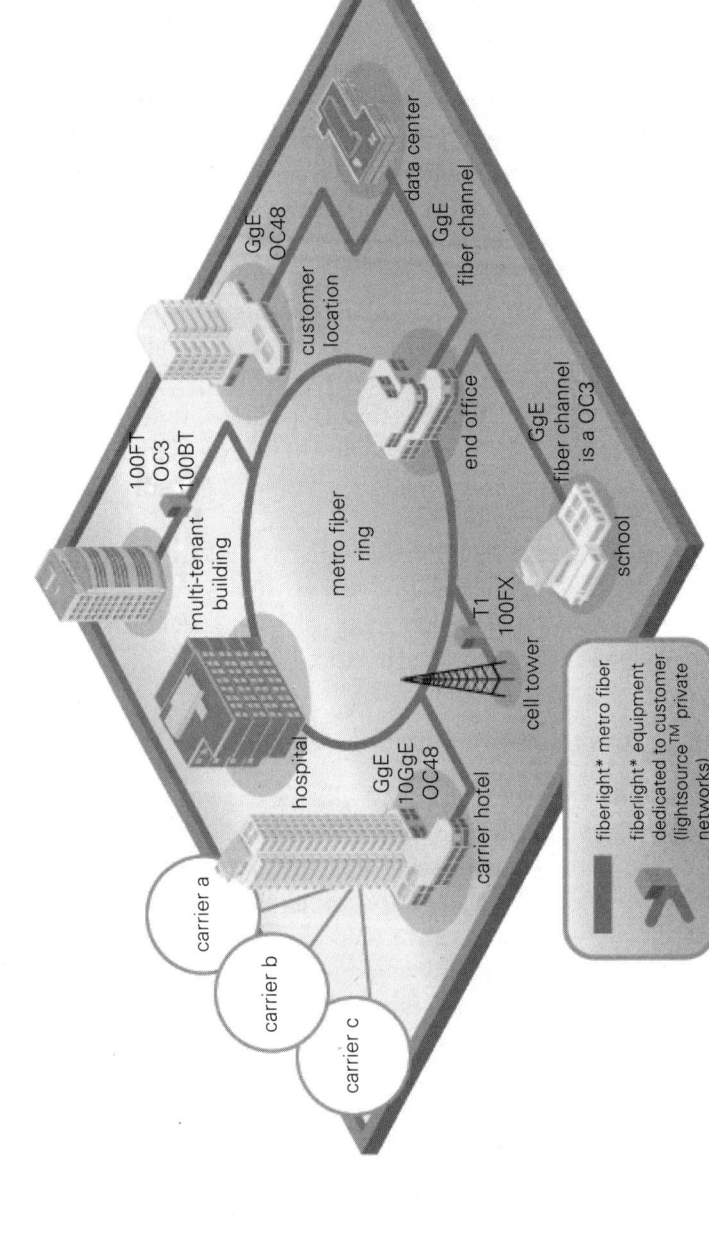

- data center
- GgE OC48
- GgE fiber channel
- customer location
- 100FT OC3 100BT
- multi-tenant building
- metro fiber ring
- end office
- GgE fiber channel is a OC3
- school
- hospital
- T1 100FX
- cell tower
- GgE 10GgE OC48
- carrier hotel
- carrier a
- carrier b
- carrier c
- fiberlight* metro fiber
- fiberlight* equipment dedicated to customer (lightsource™ private networks)

Source: Fiberlight, http://www.fiberlight.com/wp-content/uploads/2010/05/diag_opttrans1.jpg.

broadband widely with the reality that not all areas can be served right away. In Australia in 2009, for example, the government announced a $A 250 million "blackspots" program designed to bring high-capacity links to regional centers without adequate connectivity, holding a consultation to determine which regions should receive new links (Australia, Department of Broadband, Communications, and the Digital Economy n.d.).

Network design issues can also be difficult. In most developed countries and in countries with a liberalized telecommunications framework, competing alternative carriers use the dominant carrier's network through leases or open-access requirements and build their own networks around the dominant carrier's physical facilities. But as new broadband links are installed at the metropolitan level, an important issue to resolve is determining how many points of interconnection will be offered to the new broadband facilities and where the points will be located. In Australia, there has been a strong debate over how many points of interconnection should be offered, with the government and the National Broadband Network Company originally suggesting 14, while the competition authority states that 120 interconnection points are needed.

Local Connectivity

In the broadband supply chain, local access networks are those that directly connect end users to broadband services, the so-called "last mile." Several wireline and wireless broadband technologies are used today to support local access networks. Having multiple broadband access options in a country increases consumer choice, stimulates intermodal competition, enhances quality and innovation, and is generally associated with lower retail prices. However, countries may not be able to use all possible technological choices for historical, technical, regulatory, or financial reasons. As governments seek ways to promote broadband development, they will need to recognize the strengths and limitations of their existing level of infrastructure development—both for its upgrade possibilities as well as for developing appropriate incentive and competition policies.

Wireline Access Technologies

This section examines wireline broadband access technologies, including digital subscriber line (DSL), cable modem, fiber to the premises (FTTP), and other options. The first three account for almost all wireline local access technologies worldwide (figure 5.8).

Figure 5.8   Number of Broadband Subscribers Worldwide, 2007–09, by Type of Wireline Technology

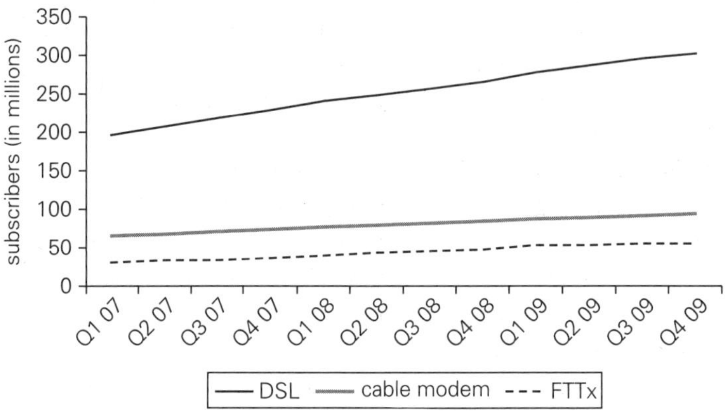

Source: Point-Topic.

Digital Subscriber Line

The public switched telephone network (PSTN) line running to the subscriber's premise has traditionally been copper wire, with a bandwidth of 3 to 4 kilohertz (kHz). This narrowband channel offers an analog carrier originally configured to provide a single telephone call. Two "twisted-pair" copper wires are used to support duplex communications (that is, the ability to send and receive at the same time). The PSTN has also supported the capability for narrowband Internet access, with subscribers using a modem to dial up an ISP.

DSL technologies use special conditioning techniques to enable broadband Internet access over that same PSTN copper wire. Transmission speeds vary as a function of the subscriber's distance from the telephone company switching facilities, the DSL version, the extent of fiber in the network, and other factors. DSL requires that the bandwidth over the copper line be separated between voice and data. A quartz crystal splitter is used to filter the data channel when using the shared copper local loop for telephone service. Similarly the voice channel must be filtered when the line is used for broadband Internet access. Nonetheless, users can continue to make and receive PSTN telephone calls when using DSL data services. As is the case with dial-up access to the Internet, subscribers must have a modem installed between their computer and the copper wire. A DSL modem modulates upstream signals to the Internet and demodulates downstream traffic to the subscriber.

In addition to retrofitting their copper lines, telephone companies also have to upgrade their switching facilities in order to split traffic into voice and data streams and to route data traffic between subscribers and the Internet. Traffic exchanged with the Internet is routed through a digital subscriber line access multiplexer (DSLAM). This device aggregates (multiplexes) upstream traffic from DSL subscribers onto high-speed trunk lines to be delivered to the Internet. Similarly, the DSLAM disaggregates (demultiplexes) traffic arriving from the Internet and routes it to the intended subscriber.

DSL has gone through several evolutions supporting increasing speeds and distances (table 5.2). The technology is standardized within the International Telecommunication Union (ITU) under Study Group 15 and the G series of ITU-T recommendations.[22] An asymmetric digital subscriber line (ADSL) maintains the frequency bandwidth of voice (that is, below 4 kHz) for telephony service. Broadband is transmitted on two other frequency bands; one is allocated to a low-speed upstream channel (25 to 138 kHz), and the other is allocated to a high-speed downstream channel (139 kHz to 1.1 MHz). The theoretical maximum downstream bit rate of 6 Mbit/s and maximum upstream rate of 640 kbit/s are defined by the standard.

In the ADSL2 standard, more efficient modulation and coding are implemented to improve the bit rate, quality, and, to a lesser degree, coverage. The standard defines maximum bit rates of about 8 Mbit/s downstream and 800 kbit/s upstream. The data rate is increased with ADSL2+ through doubling the frequency bandwidth by including the frequency band between 1.1 and 2.2 MHz. This results in a standard of 16 Mbit/s downstream and 800 kbit/s upstream.

Very high-speed digital subscriber line (VDSL) allows for much greater symmetrical data rates accomplished by using improved modulation techniques and adding more frequency bandwidth to the copper wire. However, the distances from the switch to the end user must be short or fiber must be installed to the curb. The VDSL2 standard overcomes some of these

Table 5.2 DSL Connection Speeds, by Type of Line

Type of line	Downstream speed	Upstream speed	ITU-T standard
Asymmetric DSL (ADSL)	6 Mbit/s	640 kbit/s	G.992.1
ADSL2	8 Mbit/s	800 kbit/s	G.992.3
ADSL2+	16 Mbit/s	800 kbit/s	G.992.5
Very high-speed DSL (VDSL)	52 Mbit/s	52 Mbit/s	G.993.1
VDSL2	100 Mbit/s	100 Mbit/s	G.993.2

Source: Adapted from ITU 2008.

Note: The speeds shown are those specified in the standard, not necessarily those experienced by end users.

challenges by extending distances and reducing interference, while increasing bit rates up to 100 Mbit/s for distances less than 300 meters.

In Israel, incumbent operator Bezeq has been rolling out VDSL2 as part of its NGN deployment, with coverage to around half the households by the end of 2010 (Bezeq Group 2011). It was advertising bandwidth of 100 Mbit/s for DSL connections on its website in March 2011. Bezeq plans to offer up to 200 Mbit/s through VDSL bonding, which uses two copper pairs per subscriber.[23]

While DSL technology has evolved with ever-increasing data rates and remains the most popular wireline broadband technology in terms of subscriptions, its biggest constraint is bandwidth deterioration as the distance from the exchange increases, as shown in figure 5.9.

Cable Modem

Cable modems provide subscribers with access to broadband services over cable television (CATV) networks. CableLabs developed standards for cable modem technology in the late 1990s.[24] The technical guidelines are called Data over Cable Service Interface Specification (DOCSIS). The DOCSIS guidelines have been progressively enhanced in terms of functionality (for example, support for IPv6) and speed (figure 5.10). The latest version is 3.0, with a slightly different European implementation (EuroDOCSIS). DOCSIS has been approved as an ITU recommendation.[25]

The first DOCSIS specification was version 1.0, issued in March 1997, which uses the subscriber's copper wire telephone line for upstream traffic. Beginning in April 1999 with the DOCSIS 1.1 revision, cable operators

Figure 5.9   Speed of DSL and Distance from Exchange

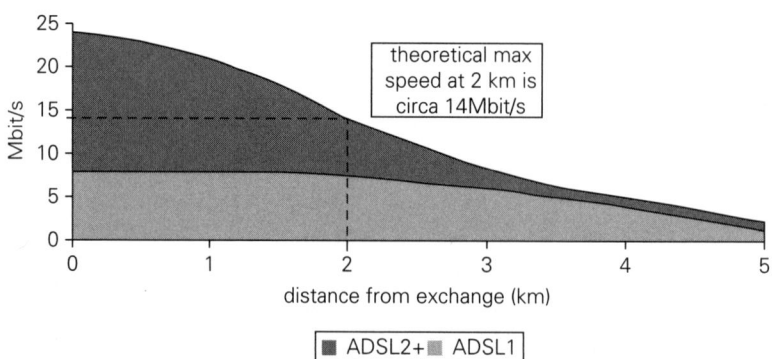

Source: Ofcom.

Note: Theoretical maximum speed at 2 kilometers is about 14 Mbit/s.

Figure 5.10   Cable Modem Connection Speeds, by Specification

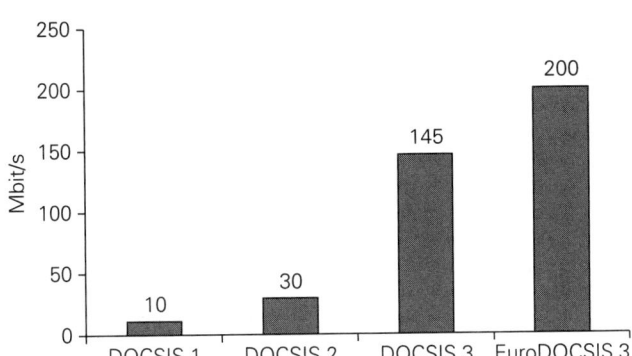

Source: Adapted from Motorola, "Planning an Effective Migration to DOCSIS® 3.0," http://www.motorola
.com/staticfiles/Video-Solutions/ultrabroadbandsolutions/pdf/Migration_to_DOCSIS30.pdf.

added quality of service capabilities and began installing fiber optic cables originating at the cable operator's switching facility (that is, the head end) and terminating at a junction box near the subscriber. This combination of coaxial cable and fiber optic is referred to as a hybrid fiber coaxial (HFC) network. Due to increased demand for symmetric services such as IP telephony, DOCSIS 2.0 was released in December 2001 to enhance upstream transmission speeds. Most recently, the specification was revised to increase transmission speeds significantly (DOCSIS 3 and EuroDOCSIS 3).

Older CATV networks cannot sustain higher bandwidths without significant upgrades. CATV operators that have recently built out their networks generally have a high-capacity bandwidth network from which they can partition a portion for broadband data service. Internet access via CATV networks uses a modem, and broadband access is typically called cable modem service. Television content is separated from Internet traffic at the head end. A cable modem termination system (CMTS) exchanges digital signals with cable modems and converts upstream traffic into digital packets that are routed to the Internet. The CMTS receives traffic from the Internet and routes it to the appropriate cable modem of the subscriber. Because CATV networks use a cascade of amplifiers to deliver video programming, cable modem service has fewer limitations than DSL with regard to how far subscribers can be located from the head end.

Allocating additional frequency has enabled bandwidth increases for cable modem broadband. For example, adding a 6 MHz channel for

Internet access provides download speeds typically between 1.5 and 15 Mbit/s and upload speeds of 384 kbit/s to 3 Mbit/s. Channel bonding adds an additional 6 MHz channel to increase speed. However, unlike DSL where subscribers are provided a dedicated connection between their home and the provider's switch, cable modem broadband capacity is shared among nearby users, which can cause a marked deterioration in service at peak times.

Until recently, the world's fastest cable broadband network was in Japan, where J:Com offers speeds of 160 Mbit/s based on DOCSIS 3 (Hansell 2009). It achieved this rate through a US$20 per subscriber upgrade, considerably cheaper than building out a new fiber to the home (FTTH) network. In 2011, however, several companies began rolling out EuroDOCSIS services at speeds up to 200 Mbit/s (Nastic 2011).

Although some countries have a significant number of CATV subscribers, cable broadband penetration on a worldwide basis remains relatively low, particularly in developing countries. A main reason is that cable operators have not made the necessary investment in HFC networks. Another factor is that regulatory restrictions in some countries forbid cable operators from providing Internet or voice services. In many countries, however, cable has never achieved significant market penetration, and satellite TV or digital terrestrial TV offers a substitute for multichannel television distribution.

Fiber to the Premises

Fiber to the premises refers to a complete fiber path linking the operator's switching equipment to a subscriber's home (FTTH) or business (FTTB). This distinguishes FTTP from fiber to the node (FTTN) and fiber to the curb (FTTC), which bring fiber optic cable part of the way to a subscriber's premises (figure 5.11). FTTN and FTTC are therefore not subscriber access technologies like FTTP, but are used to extend the capabilities of DSL and cable modem networks by expanding fiber optic cable deeper into the network. Again, the exact technology a company or government chooses to deploy or promote will depend on the unique circumstances in each country. FTTP offers the highest speeds of any commercialized broadband technology. However, it is not widely available around the world, with the FTTH Council reporting that only 26 economies had at least 1 percent of their households connected.[26]

FTTP sometimes replaces existing copper wire or coaxial cable connections but is also increasingly popular for greenfield building projects (where a new housing or commercial development is being built and no telecommunications infrastructure presently exists). FTTP can be

Figure 5.11 Diagram of Various FTTx Systems

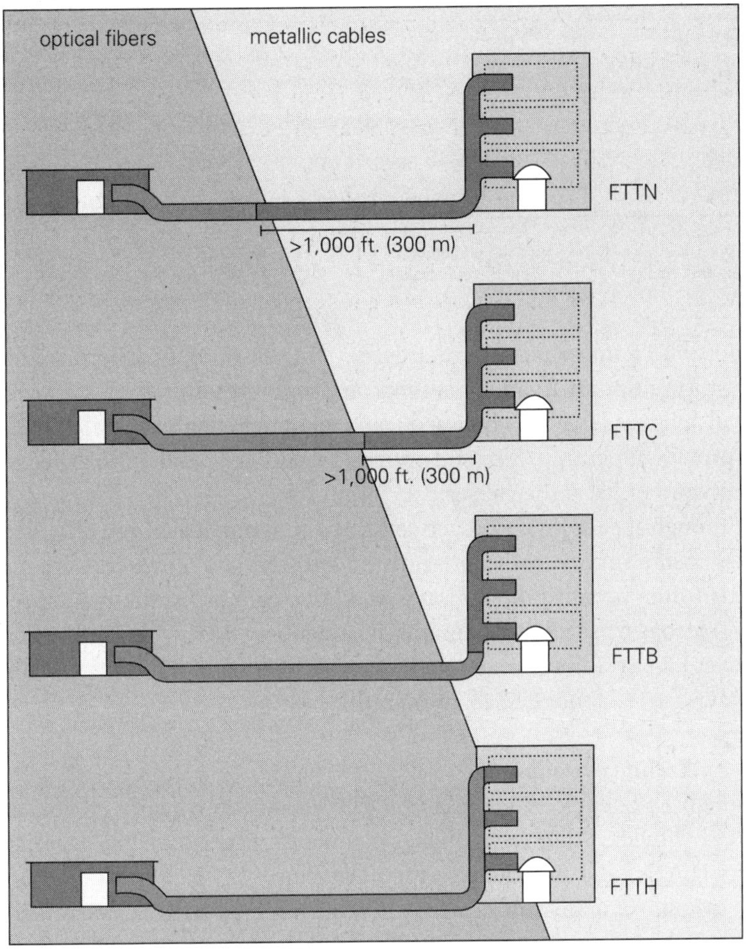

Source: Wikipedia, http://upload.wikimedia.org/wikipedia/commons/3/32/FTTX.png.

Note: The building on the left represents the central office; the building on the right represents one of the buildings served by the central office. The dotted rectangles represent separate living or office spaces within the same building.

designed with various topologies: point-to-point, where the optical fiber link is dedicated to traffic from a single subscriber; point-to-multipoint, where fiber optic cables branch to more than one premise and thus share traffic; and a ring, where the fiber optic cable is designed in a closed loop that connects various premises. The information flowing over the fiber optic cable is guided by protocols that have been standardized by the IEEE or the ITU (table 5.3).

Table 5.3   FTTP Access Protocols

Access protocol	Name	Standard
EFM	Ethernet in the first mile	IEEE 802.3ah
EP2P	Ethernet over point-to-point	IEEE 802.3ah
EPON	Ethernet passive optical network	IEEE 802.3ah
BPON	Broadband passive optical network	ITU-T G.983
GPON	Gigabit passive optical network	ITU-T G.984

Source: Fiber to the Home Council, "Definition of Terms," January 9, 2009, http://www.ftthcouncil
.eu/documents/studies/FTTH-Definitions-Revision_January_2009.pdf.

Most FTTP implementations are based on passive optical network using point-to-multipoint topology serving multiple premises with unpowered optical splitters. Traffic is handled using an optical line terminal at the service provider's central office and optical network terminals, also called optical network units, at the subscriber's premises.

Although speeds on FTTP networks can be symmetrical and offer up to 1 Gbit/s, many service providers provide substantially lower asymmetrical speeds (often because the national backbone cannot handle high speeds). City Telecom, a broadband operator in Hong Kong SAR, China, for example, has over half a million homes connected to a fiber network. It offers 1 Gbit/s fiber service for about US$25 per month.[27]

Other Wireline Broadband

Although DSL, cable modem, and FTTP account for nearly all subscriptions worldwide, other technologies include Ethernet-based local area networks (LANs) and broadband over powerline (BPL). Wireline LANs are used to connect many subscribers in a large building such as apartments or offices. Subscribers are typically connected directly to a fiber or Ethernet backbone where broadband access is distributed through the LAN. Some countries report LAN subscriptions as a separate wireline broadband access category. LANs can be wireline (using coaxial cable or twisted pair [Cat3 or 10Base-T]) or wireless, based on the IEEE 802.3 or 802.11 standards. They are typically used within a home or a public access facility.

BPL uses the electricity distribution network to provide high-speed Internet access. BPL operates by differentiating data traffic from the flow of electricity. This separation occurs by using a much higher frequency to carry data through the copper wires, coupled with encoding techniques that subdivide data traffic into many low-power signals or that spread the bitstream over a wide bandwidth. The former encoding scheme is known as

Orthogonal Frequency Division Multiplexing (OFDM), and the latter is a type of spread spectrum technology. In both technologies, digital signal processing integrated circuits help to keep data traffic intact, identifiable, and manageable.

BPL has so far failed to achieve wide-scale commercial success, partly because of interference issues and partly because of uncertainty over whether and how data transmission can take place at significant volumes over an entire electricity distribution grid. The problem stems from when transformers are used to reduce the voltage of electricity to that used by residential and business users. Because a BPL distribution grid requires repeaters that amplify data signals, such networks can be costly to build. In addition, BPL reportedly can interfere with some radio transmissions, and there is no international standard for BPL. Finally, a big barrier in many low-income nations is the lack of a reliable electrical grid to carry the data signals.

A building's internal electrical wiring can also be used as a type of LAN. Devices with Ethernet ports can be interconnected using plug-in adapters over electrical wiring to create home and office networks. The HomePlug Powerline Alliance has created an adapter standard and reports that it had sold over 45 million such devices by March 2010, accounting for 75 percent of the market.[28] The ITU covers the use of electrical wiring for home networking in its G.hn Recommendation.[29]

Wireless Access Technologies

The immense success of cellular telephone service attests to the attractiveness of wireless technologies as a local access solution. Factors in their success include being generally easier and cheaper to deploy than wireline solutions and consumers' fondness for mobility. Technological innovations offer the near-term opportunity for widespread mobile access to the Internet, as next-generation wireless networks have the technological capability to offer bit rates at near parity with current wired options, though not yet at the same price points. The ability of carriers to offer such services will depend on whether sufficient radio spectrum can be allocated for mobile broadband services and whether innovations in spectrum conservation techniques can help operators to meet consumer demand.

Early Wireless Broadband Standards

EDGE. Although an International Mobile Telecommunications-2000 (IMT-2000) standard, EDGE initially offered less than broadband speeds

(120 kbit/s, according to the GSMA).[30] A newer version of EDGE (Evolution) can achieve top speeds of up to 1 Mbit/s, with average throughput of around 400 kbit/s (Ericsson 2007), but EDGE is not considered a true mobile broadband solution. It can be attractive since it provides an upgrade path for global system for mobile (GSM) communications networks, allowing higher speeds than GPRS,[31] particularly where investment is constrained, regulators have not released mobile broadband spectrum, or gaps in coverage need to be filled.

CDMA2000 1X. Code Division Multiple Access (CDMA) 2000 refers to the CDMA2000 1X and CDMA2000 Evolution Data Optimized (EV-DO) technologies that are part of the IMT-2000 standards. CDMA2000 builds on second-generation (2G) CDMA technologies, known as ANSI-95 or cdma-One, and uses a 1.25 MHz channel size. CDMA2000 attractions include backward compatibility with earlier standards, use for either wireline or mobile wireless, and spectrum flexibility due to small channel size and availability in a range of frequencies including 450 MHz, the only IMT-2000 standard commercially available in that band (figure 5.12; CDMA Development Group 2000).

CDMA2000 1X supports circuit-switched voice up to and beyond 35 simultaneous calls per sector and high-speed data of up to 153 kbit/s in both directions. Although it was the first IMT-2000 technology to be commercially adopted, it is not considered mobile broadband due to the low speed. However, CDMA2000 EV-DO uses packet-switched transmission specifically designed and optimized for mobile broadband networks. There have been three revisions to the EV-DO standard (Rel. 0, Rev. A, and Rev. B), each offering higher speeds than its predecessor (table 5.4). In December 2011, there were 122 Rel. 0 networks in 67 countries, 138 Rev. A networks in 60 countries, and 10 Rev. B networks in 9 countries, serving 186 million subscribers around the world.[32] One of the fastest EV-DO networks is in Indonesia, where operator Smart Telecom is using Rev. B to achieve an average download speed of 8.6 Mbit/s and a peak download speed of 9.3 Mbit/s.[33] Box 5.2 describes the experience of Mexico and Sweden with CDMA 450 MHz.

IMT-2000

The first two generations of mobile networks were characterized by analog and then digital technology. There were no global standards, and a variety of technologies evolved. In an effort to standardize third-generation (3G) mobile systems expected to be commercialized around the year 2000, the ITU developed the International Mobile Telecommunications (IMT)

Figure 5.12 Frequency Bands Used by CDMA2000

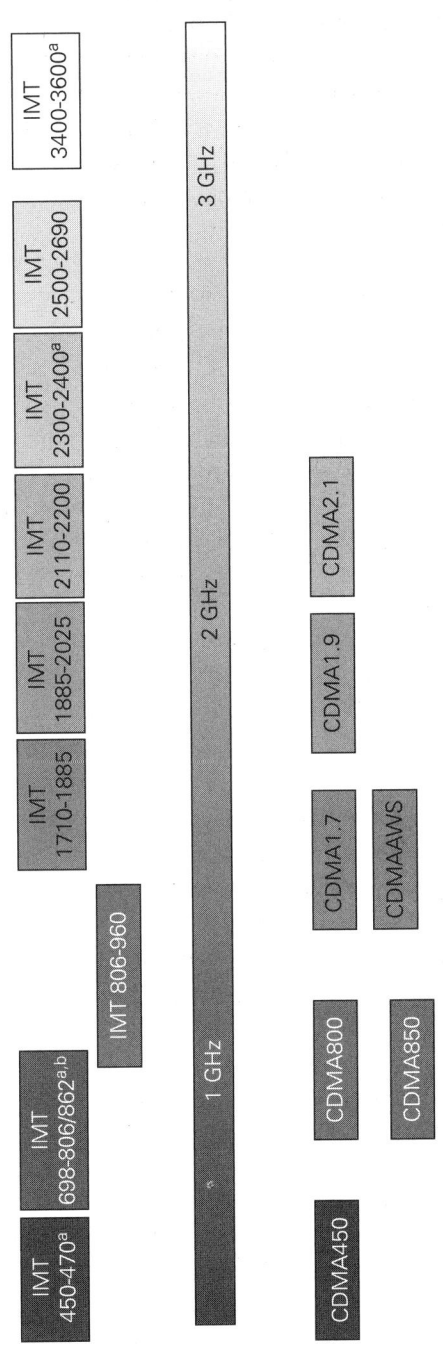

Source: CDMA Development Group, http://www.cdg.org/technology/cdma2000/spectrum.asp.

a. Identified at WRC-07.

b. Includes 698–862 MHz band in region 2 (Americas), 790–862 MHz band in Region 1 (Europe, Middle East, Africa, Russia and CIS), and 790–960 MHz identified for IMT in Region 3 (Asia-Pacific).

Note: CDMA = Code Division Mutiple Acess; IMT = International Mobile Telecommunications.

Table 5.4 EV-DO Peak and Average Speeds

EV-DO version	Peak speeds		Average user speeds		Number of countries
	Download	**Upload**	**Download**	**Upload**	
Rel. 0	2.4 Mbit/s	153 kbit/s	300–700 kbit/s	70–90 kbit/s	67
Rev. A	3.1 Mbit/s	1.8 Mbit/s	600–1,400 kbit/s	500–800 kbit/s	60
Rev. B	14.7 Mbit/s	5.4 Mbit/s	1.8–4.2 Mbit/s	1.5–2.4 Mbit/s	9

Source: CDMA Development Group 2011.

■

Box 5.2: CDMA 450 MHz for High-Speed Rural Internet Access

One of the attractions of 450 MHz spectrum is its use for rural communications. Because of the lower frequency range, coverage is wider, and fewer base stations are required so that investment costs are significantly lowered. CDMA2000 1X and EV-DO operate in 450 MHz, and their use is helping to extend high-speed connectivity to rural areas. Although the number of subscriptions may not be high, they are often the only high-speed networks available in small rural communities, where they can have an important socioeconomic impact.

In Mexico, the incumbent Telmex won the government's Fund for Telecommunications Social Coverage with its bid to provide services in some 8,500 rural communities with around 7 million low-income inhabitants. It is using CDMA450 where each base station covers more than 80 kilometers, providing 150 kbit/s Internet connections. In addition to regular post- and prepaid subscriptions (around 180,000 by late 2009), Telmex also set up some 500 "digital agencies," which offer personal computers, printers, and Internet access to the public.

In Sweden, CDMA2000 1xEV-DO in the 450 MHz band is attributed with reducing by half the number of people with no access to broadband between 2009 and 2010. Over 99 percent of Swedes living in sparsely populated regions have access to the CDMA 450 MHz network. Service is provided by Net 1, which has built a nationwide CDMA network in the 450 MHz frequency band, providing up to 25 times more coverage per transmitter than Universal Mobile Telecommunications System (UMTS) networks using the 900 MHz, 1,800 MHz, and 2,100 MHz bands. As a result, the 450 MHz network is available in places where it is not economically viable for competitors to provide coverage. Net 1 is using EV-DO Rev. A, offering download speeds of 3.1 Mbit/s for SKr 229 (US$32) per month.

According to the CDMA Development Group, CDMA450 can be profitable at average revenue per user of less than US$8 per month, and handsets are available for less than US$25.

Sources: CDMA Development Group 2009, 2011; Swedish Post and Telecom Agency 2011; Swedish Post and Telecom Agency, "Broadband Survey, PTS Statistics Portal," http://www.statistik.pts.se/broadband; Net1, "Teknik," http://www.net1.se/omnet1/teknik.aspx; Net 1, "Mobilt Bredband," http://www.net1.se/privat/bredband.aspx.

Note: Swedish kroner converted to U.S. dollars using 2010 annual average exchange rate.

family of standards. Despite the goal of standardization, five significantly different radio interfaces for IMT-2000 were approved in ITU-R Recommendation M.1457 in 1999. WiMAX was added to M.1457 in 2007 (table 5.5).[34]

W-CDMA/UMTS. Wideband CDMA (W-CDMA), also referred to as UMTS, is characterized by the use of Frequency Division Duplexing (FDD). It uses paired spectrum in 5 MHz wide radio channels. W-CDMA is often marketed as an upgrade from GSM, although it requires new base stations and initially new frequency allocation. However, since W-CDMA handsets are generally dual-mode to support GSM, roaming between the two networks is typically seamless. Given its ties to the dominant GSM standard, W-CDMA has been the most successful of the IMT-2000 technologies in terms of subscriptions.

Table 5.5   IMT-2000 Radio Interfaces

Radio interface technology	Common name	Comment
CDMA direct spread	W-CDMA/UMTS	Original frequencies in standard: 1,920–1,980 MHz as uplink and 2,110–2,170 MHz as downlink
		Later added: 2.6 GHz, 1,900 MHz, 1,800 MHz, 1,700 MHz, 1,500 MHz, 900 MHz, 850 MHz, and 800 MHz bands as well as a pairing of parts, or whole, of 1,710–1,770 MHz as uplink with whole, or parts, of 2,110–2,170 MHz as downlink
CDMA multicarrier	CDMA 2000	Including 1X and EV-DO. As the 3G-evolution path for 2G TIA/EIA-95-B standards, assumption is that 3G would use the same 2G frequencies
CDMA TDD	TD-SCDMA	Original frequencies in standard: 1,900–1,920 MHz and 2,010–2,025 MHz for both uplink and downlink operation. Added later: 1,850–1,910 MHz, 1,910–1,930 MHz, and 1,930–1,990 MHz
TDMA single-carrier	EDGE	Provides an evolution path for GSM/GPRS so assumption is that implementation would use the same 2G frequencies
FDMA/TDMA	DECT	Not widely used as a mobile cellular technology
OFDMA TDD WMAN	WiMAX (IEEE 802.16)	Frequencies not mentioned in standard. Generally commercially implemented in the 2.3, 2.5/2.6, and 3.5 GHz bands

Source: ITU, Radio Communication Sector, Recommendation M.1457.
Note: FDMA = Frequency Division Multiple Access; OFDMA = Orthogonal Frequency Division Multiplexing; TDD = Time Division Duplexès; TDMA = Time Division Multiple Access; TD-SCMA = Time Division–Synchronous Code Division Multiple Access; TIA/EIA = Telecommunications Industry Association/Electronics Industry Association; WMAN = Wide area Metropolitan Access Network.

High-Speed Packet Access refers to the various software upgrades to achieve higher speeds on W-CDMA networks (table 5.6).[35] Initial speed improvements are listed below, although some operators have been able to achieve even higher data rates through various enhancements:

- High-Speed Downlink Packet Access (HSDPA) increases download data rates. Speeds achieved top 14.4 Mbit/s, with most operators offering speeds up to 3.6 Mbit/s. Upload speeds are 384 kbit/s.

- High-Speed Uplink Packet Access (HSUPA) increases upload rates. Upload speeds are increased to a maximum of 5.7 Mbit/s.

- HSPA+ (also known as HSPA Evolved) offers significant speed improvements. HSPA+ enables speeds up to 42 Mbit/s in the downlink and 11 Mbit/s in the uplink. In March 2011, there were 128 HSPA+ networks in 65 countries, including 95 HSPA+ networks offering peak rates of 21 Mbit/s, 11 offering peak rates of 28 Mbit/s, and 22 offering peak rates of 42 Mbit/s.[36]

TD-SCDMA. Some of the key characteristics of Time Division–Synchronous Code Division Multiple Access (TD-SCDMA) are that it uses Time Division Duplexing (TDD), unlike W-CDMA, which uses FDD, and does not require paired spectrum, increasing spectrum flexibility. The word "synchronous" refers to the fact that the base station synchronizes upstream signals. Interference is reduced and capacity is increased; however, there is reduced coverage compared to other technologies. China is the only country where TD-SCDMA has been deployed on a significant scale (box 5.3). Launched by China Mobile on January 7, 2009, the network covered 656 cities by the end of 2010, with 20,702,000 subscribers (China Mobile 2011).

WiMAX. WiMAX consists of several products based on IEEE 802.16 standards for wireless broadband. Originally designed as a wireline backbone

Table 5.6 W-CDMA and HSPA Theoretical Data Rates

Technology	Download speed	Upload speed
W-CDMA	384 kbit/s	384 kbit/s
HSDPA	14.4 Mbit/s	384 kbit/s
HSUPA	Specification for upload and not download	5.7 Mbit/s
HSPA	42.0 Mbit/s	11.0 Mbit/s

Source: GSMA, "About Mobile Broadband," http://www.gsmamobilebroadband.com/about/.

Box 5.3: Three 3G Technologies in China

China is one of the few countries in the world with three kinds of mobile broadband networks. In early January 2009, the Ministry of Industry and Information Technology awarded 3G licenses to three different operators in China for three different IMT-2000 technologies. China Mobile received permission to use the homegrown TD-SCDMA technology, becoming the world's first implementation of this standard. China Unicom was approved to operate 3G using W-CDMA, which has been widely deployed in many countries.

Meanwhile, China Telecom was awarded a 3G license using CDMA2000 technology. It already operated a CDMA2000 network, and the new license allowed it to upgrade to faster EV-DO speeds. Competition between these three technologies has rapidly boosted the take-up of 3G: from no subscribers in 2008 to 10 million in 2009 and to 47 million by the end of 2010. Although the networks are incompatible for now, it is hoped that they will evolve to the next-generation mobile standard, LTE.

Source: China Mobile.

technology, the mobile version of WiMAX (802.16e) is a more recent incarnation that was approved by the ITU as an IMT-2000 standard in 2007.[37] Distinguishing features of WiMAX include IP packet switching, the use of Scalable Orthogonal Frequency Division Multiple Access (SOFDMA), unpaired spectrum using TDD, and operation in the 2.3, 2.5/2.6, and 3.5 GHz bands. Top theoretical speeds for wireless WiMAX are 46 Mbit/s on the uplink and 7 Mbit/s on the downlink, roughly equivalent to HSPA+ networks (Pinola and Pentikousis 2008).

Although mobile WiMAX is standardized as an IMT-2000 technology by the ITU, it is often used as a fixed wireless access technology (IEEE 802.16; Marks 2010). One of the early implementations was the Korean variation called WiBro (WiMax Forum 2008). The government issued spectrum in the 2.3/2.4 GHz band in 2005, and WiBro was commercially launched in April 2007. By the end of 2010, WiMAX networks were used in 149 countries covering more than 823 million people.[38] The number of WiMAX subscribers around the world was estimated at 13 million in December 2010 (Maravedis 2011).

IMT-Advanced

The ITU has been working on standards for the next generation of wireless systems for several years. In March 2008, it issued a circular letter specifying the provisions for International Mobile Telecommunications-Advanced (IMT-Advanced) networks, which are generally defined as systems "that go beyond those of IMT-2000" (Blust 2008). One of the most significant requirements is peak data rates of 100 Mbit/s for high mobility and 1 Gbit/s for low mobility. In October 2010, the ITU announced that two technologies met the requirements for IMT-Advanced: LTE-Advanced and Wireless-MAN-Advanced (ITU 2010).

LTE and LTE-Advanced. Development of the LTE mobile network standard started in 2004. One goal was to achieve higher data speeds to support the rising growth of Internet access over mobile phones. Targeted speeds were initially 100 Mbit/s for downloads and 50 Mbit/s for uploads. LTE uses OFDM for downloads and Single Carrier-Frequency Division Multiple Access (SC-FDMA) for uploads. LTE is designed for frequency flexibility, with bandwidth requirements ranging from 1.25 and 20 MHz and support for both paired (FDD) and unpaired (TDD) bands.

LTE standards have been developed under the auspices of the 3G Partnership Project (3GPP). The 3GPP Release 8, issued in December 2008, forms the basis for initial LTE deployments. It has theoretical maximum download speeds of 300 Mbit/s and upload speeds of 75 Mbit/s. In order to meet global requirements for fourth-generation (4G) mobile networks, 3GPP developed LTE Release 10 and Beyond (LTE-Advanced), which was submitted to the ITU in October 2009.

Although LTE was developed within the auspices of the 3GPP, whose work includes technical specifications for GSM, W-CDMA, and HSPA technologies, there is no straightforward migration path. So far, LTE deployments have required the purchase of new equipment by operators and new devices by users.

The world's first LTE deployment was by TeliaSonera when it simultaneously launched networks in Stockholm, Sweden, and Oslo, Norway, at the end of 2009 using the 2.6 GHz frequency band.[39] Verizon's LTE network launch in the United States in December 2010 is noteworthy for using the 700 MHz frequency band.[40] Verizon reported that speeds were 5–12 Mbit/s download and 2–5 Mbit/s upload. According to 4G Americas, 19 commercial LTE networks were operating worldwide in 14 countries in March 2011.

WirelessMAN-Advanced. WirelessMAN-Advanced is standardized as IEEE 802.16m and offers backward compatibility with IEEE 802.16e, an IMT-2000 technology. It meets the IMT-Advanced data rate requirements with a theoretical 180 Mbit/s downlink using a 20 MHz TDD channel (WiMax Forum 2010). Multiple channels can be aggregated to support 1 Gbit/s speeds (Jiaxing and Guanghui 2010).

Wi-Fi

Wi-Fi refers to the IEEE 802.11 family of standards specifying wireless local area networking over 2.4 and 5 GHz frequency bands. Wi-Fi is not typically deployed as a commercial local access network; it is used most often to redistribute a broadband connection to a wider group of users in homes, offices, and "hotspots." Wi-Fi technology has gone through several updates that provide varying speeds depending on the frequency and version used (table 5.7).

Reportedly, one in 10 people around the world uses Wi-Fi.[41] Its success is attributed to several factors, including embedding Wi-Fi chips in portable computers and smartphones, the fact that it operates on a license-exempt (unlicensed) basis,[42] and the relative ease of installation compared to wired networks, with the majority of the upgrade costs lying with the consumer rather than the operator.

In addition to sharing broadband connectivity with devices in home and office networks, Wi-Fi is being used for the following significant applications:

- *Subscription-based access to broadband.* Many wireline ISPs around the world offer Internet access through Wi-Fi hotspots at airports, coffee shops, and other locations. This is seen as a complement to their traditional service.

- *Municipal Wi-Fi networks.* Large-scale Wi-Fi networks have been deployed in some urban areas around the world to provide free Internet access. Wireless@KL in Kuala Lumpur, Malaysia, provides free 512 kbit/s access throughout the city; faster speeds are enabled through payment.[43] The Kuala Lumpur City Hall and the Malaysian Communications and Multimedia Commission sponsor the KL Wireless Metropolitan project in collaboration with Packet One Networks, an ISP. Some 1,500 hotspots have been deployed in the city.

Table 5.7 Wi-Fi Speeds

Wi-Fi technology	Frequency band (GHz)	Maximum data rate (Mbit/s)
802.11a	5	54
802.11b	2.4	11
802.11g	2.4	54
802.11n	2.4, 5	450

Source: Wi-Fi Alliance, "Discover and Learn," http://www.wi-fi.org/discover_and_learn.php.

- *Relief for congested mobile networks.* Mobile operators were initially luke-warm about handsets with Wi-Fi capability, since users could bypass more expensive cellular network data offerings. That view is changing due to the rapid growth in demand for data over mobile cellular networks and consequent capacity constraints. Today, many mobile operators embrace Wi-Fi as a way to offload 3G-network traffic as a complement to their regular commercial service. For example, AT&T in the United States is automatically switching smartphone users to Wi-Fi when they are within range of a hot spot (Fitchard 2010).

Satellite

Aside from its role in the international and backbone segments of the broadband supply chain, satellites are also used to provide direct sub-scriber access to broadband services, particularly in remote areas where wireline broadband is not available and there is no terrestrial high-speed wireless coverage.[44] The subscriber uses a satellite antenna or dish that is connected to a satellite modem. Speeds vary depending on the satellite technology, antenna, and the weather. Latency can be an issue for some applications (for example, gaming). Although they serve specific niches, satellites do not offer the same price to quantity ratio as other broadband solutions. For example, in March 2011, the highest speed available from a leading retail broadband satellite provider in the United States was 5 Mbit/s for US$300 per month.[45]

Implementation Issues for Local Connectivity

Countries face numerous challenges in deploying local broadband access networks, including whether and how physical infrastructure can be shared, quality of service, and spectrum.

Local Loop Unbundling

In many countries, an incumbent, former monopoly wireline provider often controls the only extensive local access network. In such cases, regulators have sought ways to introduce more competition and innovation into the local access market. Local loop unbundling (LLU) has been one of the main methods implemented in developed nations for service providers to gain access to the incumbent's switched telephone network in order to provide DSL service. There are three main types of implementation:

- *Full unbundling.* The entire copper local loop is leased to a service provider. The service provider installs its own broadband equipment either in, or close to, the incumbent's site.

- *Line sharing.* The copper local loop is shared between the incumbent and the other service provider. The incumbent provides voice telephony over the lower-frequency portion of the line, while the other provider offers DSL services over the high-frequency portion of the same line.

- *Bit stream access.* DSL service is essentially sold at wholesale prices to the service provider, who in turn resells it to customers. The incumbent operates all of the key infrastructure components in the loop.

Quality of Service

There is often a significant difference between advertised speeds and actual speeds achieved by users (figure 5.13). The problem is that the advertised speeds are usually based on the theoretical capability of the technology or standard. In reality, however, numerous factors make such speeds very difficult or even impossible to achieve, including network congestion or (for wireless networks) radio interference.

In an effort to manage network quality, many providers are moving away from unlimited broadband packages and adopting so-called "fair use policies" in order to control and regulate traffic. One practice is to use data caps where providers establish a threshold on the amount of data that can be downloaded per month. Once the cap is exceeded, either the subscriber

Figure 5.13 Difference between Advertised and Actual Speeds in the United Kingdom, 2009 and 2010

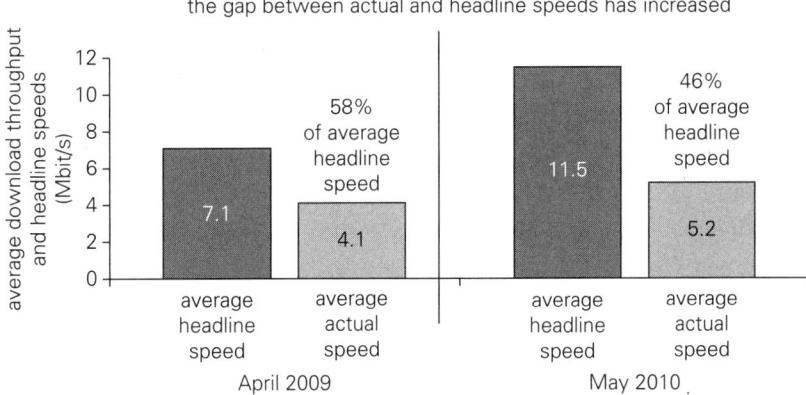

Source: Ofcom, "The Communications Market 2010: UK," http://www.ofcom.org.uk/static/cmr-10/UKCM-5.10.html.

Note: Headline speeds are based on data from the operator, while actual speed are based on measurement data from SamKnows for all panel members with connections in April 2009 and May 2010 (single-thread tests).

has to purchase additional download volume or the subscriber's speed is reduced or, in the worst-case scenario, service is terminated for that month. Some operators establish different caps for domestic and international traffic. Another practice is to control the use of high-bandwidth applications or access to traffic-intensive sites by restricting or degrading service. This practice has been banned in some countries as a violation of network neutrality. Providers have been known to "throttle" service by limiting the subscriber's bandwidth when they have exceeded data caps or tried to access traffic-intensive sites.

These network management practices have been contentious since they are often covered by the "small print" of customer contracts and many users are not aware of them. In an effort to alleviate consumer concerns about service quality, some governments monitor and compile reports on service quality. The Telecommunications Regulatory Authority (TRA) in Bahrain, for example, publishes data on wireline broadband performance (Bahrain, Telecommunications Regulatory Authority 2011). The TRA measures upload and download speeds for different broadband packages, domain name system (DNS) response (time taken in milliseconds to translate a domain name to its IP address), and ping (an echo request sent to a server to test latency). In other countries, although governments do not publish quality of service reports, they offer sites where consumers can check their speed.[46]

Spectrum

One of the biggest constraints on wireless broadband deployment and usage is the availability of spectrum. Some countries have yet to allocate mobile broadband spectrum, have not allocated certain frequencies, or have not allocated sufficient spectrum.

Although the number of frequency bands in which mobile broadband operates has increased, not every technology operates in every band. Therefore, by not licensing certain bands, countries prevent the availability of some mobile broadband technologies. Another issue is that even slight differences in frequency assignments can make a difference in equipment compatibility, affecting prices and roaming. Growing mobile broadband demands are placing increasing pressure on spectrum availability. Providers use several techniques to increase capacity, including splitting cells, upgrading to more efficient technology, and offloading some uses onto other networks like Wi-Fi. However, there may come a point where technology cannot fix the capacity shortage and additional spectrum is required. Some countries have already begun examining how to use the various bands identified for broadband, including the so-called "digital dividend" spectrum that can be made available as the result of the transition from analog to digital television. One promising solution could be *cognitive radio,* where devices reconfigure themselves according to whatever spectrum is available, while avoiding interference. The first call in the world using cognitive radio was made in Finland in 2010 (Centre for Wireless Communications 2010).

In looking at spectrum, regulators need to determine the best procedure to follow in awarding spectrum, whether to impose limits on the amount of spectrum a single operator can hold, and whether to allow operators to engage in secondary trading. These issues are discussed in more detail in chapter 3.

Notes

1. MPLS packetizes and labels information coming from different network protocols so that the underlying architecture does not have to be changed; it then routes the information to its destination.
2. See ITU, "ITU-T's Definition of NGN," http://www.itu.int/en/ITU-T/gsi/ngn/Pages/definition.aspx.
3. KPN, "Facts and Figures Q4 2010," http://www.kpn.com/corporate/aboutkpn/investor-relations/publications/Financial-publications.htm.
4. Intelsat, "Satellite Basics," http://www.intelsat.com/resources/satellite-basics/how-it-works.asp.
5. ViaSat, "Meeting the Demand for Media-Enabled Satellite Broadband Satellite Services," http://www.viasat.com/files/assets/Broadband%20Systems/MediaEnabledSatellite9-09.pdf.
6. Telesat, "Satellite's Growing Role in Data Networking," http://www.telesat.ca/en/Satellites_Growing_Role_in_Data_Networking.
7. RADWIN, "IP Backhaul," http://www.radwin.com/Content.aspx?Page=ip_backhaul.
8. NEC, "Pasolink (Egypt)," http://www.nec.com/global/onlinetv/en/business/pasolink_l.html#NF-project.
9. TeleGeography, "Submarine Cable Map," http://www.telegeography.com/telecom-maps/submarine-cable-map/index.html.
10. EASSy, "EASSy Ownership," http://www.eassy.org/ownership.html. EASSy's largest shareholder is WIOCC, which is owned by 14 African telecommunications operators and partially funded by several development financial institutions, including the World Bank.
11. In this context, ISPs include wireless operators that offer Internet connectivity.
12. Tata Communications, "IP," http://www.tatacommunications.com/providers/ip/.
13. SLT Hong Kong, "SLT Hong Kong Is Gateway to East Asia and US," http://www.slthkg.com/Company.htm.

14. Internet Service Providers' Association, "CINX Users," http://www.ispa.org.za/inx/cinx-users/.

15. Technology Zimbabwe, "TelOne's Fibre Connection on EASSy Now Live, Total 2.48 Gbps Lit," March 31, 2011, http://www.techzim.co.zw/2011/03/telone%E2%80%99s-fibre-connection-on-eassy-now-live-total-2-48-gbps-lit/.

16. "Sri Lanka to Set up a National Backbone Network." *Sri Lankan News*, February 13, 2011, http://firstlanka.com/english/news/sri-lanka-to-set-up-a-national-backbone-network/.

17. Pakistan Universal Service Fund, "Optic Fiber Project," http://www.usf.org.pk/project.aspx?pid=6.

18. Kenya Power and Lighting Company, "KDN Leases KPLC Dark Fibres," Press Release, March 18, 2010, http://www.kplc.co.ke/fileadmin/user_upload/kplc09_files/UserFiles/File/Press%20Release%20-%20KDN%20leases%20KPLC%20dark%20fibres.pdf.

19. Ventelo, "Capacity Product Line," http://www.ventelo.no/wholesale-english/capacity.html.

20. See, for example, numerous comments filed generally in a U.S. Federal Communications Commission proceeding that was established to develop the National Broadband Plan (United States, FCC 2009).

21. IEEE Standards, "IEEE 802.17™: Resilient Packet Rings," http://standards.ieee.org/about/get/802/802.17.html.

22. ITU, "Study Group 15 at a Glance," http://www.itu.int/net/ITU-T/info/sg15.aspx.

23. Nokia Siemens Networks, "Bezeq to Offer World-First 200 Mbit/S Broadband While Cutting OPEX," http://www.nokiasiemensnetworks.com/pt/portfolio/customer-successes/success-stories/bezeq-to-offer-world-first-200-Mbit/s-broadband-while-cutting-OPEX.

24. CableLabs, "Home Page," http://cablelabs.com/.

25. ITU, "J.112: Transmission Systems for Interactive Cable Television Services," http://www.itu.int/rec/T-REC-J.112/en.

26. Fiber to the Home Council, "Global FTTH Councils' Latest Country Ranking Shows Further Momentum on All-Fiber Deployments," February 10, 2011, http://www.ftthcouncil.org/en/newsroom/2011/02/10/global-ftth-councils-latest-country-ranking-shows-further-momentum-on-all-fiber-.

27. Global Telecoms Business, "How to Be a Fat Dumb Pipe at $25 a Month for One Gigabit," February 2, 2011, http://www.globaltelecomsbusiness.com/Article/2760589/Interview-NiQ-Lai-and-Ivan-Tam-of-City-Telecom.html.

28. HomePlug Powerline Alliance, "HomePlug® Powerline Alliance Announces Milestones on 10th Anniversary as Powerline Technology Leader," March 22, 2010, http://www.homeplug.org/news/pr/view?item_key=a633eafa198466341aa340327092bc76f8169135.

29. ITU, "New ITU Standard Opens Doors for Unified 'Smart Home' Network," Press Release, November 15, 2009, http://www.itu.int/newsroom/press_releases/2009/46.html.

30. According to the GSMA, "GPRS offers throughput rates of up to 40 kbit/s. . . . Using EDGE, operators can handle three times more subscribers than GPRS, triple their data rate per subscriber." See "GPRS" and "EDGE" on the GSMA website, http://gsmworld.com/technology/index.htm.

31. In theory, the speed limit of GPRS is 115 kbit/s, but in most networks it is around 35 kbit/s.

32. CDMA Development Group, "CDMA Statistics," http://www.cdg.org/resources/cdma_stats.asp.

33. ZTE, "ZTE Launches the World's First Commercial EV-DO Rev.B Network in Indonesia," Press Release, January 18, 2010, http://wwwen.zte.com.cn/en/press_center/news/201001/t20100118_179633.html.

34. There have been 10 revisions of Recommendation ITU-R M.1457. The latest is M.1457-9 of May 2010. See ITU (2010).

35. GSMA, "About Mobile Broadband," http://www.gsmamobilebroadband.com/about/.

36. 4G Americas, "HSPA+ and LTE: Fastest Speeds for Mobile Broadband Today," March 18, 2011, http://www.3gamericas.org/index.cfm?fuseaction=pressrelease display&pressreleaseid=3084.

37. ITU, "ITU Defines the Future of Mobile Communications," Press Release, October 19, 2007, http://www.itu.int/newsroom/press_releases/2007/30.html.

38. WiMAX Forum, "WiMAX™ on Track to Cover One Billion by EOY 2011," February 15, 2011, http://www.wimaxforum.org/news/2761.

39. TeliaSonera, "TeliaSonera First in the World with 4G Services," Press Release, December 14, 2009, http://www.teliasonera.com/News-and-Archive/Press-releases/2009/TeliaSonera-first-in-the-world-with-4G-services/.

40. Verizon, "Blazingly Fast: Verizon Wireless Launches the World's Largest 4G LTE Wireless Network," December 4, 2009, http://news.vzw.com/news/2010/12/pr2010-12-03.html.

41. Wi-Fi Alliance, "Organization," http://www.wi-fi.org/organization.php.

42. The ITU has designated the 2,450 MHz and 5,800 MHz bands for industrial, scientific, and medical applications that "must accept harmful interferences." This is often interpreted to mean that they are considered unregulated. See ITU-R, "Frequently Asked Questions," http://www.itu.int/ITU-R/terrestrial/faq/index.html#g013.

43. Wireless@KL, "About," http://www.wirelesskl.com/?q=about.

44. In the United States, users in remote areas without wireline broadband availability were offered a discount for satellite broadband Internet access (including no installation or equipment charges) through the American Recovery and Reinvestment Act. See HughesNet, "Frequently Asked Questions," http://consumer.hughesnet.com/faqs.cfm.

45. HughesNet, "Business Solutions," http://business.hughesnet.com/explore-our-services/business-internet/business-internet-high-speed.

46. United States, FCC, "About the Consumer Broadband Test (Beta)," http://www.broadband.gov/qualitytest/about/.

References

Australia, Department of Broadband, Communications, and the Digital Economy. n.d. "Regional Backbone Blackspots Program: Fast Facts." Department of Broadband, Communications, and the Digital Economy, Canberra. http://www.dbcde.gov.au/__data/assets/pdf_file/0017/123605/DBCDE-NBN-Blackspots-Program-Fact-Facts.pdf.

Bahrain, Telecommunications Regulatory Authority. 2011. *Fixed Broadband Analysis Report 01 Oct 2010–31 Dec 2010*. Manama: Telecommunications Regulatory Authority. http://www.tra.org.bh/en/pdf/FixedBroadbandAnalysisReport Q42010.pdf.

Bezeq Group. 2011. "Investor Presentation Full Year 2010." Bezeq Group, Jerusalem, March. http://ir.bezeq.co.il/phoenix.zhtml?c=159870&p=irol-presentations.

Blust, Stephen. 2008. "Development of IMT-Advanced: The SMaRT Approach." *ITU News* 10: 39. http://www.itu.int/itunews/manager/display.asp?lang=en &year=2008&issue=10&ipage=39&ext=html.

Bressie, Kent. 2010. "Eight Myths about Undersea Cables." Paper presented at the "SubOptic2010" conference, Wiltshire and Grannis LLP, Washington, DC. http://www.wiltshiregrannis.com/siteFiles/News/0D12892E4C32D99099222C 3A323EE38C.pdf?CFID=9388181&CFTOKEN=68748222.

Cavalli, Olga, Jorge Crom, and Alejandro Kijak. 2003. "Desarrollo de NAPS en Sudamérica." Instituto para la Conectividad en las Américas, Montevideo. http://www.idrc.ca/uploads/user-S/11660380021NAPs-Sp.pdf.

CDMA Development Group. 2000. "CDMA2000." CDMA Development Group, Chelmsford, MA. http://www.cdg.org/technology/cdma2000.asp.

———. 2009. "TELMEX: Latin America's Largest CDMA450 Operator." CDMA Development Group, Chelmsford, MA. http://www.cdg.org/resources/files/ case_studies/CDG_Telmex_Final_101609.pdf.

———. 2011. "CDMA450 Market Facts." CDMA Development Group, Chelmsford, MA, December. http://www.cdg.org/resources/market_technology.asp.

Centre for Wireless Communications. 2010. "Public CWC GIGA Seminar Programme 2010." Centre for Wireless Communications, University of Oulu, Oulu, Finland, January 11. http://www.cwc.oulu.fi/public-cwc-giga-2010/ programme.html.

China Mobile. 2011. *Annual Results 2010*. Beijing: China Mobile, March 16. http://www.chinamobileltd.com/images/present/20110316/pp01.html.

Cisco. 2009. "IP NGN Carrier Ethernet Design: Powering the Connected Life in the Zettabyte Era." White Paper, Cisco, Petaluma, CA. http://www .cisco.com/en/US/solutions/collateral/ns341/ns524/ns562/ns577/net _implementation_white_paper0900aecd806a7df1_ns537_Networking _Solutions_White_Paper.html .

Ericsson. 2007. "The Evolution of EDGE." White Paper, Ericsson, Stockholm. http://www.ericsson.com/res/docs/whitepapers/evolution_to_edge.pdf.

——. 2010. "Annual Report 2010: Highlights." Ericsson, Stockholm. http://www
.ericsson.com/thecompany/investors/financial_reports/2010/annual10/sites/
default/files/2010_highlights_EN_0.pdf.

Fitchard, Kevin. 2010. "AT&T Tests Free Wi-Fi for Mobile Offload in Times
Square." *Connected Planet*, May 25. http://blog.connectedplanetonline.com/
unfiltered/2010/05/25/att-tests-free-wi-fi-for-mobile-offload-in-times-
square/.

Gallagher, Ian. 2010. "Daylight Coppery as Thieves Tear up Manhole Covers and
Steal £70,000 of Phone Cable." *Mail Online*, March 21. http://www.dailymail
.co.uk/news/article-1259460/Daylight-coppery-thieves-tear-manhole-covers-
steal-70-000-phone-cable.html.

Guzmán, José Miguel. 2008. "Google Peering." Google, Mountain View, CA.
http://lacnic.net/documentos/lacnicxi/presentaciones/Google-LACNIC-
final-short.pdf.

Hansell, Saul. 2009. "World's Fastest Broadband at $20 per Home." *New York
Times*, April 3. http://bits.blogs.nytimes.com/2009/04/03/the-cost-to-offer-the-
worlds-fastest-broadband-20-per-home/.

Hansryd, Jonas, and Per-Erik Eriksson. 2009. "High-Speed Mobile Backhaul
Demonstrators." *Ericsson Review* (February): 10–16. http://www.ericsson.com/
ericsson/corpinfo/publications/review/2009_02/files/Backhaul.pdf.

Ibrahim, Malika, and Ilyas Ahmed. 2008. "Maldives." In *Digital Review of Asia
Pacific 2007–2008*. Ottawa: International Development Research Centre.
http://www.idrc.ca/openebooks/377-5/#page_204.

IDA (Info-communications Development Authority of Singapore). 2008. "International Sharing: International Gateway Liberalization." IDA, Mapletree Business
City.

——. 2010. "What Is Next Gen NBN?" IDA, Mapletree Business City, September 9.
http://www.ida.gov.sg/Infrastructure/20090717105113.aspx.

ITU (International Telecommunication Union). 2008. "DSL: Digital Subscriber
Line." ITU, Geneva, May. http://www.itu.int/dms_pub/itu-t/oth/1D/01/
T1D010000040003PDFE.pdf.

——. 2010. "Recommendation ITU-R M.1457-9: Detailed Specifications of the
Terrestrial Radio Interfaces of International Mobile Telecommunica-
tions-2000." ITU, Geneva, May. http://www.itu.int/rec/R-REC-M.1457/e.

Jensen, Mike. 2009. "Promoting the Use of Internet Exchange Points: A Guide to
Policy, Management, and Technical Issues." Internet Society, Reston, VA.
http://www.isoc.org/internet/issues/docs/promote-ixp-guide.pdf.

Jiaxing, Xiao, and Fan Guanghui. 2010. "802.16m: Ready for 4G." *Communicate*,
June. http://www.huawei.com/en/static/hw-076500.pdf.

Kisambira, Edris. 2008. "Infocom Uganda Leases Fiber from Power Utility."
Network World, June 6. http://www.networkworld.com/news/2008/060608-
infocom-uganda-leases-fiber-from.html?inform?ap1=rcb.

Maravedis. 2011. *4G Counts Quarterly Report*. Issue 13. Miami: Maravedis,
January. http://www.maravedis-bwa.com/en/reports.

Marks, Roger. 2010. "IEEE 802.16 WirelessMAN Standard: Myths and Facts." IEEE 802.16 Working Group on Broadband Wireless Access Standards, June 29. http://wirelessman.org/docs/06/C80216-06_007r1.pdf.

Muwanga, David. 2009. "Telecom Costs to Drop as Fibre Optic Cable Lands in Kampala." *New Vision*, July 5. http://www.newvision.co.ug/D/8/220/686891.

Nastic, Goran. 2011. "Get Launches EuroDOCSIS 3." *CSI Magazine*, January 19. http://www.csimagazine.com/csi/Get-launches-EuroDOCSIS-3.php.

OECD (Organisation for Economic Co-operation and Development). 2008. "Public Rights-of-Way for Fibre Deployment to the Home." OECD, Paris. http://www.oecd.org/dataoecd/49/9/40390753.pdf.

Pinola, Jarno, and Kostas Pentikousis. 2008. "Mobile WiMAX." *Internet Protocol Journal* 11 (2, June). http://www.cisco.com/web/about/ac123/ac147/archived_issues/ipj_11-2/112_wimax.html.

Swedish Post and Telecom Agency. 2011. "The Broadband Survey 2010." Swedish Post and Telecom Agency, Stockholm. http://www.pts.se/en-gb/Documents/Reports/Internet/2011/The-Broadband-Survey-2010/.

TeliaSonera. 2010. "TeliaSonera International Carrier Global Peering Policy." TeliaSonera, Stockholm, January. http://www.teliasoneraic.com/Ourservices/IP/IPTransit/index.htm.

Uganda, Parliament of. 2009. "Government to Borrow US$ 61,059,125 and US$15,391,511 from China Export and Import Bank (EXIM) for Phase II and III of the National Data Transmission Backbone Infrastructure and E-Government Project." Parliament of Uganda, July 30. http://www.parliament.go.ug/index.php?option=com_docman&task=doc_details&gid=58&Itemid=102.

United States, FCC (Federal Communications Commission). n.d. "Appendix C, Glossary." In *National Broadband Plan*. Washington, DC: FCC. http://www.broadband.gov/plan/appendices.html#s18-3.

———. 2009. "Notice of Inquiry, in the Matter of a National Broadband Plan for Our Future." FCC 09-31 (rel. April 8). FCC, Washington, DC.

Van der Berg, Rudolf. 2008. "How the 'Net Works: An Introduction to Peering and Transit." *Ars Technica*, September. http://arstechnica.com/old/content/2008/09/peering-and-transit.ars/4.

Williams, Mark. 2010. "Broadband for Africa: Developing Backbone Communications Networks." World Bank, Washington, DC.

WiMAX Forum. 2008. "WiMAX Enables Wireline Incumbent to Become Leading Provider of Broadband Wireless Data Services in Korea." WiMAX Forum. http://www.wimaxforum.org/sites/wimaxforum.org/files/document_library/kt_wibro_v1.6.pdf.

———. 2010. "IEEE 802.16m Approved as IMT-Advanced Technology." WiMAX Forum, October 25. http://www.wimaxforum.org/printpdf/2650.

World Bank. 2011. "Kenya Broadband Case Study." World Bank, Washington, DC.

CHAPTER 6

Driving Demand for Broadband Networks and Services

In general terms, demand for broadband services, applications, and content is thriving and may not appear to need a large amount of government effort to spur adoption by those who have broadband access. In 2010, for example, 40 percent of all consumer Internet traffic was video,[1] which was 1.6 times the amount of video traffic in the previous year and consisted mostly of private sector–created or user-generated video. Broadband use is growing quickly and is heavily driven by private sector content. Nevertheless, governments have sought to complement supply-side policies that focus on building infrastructure with demand-side efforts that seek to drive demand for broadband access and services. Although demand stimulation is particularly relevant in the early stages of broadband market development, it is also important in more mature broadband markets, where some potential users, such as elderly and less-educated persons, may not be taking advantage of the benefits offered by broadband.[2]

Demand facilitation or stimulation refers to efforts to boost the use of broadband by raising awareness of its possible benefits as well as making it affordable and more attractive to users. As discussed in chapter 1, supply-side strategies focus on the "availability" of broadband by promoting investment in broadband technologies and infrastructure, based on the assumption that there is unsatisfied demand or that demand will grow to justify those investments. Demand-side strategies focus on expanding the market through programs designed to encourage broadband Internet access and adoption. With more visible demand, infrastructure providers are more likely to make the investments needed to spur greater broadband development.

Demand facilitation strategies can be included in top-down national plans, can originate from grassroots efforts, or can involve the public and private sectors as well as civil society.[3] The scope of such strategies may be targeted at one particular obstacle to access, such as the high cost of connections or computer ownership, or may be broader, resulting in more comprehensive programs that attempt to address multiple barriers (Hauge and Prieger 2009). The Dominican Republic, for example, established legislation to address not only the financing mechanisms needed to achieve universal broadband, but also the deployment of infrastructure and the acquisition and installation of terminal equipment such as computers, personal digital assistants (PDAs), smartphones, and other devices that enable consumers to use a broadband connection (San Román 2009). Demand facilitation may also involve packaging broadband with applications that appeal to specific sectors of the economy or groups within society.

This chapter analyzes the various approaches that can be used to facilitate additional demand for broadband services. These approaches can be roughly characterized into three categories—awareness, affordability, and attractiveness. Governments seeking to promote broadband services will need to address all three of these issues. In Malaysia, for example, a new National Broadband Initiative (NBI) was launched in March 2010. With regard to spurring demand for broadband, the NBI focuses specifically on these three issues. As stated by the Malaysian Communications and Multimedia Commission (MCMC),

> The approach for creating awareness will be through continuous government and private sector involvement in the awareness programs and capacity building initiatives. In order to improve the attractiveness of the online content, efforts will be focused to enhance and promote e-government, e-education, and e-commerce. Efforts are also on the way to digitalize the traditional information resources such as library, archive, etc. to be available online. The affordability factor and bridging the digital divide is being improved by developing various incentives to reduce the broadband access costs and widening the community access.[4]

Figure 6.1 summarizes the mechanisms that can be used to spur demand. See chapter 2 for an introduction to the three pillars of awareness, affordability, and attractiveness.

Figure 6.1 The Three Pillars of Facilitating Broadband Demand

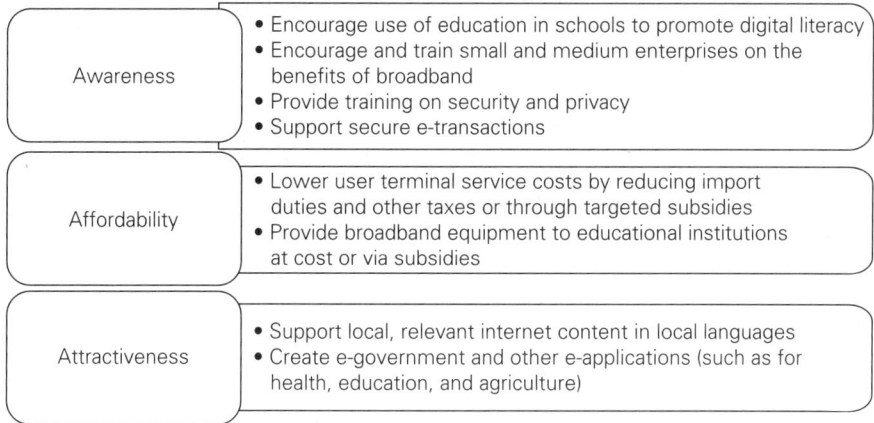

Awareness	• Encourage use of education in schools to promote digital literacy • Encourage and train small and medium enterprises on the benefits of broadband • Provide training on security and privacy • Support secure e-transactions
Affordability	• Lower user terminal service costs by reducing import duties and other taxes or through targeted subsidies • Provide broadband equipment to educational institutions at cost or via subsidies
Attractiveness	• Support local, relevant internet content in local languages • Create e-government and other e-applications (such as for health, education, and agriculture)

Source: World Bank.

Awareness

Awareness of the benefits of broadband and the capability to use broadband are critical first steps in building demand for broadband services. In order for people to use broadband successfully, they must have the necessary interest and competency. This is sometimes referred to as digital literacy, which has been defined as "using digital technology, communications tools, and/or networks to access, manage, integrate, evaluate, and create information in order to function in a knowledge society" (Educational Testing Service 2002). Digital literacy ideally makes users aware of and capable of accessing broadband applications and services. This, in turn, widens the information available to them, provides new ways of learning, and creates new employment opportunities.

There is a spectrum of digital skills that increase in complexity as users gain expertise. Therefore, competency in information and communication technology (ICT) skills can range from a basic understanding, which enables users to access information using broadband, to deeper technical knowledge, which enables them to create and disseminate their own information, including new applications and services. This is acknowledged in definitions of the different stages of digital literacy (figure 6.2).

People learn digital literacy skills in various ways and institutional settings. These range from watching friends, to being taught in schools,

Figure 6.2 Elements of Digital Literacy

	Elements	Definitions
increasing complexity of knowledge and expertise ↓	Access	Knowing about and knowing how to collect and retrieve information
	Manage	Applying an existing organizational or classification scheme
	Integrate	Interpreting and representing information: summarizing, comparing, and contrasting
	Evaluate	Making judgments about the quality, relevance, usefulness, or efficiency of information
	Create	Generating information by adapting, applying, designing, inventing, or authoring information

Source: Educational Testing Service 2002.

to participating in special programs (figure 6.3). The range of skills and settings vary and overlap. For example, some people may choose simply to acquire basic skills in a formal academic environment, while others may choose to pursue a higher degree of ICT knowledge. Although there are a variety of institutional settings for gaining knowledge about the use of broadband networks, self-training plays an ongoing lifetime role. This is particularly important since the services and applications available over broadband networks continually evolve.

There are several challenges to ensuring that people are digitally literate. Some studies suggest that the main way people learn about ICTs is through self-study (that is, through their own initiative and assistance from friends, family, and colleagues) rather than through formal courses. Motivating people to continue to learn on their own is essential in order for them to adapt to the constant evolution in broadband services and applications without always having to resort to more formal training. This is related to the interaction of digital literacy with "value addition." Although training is important, it does not necessarily build peoples' understanding of how broadband and associated technologies can transform their lives. This lack of understanding risks creating a "value divide" in which the people who have broadband diverge widely in their ability to derive value from it. As broadband spreads to other platforms, particularly mobile phones in developing countries, the notion of digital literacy, which has typically been associated with learning on personal computers (PCs), must be adapted to entail familiarity with using applications and services delivered via various mobile devices such as smartphones and tablets.

Broadband Strategies Handbook

Figure 6.3   How People Obtain ICT Training in Europe, 2007

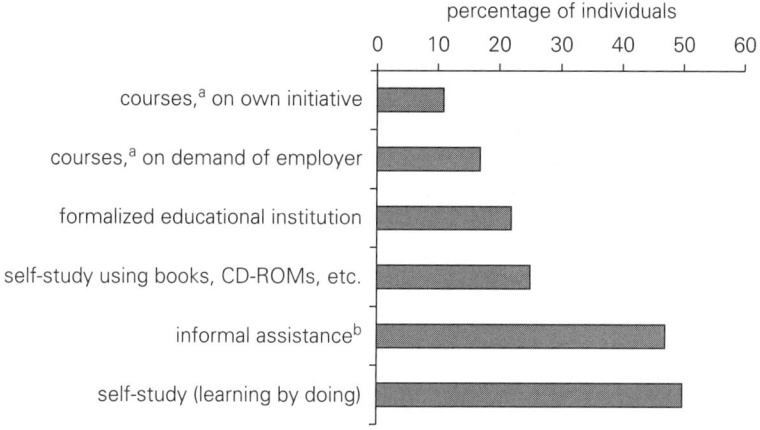

Individuals who have obtained ICT skills through

Source: Telecommunications Management Group, Inc., adapted from the Eurostat Information Statistics database, http://epp.eurostat.ec.europa.eu/portal/page/portal/information_society/data/database.
a. Training courses and adult education centers.
b. Informal assistance from colleagues, relatives, and friends.

Basic Digital Literacy

People must have basic skills (one of the components of "access" in figure 6.2) if they are to reach the point where they can teach themselves.[5] Basic and some advanced skills are increasingly offered to students in primary and secondary educational institutions, while adults or other potential users can obtain skills through community learning centers or similar institutions. Advanced skills are typically developed in postsecondary environments, including training provided by the private sector and through more informal methods. In addition, remedial basic skills development may be needed by those who have been bypassed in the acquisition of ICT skills through formal primary and secondary education—either because they did not complete their schooling or because ICT training was not available.

Digital Literacy through Education

To enhance awareness of the benefits of broadband, countries may need to impart basic digital literacy skills to their people as part of, or associated with, their general educational programs. The extent of such need varies depending on the level of sophistication of the ICT sector and overall educational background of a country's inhabitants. Quite often, those lacking basic digital literacy tend to be the "at risk" groups, such as the elderly, women, the uneducated, people with disabilities, and the unemployed. These groups need to be included in plans to enhance digital literacy. This is particularly critical given that, as the average level of broadband penetration in a country grows, the social and economic costs of being excluded from access also increases.

The European Union (EU) has acknowledged the importance of digital literacy through various programs over a number of years. A key thrust of the EU's i2010 Strategy is "e-Inclusion"—the ability and willingness of individuals and communities to participate in the information society. In 2005, for example, there were large disparities in Internet use between the average population and persons over 65, those with low education, and the unemployed.[6] The EU set a target of cutting in half the gap in digital literacy between the average population and those groups as well as immigrants, people with disabilities, and marginalized young people. It proposed the following actions to improve digital literacy:

- Offering digital literacy courses through formal or informal education systems tailored to the needs of groups at risk of exclusion
- Undertaking digital literacy actions through partnerships with the private sector and in conjunction with other related educational initiatives and regularly upgrading skills to cope with technical and economic developments
- Supporting qualification methods measuring digital literacy achievement.

A review of progress since a 2006 EU conference on an inclusive information society found increases in broadband connections, use of the Internet, and digital literacy. There were advances in Internet use for disadvantaged groups, particularly the unemployed and marginalized youth (European Commission 2008). A significant factor was the number of digital inclusion initiatives launched by member countries as well as by civil society and the private sector (Council of the European Union 2008). Nevertheless, it was recognized that more efforts were needed to reduce digital exclusion; additional measures were incorporated into Pillar 6 (enhancing digital literacy, skills, and inclusion) in the Digital Agenda for Europe adopted in May 2010.[7]

Developing countries have also adopted a variety of programs to provide training on how to use computers and the Internet. As illustrated in box 6.1, Sri Lanka is enhancing the digital literacy of its people by providing training to vulnerable groups through schools and computer learning centers.

Box 6.1: Sri Lanka's Approach to Computer Literacy

In Sri Lanka, a fifth of the population was "computer literate" in 2009, according to a survey by the Census and Statistics Department, but gaps exist depending on age, education, and language fluency (figure B6.1.1). Around 60 percent of college-educated persons were computer literate compared to just over 1 percent of persons with no schooling. Over half of the country's English speakers were computer literate compared to less than a quarter of those who only speak other national languages.

Training for Sri Lanka's vulnerable groups is being supported through schools and telecenters. The Asian Development Bank helped to fund the Secondary Education Modernization project, which included a component for creating over 1,000 computer learning centers (CLCs) with Internet access in secondary schools.[8] The CLCs were open to the public after school hours to provide training and Internet access. The Ministry of Education issued a regulation for schools to keep the money earned from training and Internet access services instead of transferring it to the central treasury, allowing the CLCs to recover a portion of their operating costs. About 90 percent of schools with CLCs provide after-hour use, with 70 percent of them earning a profit. The earnings have been used to pay for access, electricity, maintenance, repairs, and equipment such as printers and scanners.

Figure B6.1.1   Computer Literacy in Sri Lanka, 2009

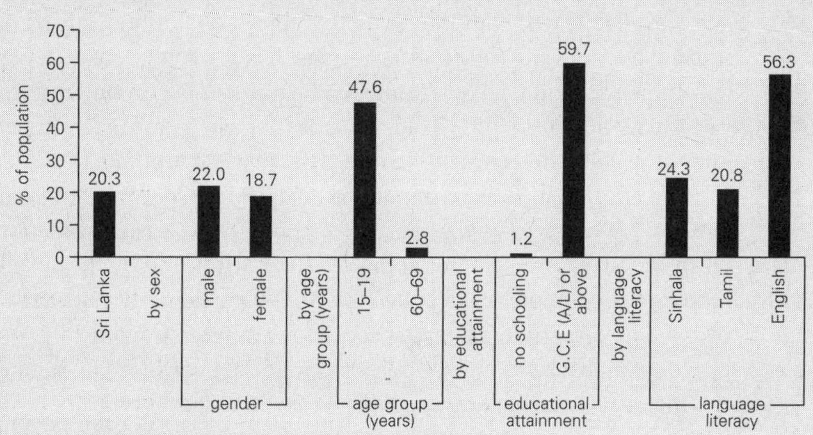

Source: Sri Lanka, Department of Census and Statistics.

Note: The GCE A/L (General Certificate of Education Advanced Level) is an exam taken by students in Sri Lanka, typically at the end of high school.

Sources: Sri Lanka, Department of Census and Statistics 2009; Dessoff 2010.

Note: Individuals are considered computer literate if they can use a computer on their own.

Broadband can also improve digital literacy through a variety of e-education services and applications, which also have the potential to increase demand for broadband services, including access to digital libraries of information, distance learning and virtual classrooms, and distance training for teachers in remote areas. For example, Colombia's National Learning Service (Servicio Nacional de Aprendizaje, or SENA) uses broadband services, along with other media, to train millions of people each year (nearly 8 million in 2009) using virtual online courses in professional and vocational subjects.[9] In large part, this has only been possible by using broadband services along with distance and online courses. SENA offers free training to all Colombians in a variety of vocational and professional subjects, including arts and sports, social sciences and education, finance and administration, manufacturing, health services, information technology, and retail services. Companies whose employees participate in SENA's training courses are shown to improve their profitability and competitiveness significantly (Colombia, Fedesarrollo 2010).

One very important way to provide digital literacy is through primary and secondary schools, particularly since enrollment is mandatory in many countries. Although many countries have installed computers and broadband access in schools, policies vary widely regarding access by students. In some cases, computers are only available to administrative staff, while in others computer labs exist but may not be accessible to all students. Adequate availability of computers, tablets, and mobile phones is an essential starting point for building digital literacy. A lack of computers, in particular, may limit educational opportunities. The number of computers per student varies widely around the world, a factor that can significantly enhance—or limit—the ability of countries to offer effective ICT training to students.

In an effort to increase computer availability for students, some countries have been moving toward a one-to-one model, where each student receives his or her own laptop.[10] This approach has been fueled by the development of low-cost computers for education and is particularly relevant in countries where few students have access to a computer at home.

However, broadband access is also essential in order to learn how to use the Internet. Internet access at school is particularly relevant in developing nations where many students come from homes without such access. The availability of Internet access in schools varies widely as well (see UNESCO n.d.).

With regard to funding, countries can pursue a range of policies in getting their schools connected, such as including broadband access in education budgets, using universal service policies or funds to have operators provide access, or working with development partners (table 6.1).

Table 6.1 Examples of Funding for School Connectivity in Three Countries

Country	Funding source	Description
Chile	Telecom operator	Under the Educational Internet 2000 project launched by the Ministry of Education, the incumbent telecom operator agreed to provide Internet service to primary and secondary schools, free of charge, for 10 years.
Namibia	Development partner	The Swedish International Development Cooperation Agency has provided ongoing financial assistance to Namibia's SchoolNet project, which provides Internet access to schools, contributing close to N$23 million (US$2.9 million) since mid-2001.
Philippines	Government	A 2009 presidential order directed the Department of Education to connect all Philippine public secondary schools to the Internet. The annual outlay for Internet subscription is ₱48,000 (US$1,115) per school or some US$6.3 million in total.[a]

Source: Adapted from ITU, "Connect a School, Connect a Community, Module 1: Policies and Regulation to Promote School Connectivity," http://www.connectaschool.org/itu-module/1/22/en/schools/connectivity/regulation/Section_3.6_funding/.

a. Based on the figure of 5,677 public secondary schools during the 2009–10 school year (Philippines, Department of Education, "Factsheet: Basic Education Statistics," September 23, 2010, http://www.deped.gov.ph/factsandfigures/default.asp).

Where countries choose to include digital training in their primary and secondary school curriculum, they should also ensure that the results of these programs are measured. As is shown in box 6.2, Australia has followed this approach (Australia, MCEECDYA 2008a).

Community Access Centers

Outside the formal educational process, additional groups of users can be targeted for digital literacy training. Turkey, for example, has opted to establish public Internet access centers. The Turkish Information Society Strategy and its annexed Action Plan endeavor to establish public Internet access centers across Turkey to provide computer and Internet access to those who do not have access at home (CIS 2009). The strategy targets libraries, public foundations, corporations, municipalities, organized industrial regions, public training centers, and volunteered foundation buildings as potential locations from which to provide access to citizens. Moreover, a public-private partnership (PPP) with Turk Telecom established 716 public

Box 6.2: Measuring Digital Literacy in Australia

ICT is incorporated into the Australian educational curriculum, and digital literacy among students is measured using a six-stage methodology in which students performing at level 1 are able to complete basic tasks using computers and software, while students performing at level 6 are able to use advanced software features to organize information and to synthesize and represent data as integrated, complete information products.[a]

In 2008, the Australian government measured the effects of the Digital Literacy Program using a standardized test for year 6 and year 10 students across Australia. As shown in figure B6.2.1, over 40 percent of year 6 students were proficient at level 3, which includes the ability to conduct simple general searches and select the best information source to meet a specific purpose. Nearly half of year 10 students were functioning at level 4, which required them to generate more complex, well-targeted searches for electronic information sources and assemble information to create new content in ways that demonstrate some consideration of audience and communicative purpose.

Figure B6.2.1 Digital Literacy in Australia, by Proficiency Level, 2008

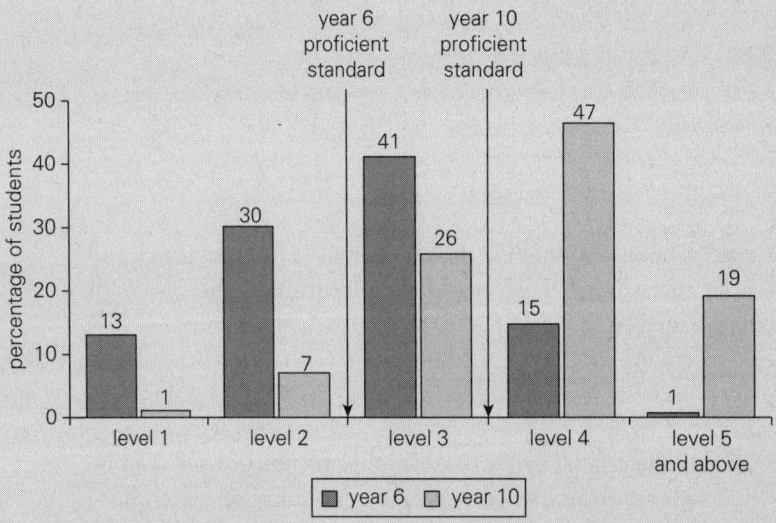

Source: Australia, MCEECDYA 2008b.

Source: Australia, MCEECDYA 2008b; Australian Curriculum, Assessment, and Reporting Authority, "General Capabilities: Literacy, Numeracy, and ICT Competence," http://www.acara.edu.au/curriculum/literacy,_numeracy_and_ict_continua.html.

a. See "Literacy, Numeracy, and ICT Continua." Australian Curriculum, Assessment and Reporting Authority, http://www.acara.edu.au/curriculum/literacy,_numeracy_and_ict_continua.html.

Internet access centers in various districts (OECD 2008, 73). The government is considering expanding the number of public Internet access centers across the country. In addition, it has extended connectivity to military conscripts through 227 public Internet access centers maintained with 4,487 computers, 227 projectors, printers, and related equipment on military campuses.

A comprehensive broadband strategy should also consider user-related issues so that available funds can finance projects to increase uptake and usage by specific groups and communities such as women, people with disabilities, and public facilities. Such projects can be part of universal broadband initiatives and funding. In many countries, community telecenters are often part of universal service programs and funded by the government or universal service fund. In the case of broadband, it is also important to ensure that programs to fund telecenters include induction programs for people who are using broadband services for the first time, including digital literacy training and broader training in how to use basic tools or available online services (such as e-government services).

Advanced ICT Training

Advanced ICT training refers to the acquisition of high-level skills necessary to support broadband networks and to develop broadband content and applications. Advanced skills are taught in two general venues: (a) specialized and more informal postsecondary schools, including training provided by the private sector, and (b) universities. Specialized postsecondary institutions include colleges, vocational schools, and courses typically taught by multinational software or hardware companies or international companies that specialize in ICT training. An example is the Cisco Networking Academy Program, which teaches network skills to almost 1 million students per year. Courses are taught at some 9,000 academies in 165 countries.[11] Countries can create similar partnerships with other hardware, software, content, and broadband services companies to fuel the development of training facilities and courses (Cooper 2010). India provides an example of the benefits of training through more informal institutions. Since the formal Indian ICT training sector through colleges and universities cannot cope with the demand for skilled ICT professionals, part of the demand is being met by India's training sector, which consists of over 5,000 private institutes offering ICT courses to over half a million students (Gupta et al. 2003).

Incorporation of ICT degrees within the formal higher education setting is important for developing highly skilled experts, fomenting a research and development culture, and addressing, understanding, and developing broadband needs within the context of national goals. Governments seeking to promote broadband in their countries should develop undergraduate, master's, and doctorate programs of study in ICTs to expand expertise in areas such as software engineering, networking, and security. A lack of domestic programs in these areas has often meant that students and professors go abroad and do not return (Rodrigues 2009). The higher education sector should forge links with industry in order to obtain funding as well as support for labs, incubators, and eventual job placement.

Privacy and Security Concerns

One obstacle to generating demand is that potential users may be afraid of using broadband services for reasons related to privacy, security, or identity theft. Training programs that address such concerns are an important part of convincing those who are not online that broadband access can be safe as well as productive. In the Republic of Korea, for example, the government created the Korea Information Security Agency and the Korea Internet Safety Commission to oversee Internet security and consumer protection as part of its efforts to get people online. The United Kingdom has a website called KidSMART that has information about safe and legal Internet use for children. Finally, Sweden has made "confidence" a cornerstone of its ICT policies since 2000. This includes not only confidence to use the technology, but confidence that personal information will be protected and secure. See chapter 3 for more information on how governments can address privacy and security concerns.

Small and Medium Enterprises

Small and medium enterprises (SMEs) are a particular group that governments may wish to focus on for purposes of demand stimulation. Such companies may not have ICT expertise or knowledge of how broadband can benefit their business functions. An Internet presence supported by broadband can help SMEs by providing them with the ability to reach new customers, reach a wider range of potential partners, and tap a wide range of resources to support their business. Concentrating on SMEs may also have important "pass-through" effects, allowing governments to reach their employees at the same time. SMEs are also likely to find e-government programs particularly helpful in interacting more efficiently with the government, whether to apply for permits, file taxes, or supply or obtain government services.

To help SMEs to use broadband networks and services most effectively, governments have adopted a variety of innovative outreach programs. The Dutch government, for example, has launched a program to stimulate and support the creation of applications for local SMEs (box 6.3). In Spain, the government is providing specific training for employees of SMEs, while Germany and Sweden have also established programs to provide training to SME employees to increase their ICT skills and increase their competitiveness. In Denmark, the government launched a program to train SMEs, providing assistance through private consultants and helping individuals to obtain the needed ICT skills to start e-businesses.[12] Providing support to SMEs to help them better use broadband is one of the important goals of the U.S. National Broadband Plan (United States, FCC 2010, sec. 13.1).

Affordability

In identifying demand-side barriers to broadband adoption, policy makers around the world have identified affordability as one of the main reasons that people do not use broadband services where they are available. The Pew Internet and American Life Project, as well as the U.S. Department of

---

Box 6.3: Stimulation of Local Applications Development for SMEs in the Netherlands

The Netherlands created a center for the development of local applications for SMEs. The center is half publicly funded, and projects require the participation of private developers. The center focuses on specific sectors of the economy (for example, hotels, restaurants, health), but also promotes cross-sector applications. Examples of applications created in this center are SME-specific solutions for customer relationship management, Internet marketplaces, and applications to manage radio frequency identification (RFID) and integrate PDAs in business processes.

The center also works as a knowledge bank that disseminates projects among SMEs through seminars and workshops. In addition, it tracks potential "breakthrough" applications on a sector-by-sector basis to disseminate them as best practices and ensure their expansion among SMEs throughout the country.

Sources: European Commission, "Information and Communication Technologies National Initiatives," http://ec.europa.eu/enterprise/sectors/ict/ebsn/national-initiatives/index_en.htm; Netherland BreedbandLand, http://www.nederlandbreedbandland.nl/.

Commerce, illustrate the importance of lack of affordability to those in the United States who do not subscribe to broadband at home (ESA and NTIA 2010). Prices for purchasing equipment and services remain a significant barrier for many consumers, especially in developing countries. Research by Ovum in 2010 showed that prices for broadband services are up to three times higher in 15 emerging markets than in developed countries, despite lower wage levels in the emerging markets.[13]

Various components affect the cost of broadband, including installation and ongoing service fees, as well as the prices of devices to access and use broadband services. In many developing countries, as well as among the low-income populations in developed nations, both the cost to acquire a broadband device and the cost of connection and service are often substantial relative to income levels. While potential users may have the necessary digital literacy skills, they may be hampered from making effective use of broadband services by the lack of affordable connections, services, and devices.

Part of the government's efforts, therefore, may also focus on supporting users who want and would benefit from broadband but cannot afford to pay prevailing commercial prices. This can apply to equipment (for example, computers), initial installation (up-front costs), connection to the network (fixed periodic charges), or use of the network to access services. One way to do this in a market context is by subsidizing providers that offer service to target population groups at less than prevailing prices. Another way is to provide subsidies directly to target users for the specific purpose of helping them to pay for broadband. Yet another approach is to include broadband in lump-sum income support to households. These approaches have been used extensively in a wide range of countries to support the use of telecommunications, electricity, transportation, and water supply, as well as to help people to pay for rent, food, health care, and other essential expenses.

The rationale for using subsidies to overcome obstacles to broadband affordability is twofold: (a) greater deployment and use of broadband services are important drivers of economic growth, and (b) the value of network services in general, and broadband services in particular, increases as more people participate. Possible measures to consider include the following:

- Subsidizing the purchase of devices or computers, by means of government financing or bulk procurements, vouchers, or distribution of devices
- Introducing tax credits for the purchase of devices or computers

- Establishing locations for shared or community access to computers and other devices to facilitate the use of broadband services
- Introducing measures that reduce or eliminate taxes on broadband service so as to reduce the final price paid by consumers.

Colombia's Plan Vive Digital, for example, addresses cost issues by making connection devices more available to the general public by eliminating customs tariffs, making access to credit for the acquisition of terminals more flexible, eliminating the value added tax for Internet service, and redirecting landline subsidies toward Internet subsidies.[14]

Device Ownership

The realization that demand for communication services, including broadband, does not generally increase if citizens do not have access to a PC or other broadband-enabled device has spurred policy makers around the world to introduce measures to facilitate ownership of devices or computers (box 6.4). The range of broadband devices includes more traditional means of access, such as PCs and laptops, as well as mobile devices, including cellular phones, smartphones, and tablets.

For many citizens in developing countries, the cost of even a discounted computer is prohibitively expensive. For example, figure 6.4 compares income levels in Sub-Saharan African countries with the cost of broadband devices. The data show that a US$400 netbook is more than the annual per capita gross domestic product (GDP) in nine Sub-Saharan African countries (Kim, Kelly, and Raja 2010). In these situations, direct distribution of low-cost devices has been used to overcome the price barrier.

Personal Computers, Laptops, and Netbooks

Programs to subsidize the purchase of laptops or computers have taken many forms, including tax breaks, government subsidies, and a reduction in price of the device itself. Some countries have provided fiscal incentives for individuals and businesses to purchase PCs, for example, by allowing pretax income to be used for these purchases. In Sweden, for example, the government established a tax rebate whereby employers could purchase computers for their employees to use at home. The program, which started in 1998, allows the purchase price of a computer to be deducted from salaries as monthly repayments over three years' time. Home computer penetration reached 90 percent by 2006. Similar programs have been used in other European countries. Governments in countries such as Korea, China, and Portugal have provided financing for the purchase of computers or are

Box 6.4: Device Price Trends

New computers. Prices have dropped more than 90 percent over the past decade for purchasing a computer capable of multimedia functions and Internet connectivity, as shown in figure B6.4.1.

Netbooks. The appearance of netbook computers in 2007, which are smaller, inexpensive laptop computers, has opened new possibilities for additional affordable devices for broadband connectivity. Prices for netbooks have fallen substantially since their introduction to the market. For example, between 2008 and 2009, the price of certain netbooks dropped dramatically in the

United States, from nearly US$500 to just over US$200 in 12 months.

Smartphones. Entry-level smartphone prices have reached the US$150 range and are expected to drop further to the US$80 level by 2015.

Refurbished computers. The purchase of refurbished computers, made possible by the donation of obsolete or malfunctioning computers, allows consumers to buy two or three computers for the price of one new model; such computers tend to come with longer warranties than their brand-new counterparts.

Figure B6.4.1   Prices of Computer Hardware in the United States, 1992–2009

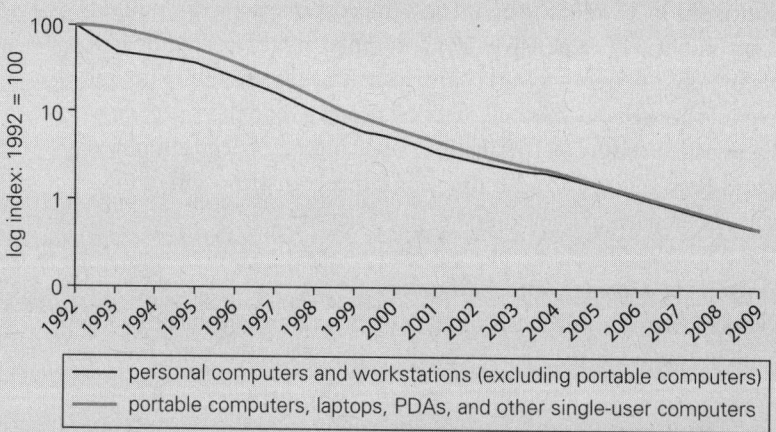

Source: U.S. Bureau of Labor Statistics, http://www.bls.gov/ppi/#tables, as cited in Kim, Kelly, and Raja 2010, 26.

Sources: "DealNews, DealWatch: Price Trends on 10" and 9" Netbooks," DealNews, July 23, 2009, http://dealnews.com/features/Deal-Watch-Price-trends-on-10-and-9-Netbooks/308433.html; Juniper Research, "Number of Entry-Level Smartphones to Reach over 185 Million by 2015, Driven by Operator Own-Brand Initiatives and Falling Prices," January 27, 2011, http://juniperresearch.com/viewpressrelease.php?pr=224; Dessoff 2010.

Figure 6.4   Cost of User Devices Relative to per Capita GDP in Selected Sub-Saharan African Countries, 2008

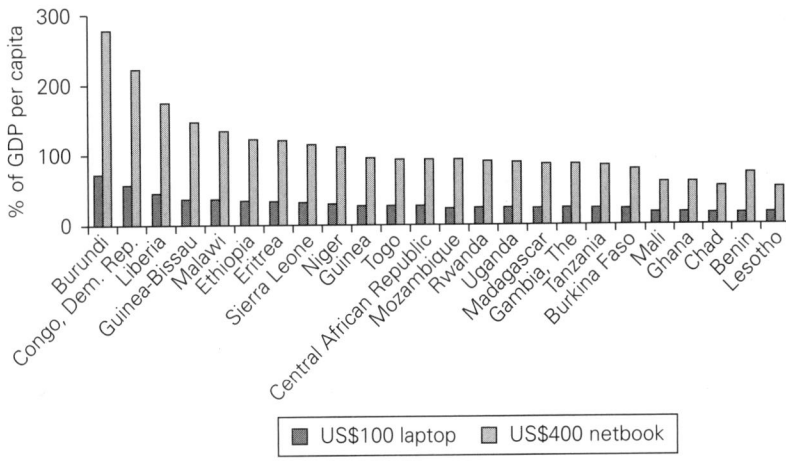

Source: Kim, Kelly, and Raja 2010, 122.

directly leasing computers to low-income families, students, or other identified groups (box 6.5).[15]

Reducing the cost of devices, particularly laptops, has also been successful in increasing device and broadband usage. One notable program to promote the spread of low-cost laptops in schools is the One Laptop per Child (OLPC) Initiative. The cost of the devices was predicted to drop to around US$100 under this program. Although the OLPC Initiative has experienced some significant setbacks, it has led to increased availability of lower-cost devices around the world. Uruguay has had some of the greatest success with the OLPC Initiative, with all of its primary students receiving their own laptop by 2009.[16] Some of the corporate participants that supported the initiative have since gone to market with their own low-cost computers, thus providing countries with additional options (Kramer, Dedrick, and Sharma 2009). The main commonalities of such devices, regardless of the brand or specific functionality, are a relatively low price (less than US$300 for the device), a flip or clamshell design, and small size (for example, screen size less than 10 inches).[17]

Mobile Devices, Smartphones, and Tablets

Mobile phones have taken the world by storm, with average mobile penetration rates in 2010 of 68 percent in developing countries and 116 percent in developed countries. Regionally, Africa has 41 percent penetration,

Box 6.5: Promoting Digital Literacy through Primary and Secondary Schools

Korea. The Korean Agency for Digital Opportunity and Promotion introduced a wide range of programs to promote digital literacy and access to computers, including subsidies for the purchase of PCs by low-income citizens. Established in 1999, this program provides low-cost PCs, partly through a purchase installment plan using the postal savings system and partly through a leasing program whereby government purchases 50,000 PCs and provides them to low-income families on a four-year lease, with free broadband for five years. Low-income students with good grades also receive free computers. Persons with disabilities and those receiving public assistance are eligible to receive free used computers.

China. China subsidizes computers for persons living in rural areas: families with a registered permanent rural residence can obtain a 13 percent subsidy if they purchase an eligible PC. Vendors compete for approval to sell computers under this program, and their maximum prices are limited under the terms of the approval. While there is a direct government outlay to pay for the 13 percent subsidy, the government's costs are at least somewhat offset by the taxes collected on all economic activity associated with the manufacturing, marketing, sale, and distribution of these computers, much of which also takes place within China.

Portugal. Portugal has launched two successful low-cost computer projects as part of its government program to promote broadband—the E-Escola (E-School) Program and the E-Escolinha Program. The E-School Program, initiated in June 2007, distributes laptops with broadband Internet access to teachers and secondary school students. By September 2010, the program had distributed over 450,000 laptops throughout the country. The laptops are sold by telecommunications providers at €150 (US$220) with a €5 discount over the basic monthly fee for 3, 5, and 7.2 megabits per second (Mbit/s) connections. Lower-income students get the laptops for free and broadband connectivity at 3 Mbit/s for between €5 and €15 per month. E-school is subsidized by the fees mobile operators paid for third-generation (3G) licenses. In July 2008, the government in partnership with Intel launched the E-Escolinha Program to produce a Portuguese version of the Intel Classmate (the "Magalhães"). The project calls for distributing these computers to 500,000 primary school students; by September 2010 over 410,000 computers had been distributed.

Sources: Atkinson, Correa, and Hedlund 2008; World Bank 2010; Escalões da Acção Social Escolar, http://eescola.pt/e-escola/

Arab states, 79 percent, Asia and Pacific, 68 percent, Commonwealth of Independent States, 132 percent, Europe, 120 percent, and the Americas, 94 percent.[18] In recent years, mobile service providers have begun to offer broadband services in addition to the original voice telephony and narrowband data services.

A business model that has contributed to the explosive growth of mobile telephony throughout much of the world is the "subsidization" of the mobile phone by revenues from subscriptions. Operators generally offer cheaper handsets subject to the consumer signing up for a one- or two-year service contract. Often, high early-termination fees are linked to such contracts to recover the remaining cost of the subsidy, if required. Besides device affordability, ease of use through prepaid services has also been one of the key benefits for low-income customers, offering them the ability to control their expenditures, the ability to switch to just receiving calls in times of economic difficulty, simple sign-up, and other features that have given mobile telephony an edge in the marketplace over traditional wireline telephone service (Oestmann 2003, 3).

It is reasonable to expect to see similar business models and programs expanded to include broadband devices, such as providing modems, smartphones, or tablet devices at reduced prices, along with contracts for broadband services. The first versions of such offers have consisted of a subsidy for the purchase of a laptop computer and modem, or a modem alone, bundled with a customer contract for Internet access.[19] In Europe and the United States, for example, network operators already subsidize other kinds of equipment in addition to phones.[20] In exchange for a two-year data contract, consumers can obtain cellular modems and sometimes even netbook computers with no up-front charge (Byrne 2009). Primarily, these offers are contingent on signing a contract for service. In Europe, studies show that the practice of bundling the cost of a laptop with an access plan is leading to robust sales in mobile access subscriptions. Thus, for example, global demand for mobile broadband pushed European operator Orange's mobile broadband customer base, including smartphone customers, to 23.2 million at the end of September 2008, which represented an 81 percent increase from the previous year. For United States–based AT&T Mobility, which started subsidizing laptops in 2008, data revenue jumped 51.2 percent in the fourth quarter of 2008 compared with the same quarter in 2007. It recorded US$3.1 billion from data revenue alone.[21]

Eventually, less expensive devices are likely to be offered, along with simplified or even no contractual commitments to purchase the broadband service, but simply with the expectation that such service will be purchased on a prepaid basis in sufficient quantities by enough customers to justify the subsidy. Already in some countries, mobile users own a USB modem enabling broadband service, but not necessarily a laptop or computer; they access the Internet at a shared computer. Throughout Africa (for example, in Tanzania, South Africa, Swaziland, Cameroon, and Kenya), operators sell subsidized modems with service contracts for 3G (or

Enhanced Data Rates for GSM Evolution [EDGE]) service, following the mobile phone subsidization business model. Since most users are prepaid, however, most of the mobile broadband uptake is prepaid as well and does not involve service contracts. In South Africa, bundled broadband products have started to emerge over the last few years, which typically include a PC, laptop, or netbook with a standard data bundle based on a 24- or 36-month contract. Incumbent operator Telkom offers its "Do Broadband" Acer netbook in a bundle, while Vodacom and MTN also have notebook and netbook offerings. *iBurst* is also selling 1 gigabyte (GB) and 2 GB notebook bundles (box 6.6).[22]

Recently, South African operators have been aggressively pursuing customers with attractive pricing of bundled mobile broadband packages. Vodacom, for example, launched a "2GB + 2GB" promotion in April 2011 for R 149 (US$22) per month, offering consumers on a 12-month contract a 2 GB per month data allowance, a 7.2Mbit/s High-Speed Packet Access (HSPA) modem, and an additional 2 GB of "night owl bandwidth" that can

Box 6.6: Trends in Low-Cost Devices

Classmate. Developed by Intel as a "mobile personal learning device for primary students in emerging markets," the Classmate was introduced in 2006. The second-generation Classmate is built around an Intel processor and has a "kid-friendly" design. Features include hardware-based theft protection, Wireless Fidelity (Wi-Fi), and a battery life of between 3.5 to 5 hours. The Classmate runs Windows XP or Linux and is available in clamshell or convertible designs. Intel has licensed the technology to various manufacturers.

Asustek. A computer manufacturer from Taiwan, China, Asustek introduced the Eee PC ("Easy, Exciting, and Economic") notebook in October 2007. Although not strictly designed for the educational environment, the Eee PC is a portable laptop that uses flash drive storage. Entry-level models are price competitive.

Mobilis. Manufactured by the Indian company Encore, Mobilis has touch-screen capabilities, a six-hour battery life, a carrying case, and a full-size, flexible, roll-up keyboard.

ITP-C. This is a touch-screen tablet computer with Wi-Fi using the Windows CE operating system. An external keyboard can be connected via a Universal Service Bus (USB) port. It is manufactured by ITP Software, based in Israel. It is being used in school projects in Argentina and Chile.

Sources: Encore Software, "Products," http://www.ncoretech.com/products/ia/mobilis/index.html; ITP Software, "ITP-C," http://www.itp-c.info; ITU, "Connect a School, Connect a Community Toolkit, Module 2: Disseminating Low-Cost Computing Devices in Schools," http://www.itu.int/ITU-D/sis/Connect_a_school/Modules/Mod2.pdf.

be used between midnight and 5 a.m. Subscribers also get free technical setup support plus a mailbox with 5 GB of storage (Muller 2011). South Africa is one of the few countries worldwide that still maintains monthly data caps on fixed-line broadband, although such caps are more common for mobile broadband.

The mobile phone subsidization business model is not without its detractors, and the practice is illegal in some countries. Concerns include whether the total cost of ownership is higher over time with subsidies and contracts versus scenarios involving unsubsidized phones and lower service prices, device locks that are used to prevent phones from being used with another operator's service, the limited variety of device models that operators are willing to subsidize, and high fees that consumers may pay if they want to terminate their contract early. Policy makers considering some type of subsidy program will need to take such concerns into account as they analyze various subsidy approaches.

Service Costs

Programs to provide affordable broadband devices to users are important, but only solve part of the problem. The longer-term issue for adoption of broadband services is the ongoing cost of receiving service. Some users may not have the means to pay for broadband access on an ongoing basis, particularly in countries where broadband service prices are still high. More information about support programs that could help users to get and keep their broadband service can be found in chapter 4, which addresses universal service funds and obligations.

In order to address the issue of service cost, in 2001 the Kenniswijk project in the Netherlands proposed a two-part subsidy program for connecting users. Notably, a year after the subsidies ended, 80 percent of subscribers were still using the service.

- *Initial connection.* A subsidy would be paid directly to the consumer and the administration would be undertaken by a government agency. This part of the subsidy would be used to encourage people to get Internet connections.
- *Ongoing support.* Ongoing support would be administered by the companies that win the contract to build the broadband infrastructure. They would receive money from the government and would then distribute the full amount to individual consumers in the region in the form of a lower connection tariff per household. It was thought that this would encourage people to adopt and keep broadband access.[23]

Shared or Community Access

In addition to using community access centers as a way to promote awareness of broadband, shared or community access can be a means of facilitating broadband affordability. Establishing locations where users are able to share broadband access is an important tool to enable broadband adoption and drive demand for otherwise willing and skilled persons who lack the financial means to purchase devices or pay long-term (contract) access charges. Public access facilities can be (a) government access facilities operated by public libraries, post offices, municipalities, or schools or (b) for-profit Internet cafés or local area network (LAN) gaming arcades operated privately. Both models are seen in abundant numbers throughout the world, including in developing countries. Public funding for access facilities may be particularly justified in localities where privately operated telecenters or Internet cafés are not yet available.

These facilities provide additional benefits, as they can also be places where training in digital skills occurs, such as those discussed earlier in this chapter. Figure 6.5 illustrates how important shared access facilities are in providing Internet access. It shows the place of access for Internet use on

Figure 6.5   Internet Use by Persons Ages 15–74 in 12 Latin American Countries, by Place of Access, 2007–09

Country and year	Household	Public access	House of another person
Brazil, 2008	60	35	19
Chile, 2009	64	22	—
Costa Rica, 2008	38	40	6
Ecuador, 2009	34	62	7
El Salvador, 2008	31	45	2
Honduras, 2007	17	77	—
Mexico, 2009	47	35	3
Panama, 2007	31	41	5
Paraguay, 2008	39	38	6
Peru, 2009	28	64	—
Dominican Republic, 2007	22	61	27
Uruguay, 2009	65	25	17

Source: Observatory for the Information Society in Latin America and the Caribbean, available at http://www.cepal.org/tic/flash/, as cited in ECLAC 2010, 32.

Note: — = Not available.

the basis of household surveys conducted in 12 countries in Latin America. In seven of the 12 countries, more persons access the Internet at public access facilities than do so at their own households.[24]

In India, the government is establishing 96,000 common service centers (CSCs) with broadband access that is configured to enable video, voice, and data in the areas of e-governance, education, telemedicine, entertainment, and other private uses.[25] E-government services from the national, state, and local governments are all available at the CSCs. One Indian state, Kerala, has implemented FRIENDS (Fast Reliable Instant Efficient Network for Disbursement of Services) as a single-window facility with at least one center in each district of the state. Currently, each center has 800 to 1,000 visitors daily. Citizens can make payments for various government-related services, obtain e-literacy training, and access a help desk to receive answers to questions or register complaints.[26] In the initial implementation of the program, 95.6 percent of participants said they lost their fear of computers because of the program, 30.5 percent felt they gained more respect in the community because of their computer knowledge, and 9.2 percent signed up a child for a computer-literacy class (Pal 2007).

Attractiveness

In order to generate demand for broadband, consumers must not only be aware of and able to afford broadband, but they must also see the relevance and attractiveness of it. This is facilitated by ensuring that the market provides sufficient choice and diversity of services, applications, and content to appeal to all consumers. Actions to boost broadband demand are generally aimed at both consumers and businesses to encourage them to produce content, services, and applications (Battisti n.d.). This section makes a distinction between services and applications, but this distinction is becoming blurred as technologies develop and services and applications begin to overlap and merge, as noted in chapter 1. While it may be arguable whether something is more appropriately classified as an "application" or a "service," for this chapter the particular category is less important than the fact that attractive services and applications both significantly increase demand.

Services to Drive Broadband Demand

Services refer to the basic connectivity function of providing access to the Internet as well as value added features that broadband operators include with the broadband subscription and that meet specific quality guidelines.

Within the broadband ecosystem, the availability of services is an important factor that influences and possibly *drives* demand. This level of demand, of course, will be affected by the attractiveness and affordability of the service offerings.

Internet

A broadband subscription provides a high-speed connection to the Internet. The way the subscription is provided can affect attractiveness and will depend on the technology and regulatory or business considerations. This includes whether the broadband subscription can be purchased on its own or requires a subscription to an underlying transport technology. For example, in the case of a digital subscriber line (DSL) broadband connection, a telephone line is required. Subscribers have typically been obligated to pay a monthly rental for the telephone line in addition to the broadband subscription even if they do not use the telephone line for anything else but broadband. This adds to costs and may require an extra bill, discouraging users from taking up the service. Some operators include the telephone line with the broadband subscription, so there is no separate bill. In a few countries, the cost of the physical broadband connection is billed separately from Internet access. In other words, the user needs to pay one bill for a broadband connection and another bill for Internet access.

Several factors make a broadband subscription more or less attractive to potential users. One important factor is speed. Although some consider all "always-on" subscriptions of at least 256 kilobits per second (kbit/s) to be broadband, in practice, speeds must be above a certain threshold to use desirable applications such as video viewing or gaming. A variety of offers with different speeds provides more choice to the user. Other factors to consider are restrictions that the broadband providers may impose on capacity (for example, data or usage caps). Some operators distinguish between domestic and international use by having no cap or a higher cap for traffic to national sites and a low cap for access to sites hosted abroad. One issue with caps is that users often do not understand the relation between volume and their usage needs. Users can easily underestimate how much data they will use, particularly if they access a lot of video services or use peer-to-peer download services (some of which may run in the background). This makes it difficult for them to know which package to select when packages vary by data caps. Some operators cap usage through time rather than data volume (for example, monthly subscription of 20 hours).

Increasingly, governments are responding to data caps and "throttling" practices by requiring service providers to disclose their network management practices clearly, in order to protect consumers and improve the overall broadband experience (see the discussion of network neutrality in chapter 3). Regulators have also instituted other measures, such as monitoring quality of service and alerting users to sites where they can test their broadband connection for speed or throttling (see chapter 5 for more discussion of quality of service issues).

Voice

Voice telephony continues to be a popular service, although it represents a declining share of revenue for public telecommunication operators. A growing number of broadband operators offer voice over broadband (VoB) service, which is a managed service (unlike voice over Internet Protocol, VoIP, which is generally considered as an application running "over the top" of the public Internet and not directly managed by the network operator). VoB provides the same quality as a traditional fixed telephone and often provides other value added features such as call waiting, voice mail, and speed dialing as well as the ability for users to monitor these features online via the provider's website. The price structure for VoB is often made attractive by including unlimited national calls for a flat rate or even including free national calls with the broadband service subscription. Since the service works through the broadband modem, users do not need to be connected to the Internet and do not even need a separate Internet subscription.

Several regulatory issues are related to VoB. The most basic is whether or not a country's laws and regulations allow it. Where VoB is legal, other regulatory considerations are often driven by the requirements placed on legacy wireline telephone networks. One is the requirement for users to be able to make emergency calls. Other regulatory requirements relating to consumers can include access for persons with disabilities and number portability.[27] The latter can be influential in encouraging users to switch from traditional telephone services to VoB.

Video

IP networks allow video services to be provided over a variety of networks. This has allowed broadband operators to provide Internet Protocol television (IPTV) or video on demand (VoD) services. The ability to provide IPTV, VoD, or both can make operators' broadband services more attractive, especially when other features are included, such as access to special programming not available elsewhere.

Television as a managed offering with a broadband subscription takes many forms. Some operators require IPTV to be bundled along with the broadband subscription, while others offer IPTV on a stand-alone basis.

Others have developed more extensive video service offerings, including BT (formerly British Telecom) in the United Kingdom, which offers its Vision service, which seamlessly integrates free-to-air digital television programs with a digital recorder and VoD feature.[28] Some operators provide additional features such as radio programming and the ability to watch programming on computers, tablets, and mobile phones in addition to the traditional television set.

The ability to bundle television with broadband Internet service is often subject to technical and regulatory considerations. In the case of IPTV, users need to have a minimum bandwidth to use the service. Some countries require companies that provide television service to obtain permission or a specific type of license. Sometimes permission is required from local authorities. Conditions vary, but in general, television service is subject to a higher level of regulatory oversight than broadband service. Regulatory limitations have sometimes meant that operators can only provide delayed service rather than live programming, making their offer less attractive.

Bundling

IP-based technology and digitalization of media allow a single network to provide a variety of voice, data, and video services. The ability to offer multiple services has led operators to bundle services together. This often includes a price reduction in the total cost of the service (that is, the bundled prices is less than the cost of buying the same services individually) and the benefit of receiving just one bill. "Double play" refers to a combination of broadband Internet and some other service, "triple play" refers to the ability to provide three services, whereas "quadruple play" also includes mobile service (see table 6.2 for bundling trends in Switzerland).

Bundling offers can be attractive to consumers because of their lower costs and a single invoice. However, some consumers may only want one service from a provider and therefore need to have an "a la carte" option and not be obligated to purchase additional services. In any case, a service provider that is only allowed to provide Internet access is at a disadvantage versus converged operators since consumers are increasingly interested in receiving multiple types of communication services offered through bundles.

Government

Government services and applications fall into the following broad categories: (a) making government information available, (b) conducting transactions with the government, and (c) participating in the political process. Governments can enhance broadband demand by acting as model

Table 6.2 Subscriptions to Bundled Services in Switzerland, 2008 and 2009

Bundle	2008	2009	Percentage change, 2008–09
Wireline telephony + broadband Internet	377,477	484,326	28
Broadband Internet + television	59,306	74,862	26
Broadband Internet + mobile telephony	42,126	66,482	58
Wireline telephony + broadband Internet + television	85,417	136,082	59
Wireline telephony + mobile telephony + broadband Internet	2,767	2,309	–17
Mobile telephony + broadband Internet + television	236	328	39
Wireline telephony + mobile telephony + broadband Internet + television	3,043	6,130	101

Source: Switzerland, OFCOM 2011.

users or anchor tenants and by promoting e-government services and broadband-related standards, putting content online, and supporting the development and distribution of digital content by other players. In addition, e-government services and broadband applications can help to organize the public sector more efficiently (in areas such as public safety, for example).

All governments collect and produce information. Applications at varying levels of sophistication can be developed to make this information available, thereby increasing demand for broadband services, as those applications are used by consumers. A 2008 study by the Organisation for Economic Co-operation and Development (OECD) contends that policy initiatives to foster more sophisticated government online services are becoming popular (OECD 2008). These initiatives include expanding secure government networks, putting administrative processes and documents online, supplying firms and citizens with more cost-effective ways to deal with the government (including once-only submission of data), and assigning firms and citizens a single number or identifier to conduct their relations with government. Government information that has been made available online in various countries includes legislation, regulations, litigation documents, reports, proposals, weather data, traffic reports, economic statistics, census reports, hearing schedules, applications for licenses and registrations, and even feeds from surveillance cameras. With always-on, high-bandwidth networks, online interactions between the government and businesses are also becoming more sophisticated, with some OECD countries offering one-stop platforms for government procurement, bidding information, and so forth.

E-services that improve openness and access to democratic institutions are also becoming feasible as a result of increases in broadband transmission capacity. Examples include Internet broadcasts of parliamentary debates and agency meetings and the use of multimedia content within the educational or cultural sector. Such applications allow citizens greater participation in the process of governance. Applications for polling, voting, campaigning, and interacting with government officials can increase the demand for broadband services. In the United States, for example, two models of e-government citizen participation are emerging. One is a deliberative model where online dialogue helps to inform policy making by encouraging citizens to scrutinize, discuss, and weigh competing values and policy options. The other is a consultative model that uses the speed and immediacy of broadband networks to enable citizens to communicate their opinions to government in order to improve policy and administration.[29] Actions to encourage citizen participation through e-government include the following:

- Connecting citizens to interactive government websites that encourage citizen feedback and participation in policy making, design, and innovation
- Encouraging library users to participate in online dialogue on topics such as health care and the economy
- Participating in government experiments with a variety of tools, including "wiki government," where citizens participate in peer review
- Educating citizens about their civic role and providing opportunities for them to interact with government agencies and officials using tools that fit individual or specific community needs
- Partnering with government officials and citizens to facilitate well-informed and productive discussions online
- Providing citizens the ability to create "my e-government" so they can personalize their interaction with government agencies and officials
- Creating "online town halls" to promote e-democracy for agenda setting and discussion of public issues as well as to promote accountability in the provision of public services.

As services and applications are developed to facilitate transactions with the government, those processes can be simplified and made more efficient, both for the government and for its citizens. A 2008 OECD study reported that since 2005 many countries have moved toward citizen-centric government (that is, measuring user satisfaction and user friendliness) with the mainstreaming of e-services via integrated multichannel service delivery strategies (OECD 2008).

A major goal of developing such services is to make government information more readily available as well as to increase transparency of government activities. The Netherlands is a leader in creating digital content and offering it via online government services (Atkinson, Correa, and Hedlund 2008, 39). In 2006, in an effort to support the development of broadband, the Dutch government decided to give all citizens a personalized webpage—the personal Internet page—where they could access their government documents and social security information as well as apply for grants and licenses. In the United States, the E-Government Act of 2002 was designed to "promote use of the Internet and other information technologies to provide increased opportunities for citizen participation in Government (Public Law 107-347, section 2b (2) Dec. 17, 2002)." These opportunities range from online tax-filing options to Social Security Administration application forms and, more recently, to electronic passport applications. In addition, the U.S. government embraced e-government as an educational tool, particularly in providing online education programs for new immigrants seeking citizenship and for school support programs within the Department of Education. In Colombia, the 2010 Plan Vive Digital aspires to create a digital ecosystem by 2014 that would achieve several demand-related goals (box 6.7).[30]

As described in chapter 3, governments may need to reform certain legal practices in order to conduct e-government transactions electronically. For example, laws may need to be modernized to define and recognize electronic signatures, electronic filings, and certification of electronic documents. These reforms will make it possible for a broad range of transactions to be conducted over broadband networks. Such reforms must also be accompanied by awareness campaigns to help users to gain knowledge of e-government services and applications. OECD governments and industry, for example, have developed campaigns to educate consumers about risks to Internet security, instructed consumers on how to protect themselves against fraudulent practices, and put into place regulatory measures to promote a culture of security.

Health

E-health involves a variety of services and tools provided by both the public and private sectors, including electronic health records and telemedicine. Broadband health care services and applications have the potential to lower

Box 6.7: Colombia's 2010 Plan Vive Digital

In Colombia, the 2010 Plan Vive Digital is set to establish a digital ecosystem by 2014 that encompasses supply (infrastructure) and demand (users, services, and applications). With respect to the latter, the plan seeks to make broadband more attractive to users and businesses in several ways:

- Have all national government entities and half of local government entities provide services online

- Support the development of applications for micro, small, and medium enterprises and enable half to use the Internet

- Assist the consolidation of the information technology and business process outsourcing industry

- Triple revenues for the creative digital industries

- Create mechanisms for public and private financial leverage for Colombian companies that develop applications and content

- Strengthen national and regional public broadcasting services incorporating the use of ICT.

Source: Colombia, Ministerio de Tecnologías de la Información y las Comunicaciones, "Vive Digital Colombia," http://vivedigital.gov.co/.

costs and lead to better health outcomes. A 2010 ITU discussion paper argues that citizens in rural areas, as well as those with limited mobility, will be able to use e-health to access specialized care that previously was not available to them (Hernandez, Leza, and Ballot-Lena 2010, 4). For example, broadband capabilities are essential to medical evaluation and other medical applications that use imaging extensively. High-definition video consultations allow rural patients and immobile patients (for example, incarcerated individuals or nursing home residents) to be seen by specialists in a timely manner when urgent diagnosis is needed and the specialists are not able to travel to where the patients are located. Other e-health services and applications include digital patient records; remote monitoring, where caregivers monitor key vital signs from a remote location, such as for diabetes or congestive heart failure patients; and access to medical information materials and advice.[31]

With the explosion of mobile devices in low-income nations and the relative lack of wireline broadband penetration, mobile health (m-health) is establishing a new frontier in health care in those countries.[32] Although basic voice and data connections are useful to improving health and medical care, broadband connectivity is necessary to realize the full potential of e-health and m-health services, particularly in rural communities.

In addition, a greater range of services becomes possible with more uniform, faster, and more affordable broadband access; greater access and coverage expand the "subscriber" base, building volume, creating incentives for players, and helping to push sustainable m-health applications beyond simple one-way data services (Vital Wave Consulting 2009, 20). As a result, improvements in telemedicine and other e-health initiatives will rely on increasing bandwidth capacity, more storage and processing capabilities, and higher levels of security to protect patient information. Cape Verde, for example, has been exploiting growing broadband connectivity by connecting two of its hospitals to the Pediatric Hospital of Coimbra, Portugal (Favaro, Melhem, and Winter 2008). The telemedicine system supports remote consultations through video conferencing. One goal is to reduce the number of Cape Verdeans who have to travel to Portugal for medical service. In addition to the Cape Verdean hospitals, two Angolan hospitals also are connected to the network, and over 10,000 remote consultations have been carried out (CVTelecom 2010, 9). In India, Ericsson and Apollo Telemedicine Networking Foundation signed a Memorandum of Understanding to "implement telemedicine applications over broadband-enabled mobile networks" in the summer of 2008 (Vital Wave Consulting 2009). The initiative is anticipated to decrease costs and improve health care outcomes, particularly for rural populations.[33]

Financial Services

Online banking has evolved considerably, with the Internet becoming an integral part of the delivery of banking services around the world. It is generally recognized that e-banking services can provide speedier, faster, and more reliable services to customers and thus also improve relationships with them. Although many types of Internet connections have online banking capabilities (for example, some m-banking transactions are conducted with narrowband short message service, SMS, messages), high-speed connectivity is essential for effective e-banking. A 2007 study, for example, found that, in the United States, banking online was performed by 66 percent of households with a home broadband connection versus 39 percent of households with a narrowband connection (DuBravac 2007, 9). Delivering financial services to low-income users through e-banking can also offer the potential to decrease operational costs dramatically, improve the quality of financial information, allow for "video chats" with bank representatives, and make banking for low-income users more profitable and less risky for mainstream financial institutions (Waterfield 2004). For these markets in particular, mobile money services have proved to be of particular importance. In countries such as Afghanistan, Bangladesh, Kenya, Indonesia,

Pakistan, the Philippines, and South Africa, various forms of m-banking services are expanding the financial services frontier. These services allow users to make payments and remittances, access existing bank accounts, conduct financial transactions, engage in commerce, and transfer balances.

Applications to Drive Broadband Demand

Among demand facilitation factors, applications (that is, function-specific software using a broadband connection to deliver content to users) have a tremendous impact on adoption. If there are no compelling applications to use on the platform, users will find no value in broadband and will not use it. Applications add value to broadband, as they provide tools and services that are tangible and valuable for both consumers and businesses. This increases the value proposition of broadband and the chances of attracting potential users to try the service. Evidence suggests that once consumers try the service, they are more willing to use it more frequently and subscribe to it. Additionally, the rise of social networking sites and the rapid increase in the amount of user-generated content being produced indicate that such applications can be strong demand drivers. Whether a user is uploading videos for friends and family or developing applications for use on a mobile device (available in various "app stores"), it is clear that individual user innovation can provide a strong incentive for people to subscribe to broadband services—whether for personal or professional reasons.

The development of local content is important, as people are more likely to be attracted to content that is developed in their local language and designed for their local culture. With greater local content, local SMEs and consumers can better understand the benefits of broadband. For example, locally developed video games played a key role in broadband diffusion in Korea, which suggests that applications that address local needs and culture are critical for broadband diffusion. Reflecting this recognition, in 2010 the Kenya ICT Board began a grant program with K Sh 320 million (US$3.7 million) to promote the development of relevant, local digital content and software by targeting developers in the film, education, entertainment, and advertising industries (Obura 2010).

The increased availability of broadband-enabled applications in government services, health care, education, and finance is also expected to boost the overall demand for broadband services. Similar to how a large merchant serves as an "anchor tenant" in a shopping center by drawing in customers who also purchase from smaller shops in the same shopping center, developing and implementing specific broadband "anchor" applications will help to attract new broadband users, who will make use of other broadband services as a result. As applications are designed and implemented, the issue of accessibility should also be kept in mind, so that those with disabilities are not excluded. The United Nation's Global Initiative for Inclusive Information and Communication Technologies offers support in this area and helps to highlight how e-government and other applications can be kept fully inclusive.[34]

Social Media and Web 2.0

Social media (for example, YouTube and Facebook) are applications that facilitate social interaction, using web and mobile technology. YouTube, for example, is one of the most widely used social media applications and requires broadband capabilities to be effective. Users generate video content, upload it, and share it with others. In 2010, some 35 hours of footage were uploaded to YouTube every minute, with over 13 million hours uploaded in total over the year (Scott 2011). Web 2.0 is closely related to social media and is a term generally associated with applications that feature user-generated content and facilitate collaboration among users (O'Reilly 2005). Web 2.0 applications—including web-based communities, hosted services, web applications, social networking sites, photo- and video-sharing sites, wikis, blogs, mashups, and folksonomies—are interoperable, user centered, and collaborative. Unlike the "traditional web," they allow users to generate, distribute, and share content in real time and typically require broadband connectivity. The availability of social media and Web 2.0 applications is stimulating demand and is an important factor to bear in mind in developing demand creation or facilitation strategies.

Social Networking

Social networking applications allow people who share interests to initiate and maintain connections, communicate with one another via various media, including text, voice, and video, interact through social games, and share user-generated and traditional media content. The highly personalized, easy, and flexible nature of social networking applications makes them some of the most-used online tools and one of the main drivers of broadband demand. Since these websites tend to offer at least limited functionality with dial-up or other low-bandwidth Internet connections, they help to drive broadband demand among users seeking to take full advantage of the website applications. Additionally, Web 2.0 applications have strong network effects, in which websites become more useful as more people participate (for example, Wikipedia entries or reviews of products on Amazon). Nonadopters who may not have found broadband to be relevant in the past may seek out broadband services in order to interact with family and friends as well as to discover and create other engaging user-generated content.

Mobility is a key component of social networking. As of September 2011, of the over 750 million active Facebook users, more than 250 million access Facebook through their mobile devices and use Facebook twice as much as users with nonmobile devices.[35] Indeed, evidence already exists that social networking applications are driving mobile broadband use in many countries. In the United Kingdom, mobile operator Hutchison 3G released traffic statistics showing the amount of data customers use when browsing social networking sites (Mansfield 2010). The operator found that social networking accounts for most mobile broadband usage in the country, with Facebook the most popular application. With the number of mobile broadband users expected to hit the 1 billion mark in 2011, the value of social networking to drive demand for ever-increasing amounts of data is substantial (Gobry 2011).

Particularly in developing countries, where mobile broadband is reaching more people than wireline broadband, social networking applications accessed through mobile devices are likely to be a major driver of demand for broadband access. In Africa, for example, the number of mobile broadband subscriptions is more than double the number of wireline broadband subscriptions.[36]

An example of the power of social media can be seen in its role in the 2011 so-called Arab Spring uprisings. Protest organizers used websites such as Facebook, Twitter, and YouTube in addition to texting and other narrowband technologies to coordinate protest activities. Social media facilitated the spread of information about citizens' grievances, through YouTube videos and conversations on social websites, when official or traditional media sources may not have given those grievances much or any coverage. These online tools also enabled the organizers to spread awareness and increase participation and attendance at demonstrations faster than more traditional media could allow.

Indonesia is another good example of mobile broadband's use in social networking. It has become the world's second-largest Facebook country, reaching 39 million users as of June 2011.[37] One of the reasons for Facebook's popularity in Indonesia is that it is used as "a way to establish social status, success, and as a platform for self-promotion" (Thia 2011). This resonates with many people in developing countries, where Facebook has emerged as the leading application.[38] Indonesia's hunger for social networking extends to Twitter, which has the world's highest penetration: around one-fifth of Indonesian Internet users access the microblogging application.[39] All of this has spurred demand for faster connectivity, with mobile broadband speeds rising to 40 Mbit/s (Indosat 2010). Since social media focuses on user-generated content, that content can be quite localized, meaning that it is in local languages and character sets and on topics that are locally relevant. Having localized content is an important part of stimulating demand.

Social Collaboration: Wikis, Mashups, and Crowdsourcing

Web 2.0 applications allow for more than simply connecting with others and sharing information—they also allow people anywhere in the world to create content through blogs and podcasts, to co-create content, such as through wikis, to link different types of content from different sources together to create new media (for example, mashups), or to create social tags to identify folksonomies. Although perhaps to a lesser extent than social networking applications, these social collaboration tools help to increase the demand for broadband services by engaging users and making the online experience more personalized and flexible. They draw on the idea of the "wisdom of the crowd," which refers to practices where opinions and information are collectively created rather than arrived at by a single or small group of experts.

Wikipedia is a well-known example of such social collaboration. The popular collaborative encyclopedia is multilingual, web-based, free to access, and written by Internet volunteers, most of whom are anonymous. Anyone with Internet access can write and make changes to Wikipedia articles, and there are currently more than 91,000 active contributors around the world, of which nearly 60 percent create and edit non-English articles.[40] Launched in 2001, Wikipedia is now available in 281 languages—the English Wikipedia contains over 3.6 million distinct articles, followed by German with 1.2 million and French with 1.1 million.[41] Users can also create mash-ups, which are interactive web applications that integrate content (for example, video, text, audio, or images) retrieved from third-party data sources in order to create new and innovative services and applications (Merrill 2009). Mashup websites tend to rely on external websites that use open-source application programming interfaces (APIs), which expose all of the instructions and operations in an application to facilitate the interaction between different software programs. Mashups may be as simple as a restaurant's website embedded with a single API, such as a Google map to make it easier for customers to find. Other mashups combine multiple APIs. For example, a web-based interactive restaurant guide could use APIs from sites with online reviews, photos, and maps to indicate the best places to eat in a given city and where to find them.

Crowdsourcing is a Web 2.0 application referring to the outsourcing of tasks to a large, undefined group or community (the "crowd") through an open call for assistance, such as via Twitter, Facebook, or a dedicated web-page. Following the 2010 earthquake in Haiti, the Crisis Map of Haiti used crowdsourcing to coordinate relief efforts on the island. Those in need could submit incident reports via the organization's website, phone, SMS, e-mail, Facebook, Twitter, and so forth and thus request aid or even report missing persons. After being reviewed by volunteers, the reports were mapped with global positioning system coordinates in near real-time on a map also showing shelter sites and hospitals. These tools helped to speed search-and-rescue efforts and provide vital supplies to those most needing them. The events in Haiti provide a model for how to deal with future disasters, both natural and man-made, as well as demonstrating a practical application of Web 2.0 technologies.

Collaborative Working Tools for Businesses and Institutions

Businesses and institutions are taking advantage of Web 2.0 applications (often referred to as Enterprise 2.0) to improve productivity and efficiency as well as lower costs. Generally, Web 2.0 applications not only are less expensive, faster to deploy, and more flexible than commercial or customized software packages, but also offer built-in collaborative workspace tools that enable people to interact across differences in time and space (Kuchinskas 2007). These tools often center around "groupware," which allows multiple people to work together on projects and share documents, calendars, and other data and to participate in video and audio conferences. Since Web 2.0 apps require large amounts of bandwidth to download and upload the various types of digital media, a broadband connection is essential.

Education and Web 2.0

Support for school connectivity programs can be strengthened through the use of Web 2.0 applications in education. Even where virtual classrooms or other e-learning tools are in use, Web 2.0 tools can replace or complement expensive virtual learning environment (VLE) software to provide a more flexible approach through the use of blogs, wikis, and other collaborative applications. For example, a classic VLE involves the teacher sharing slides and resources with students through an enabling software program. Web 2.0 applications, such as Slideshare for presentations, Google Docs for documents, Flickr for images, and YouTube for videos, however, are capable of replicating the core functions of the VLE software at no cost to educators or students (Robertson 2011). Open-source and cloud technologies also allow for more educational opportunities where fewer resources are available. For example, students without personal computers can complete assignments at a university computer lab or Internet café via Google Docs. Other services, such as Flat World Knowledge's open-source textbooks, allow professors to review, adopt, and even customize textbooks for their classes, which students can then purchase in print format or view online for free, further reducing the cost of education. Additionally, teachers can incorporate blogging and wikis to encourage student participation and interaction.

A 2009 study by the Joint Research Centre of the European Commission on e-learning initiatives in Europe found that student and teacher participation in Web 2.0 applications supports technological innovation in education and training by providing new formats for knowledge dissemination, acquisition, and management (Redecker et al. 2009). These tools increase the accessibility and availability of learning content through a range of platforms that offer a large variety of educational material. Further, Web 2.0 tools support new strategies for studying a subject matter by making available a host of dynamic tools for transforming content and displaying information in different formats as well as contribute to diversifying and enhancing teaching methods. Students are able to have more personalized and flexible lessons targeting their specific needs and are able to learn valuable networking and community-building skills. Additionally, these tools allow collaboration among geographically dispersed groups and can facilitate intercultural exchange and cross-border, cross-institutional collaboration, while reduced costs allow institutions in developing countries to compete with those in other areas.

Content to Drive Broadband Demand

Ultimately, what motivates people to buy broadband services and devices is that they believe broadband will enrich their lives, offer convenience, provide entertainment, and improve their businesses. The network infrastructure or policies in place to expand broadband access are less important to end users on a day-to-day basis than the availability of relevant and useful online services and applications that allow them to access, create, and share content. What Bill Gates said about the Internet in 1996 remains true today: "Content is King." Attractive and useful content and, increasingly, context (with the development of location-based services, which require broadband access) are perhaps the most important underlying elements of broadband adoption.

Promoting Digital Content

"Digital content" is a catch-all for the myriad websites, applications, and services available to broadband users. It can be based on text, audio, video, or a combination of these. Much of the content available on websites today can be divided into three broad categories: (a) user generated, (b) proprietary or commercial, and (c) open source. User-generated content includes social networking and things such as blogs, podcasts, Twitter updates, YouTube videos, and Flickr photos. Addressed above, these forms of social media help to drive broadband demand by engaging users and ensuring the local and personal relevance of content. Due to the "bottom-up" nature of social media, policy makers can support the development of such content by taking a more hands-off approach in regulating it (see chapter 3). They can also promote such services by becoming active users of such applications and services; more and more government agencies and even politicians are realizing the value of such tools in reaching out to citizens.

As opposed to copyrighted materials, open-source content is available free-of-charge. In addition, the source code is also freely available to allow anyone wanting to incorporate the content or application into new forms of media, such as in mashups. Open-source content has led to the creation of property rights systems that encourage collaboration by publishing source code and allowing other users to extend those applications and develop them further, with the provision that the result should also be governed by the same open-source property rights.

Promoting Local Content

Native English speakers currently account for the majority of Internet users around the world; thus, most web content is in English.[42] Figure 6.6 shows the number of Internet users by language, which is a common metric for gauging the influence of different languages on Internet content. English continues to dominate, but the number of Internet users in China is rising quickly and expected to exceed the number of English language users in the next five years (Wilhelm 2010). Despite this shift, a significant obstacle to Internet and broadband use by non-English speakers is the scarcity of content in their own languages.

Efforts to create content that is relevant and interesting, using the local language and character sets, is expected to increase the demand for broadband services in local areas. For example, in 2010, the Kenya ICT Board launched a grant of K Sh 320 million (US$3.7 million) to promote the development of relevant, local digital content and software by targeting entrepreneurs in the film, education, entertainment, and advertising industries. The goal of the project is to increase Internet penetration and promote local content, which is viewed as a potential area for new revenues in the country.

In addition to direct grants for the production of local content, governments can support the development of local content and applications in other ways, such as the development of standardized keyboards, character

Figure 6.6   Number of Internet Users Worldwide, by Language, 2010

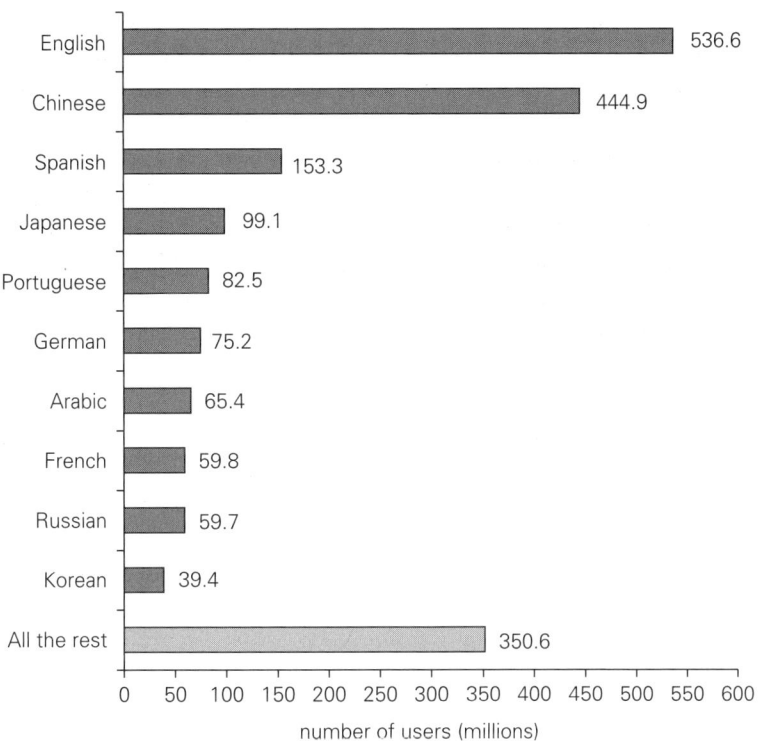

English 536.6
Chinese 444.9
Spanish 153.3
Japanese 99.1
Portuguese 82.5
German 75.2
Arabic 65.4
French 59.8
Russian 59.7
Korean 39.4
All the rest 350.6

number of users (millions)

Source: Internet World Stats, "Top 10 Languages Worldwide in Millions of Users," http://www.internet-worldstats.com/stats7.htm.

sets, and character encoding. This type of indirect intervention would affect the content available by enabling users to create content in their own languages.[43] Additionally, translation and standardization of operating systems into local languages can help to facilitate the development of local applications that are relevant and comprehensible to local users.[44] Governments can also play an important role in developing local content and local applications by directly creating local content and local applications in the form of e-government applications, as described above.

Some forms of user-generated content, such as YouTube videos, face fewer barriers to expression, as the speaker is recorded directly in his or her own language. YouTube has launched a localization system, where YouTube is available in 31 local versions as well as a worldwide version.[45] This helps to overcome some of the barriers to content reaching a possible community of interest, but not entirely, as content generated in languages other than those used in the 31 local versions or the worldwide version may encounter barriers to reaching an audience.

Nevertheless, it is likely that greater amounts of local content will continue to become available in the near term. For example, a website called d1g.com is a platform in Arabic for sharing videos, photos, and audio, a forum, and a question and answer facility. Launched in 2007, d1g.com is one of the Arab world's fastest-growing social media and content-sharing websites, with more than 13 million users and 4.8 million unique monthly visitors. It has 15 million videos and streams an extensive amount of Arabic videos—600 terabytes of data per month. Notably, nearly 100 percent of d1g.com's content is user generated, with a small amount produced in-house. d1g.com became the most popular Arab social media site (after Facebook and Twitter) when a user created the "Egyptstreet" diwan during the Egyptian revolution. During that time, unique visitors rose from 3 million to 5 million per month, and visits per month grew from 6 million to 13 million.

Notes

1. Cisco, "Visual Networking Index, Global IP Traffic Forecast 2010–2015," June 2011, http://www.cisco.com/en/US/netsol/ns827/networking_solutions_sub_solution.html#~forecast.

2. Based on the Pew Research Center's survey, almost half of the non-Internet-user adults surveyed in the United States indicated that they did not use the Internet because they did not find it relevant (they were not interested, considered it a waste of time, were too busy, or did not feel it was something they wanted or needed). See Pew Internet and American Life Project (2010).

3. Some examples of civil society organizations involved in making Internet and especially broadband services more available, accessible, and attractive include Tribal Digital Village, which works with Native American reservations in the United States (see http://www.sctdv.net), and CUWiN (Champaign-Urbana Community Wireless Network), which develops community-based wireless mesh technologies in various communities in the United States, West Africa, and South Africa (see http://www.cuwin.net).

4. MCMC, "National Broadband Initiative Overview," http://www.skmm.gov.my/index.php?c=public&v=art_view&art_id=36.

5. The level of basic literacy required to make effective use of a network may actually be lower for a broadband network than for a narrowband network, because of the greater opportunity to use visuals, sound, and icons rather than simple text. This may be important in low-literacy environments.

6. European Commission, "Information Society Thematic Portal: June 2006, Riga Ministerial Conference, ICT for an Inclusive Society," http://ec.europa.eu/information_society/activities/einclusion/events/riga_2006/index_en.htm.

7. See European Commission, "e-Inclusion," http://ec.europa.eu/information_society/activities/einclusion/index_en.htm.

8. ITU, "Connect a School, Connect a Community Toolkit, Module 1: Policies and Regulation to Promote School Connectivity," http://www.connectaschool.org/itu-html/1.

9. Servicio Nacional de Aprendizaje, "Home Page," http://www.sena.edu.co.

10. ITU, "Connect a School, Connect a Community Toolkit, Module 2: Disseminating Low-Cost Computing Devices in Schools," http://www.connectaschool.org/itu-html/8. For a discussion of a Peruvian one-to-one school laptop program, see Trucano (2010).

11. Cisco, "About Networking Academy Program Overview," http://www.cisco.com/web/learning/netacad/academy/index.html.

12. European Commission, "Connecting Europe at High Speed: Recent Developments in the Electronics Communications Sector," Europa Press Release, Brussels, February 3, 2004, http://europa.eu/rapid/pressReleasesAction.do?reference=IP/04/154&format=HTML&aged=0&language=EN&guiLanguage=en.

13. Ovum, "Emerging Markets Paying Three Times More Than Rest of the World for Broadband," September 20, 2010, http://about.datamonitor.com/media/archives/4775.

14. Colombia, Ministerio de Tecnologías de la Información y las Comunicaciones, "Ecosistema Digital: Servicios," http://201.234.78.242/vivedigital/ecosistema_2_servicios.php.

15. For a further example, see the global nonprofit organization One Economy Corporation, http://www.one-economy.com/who-we-are. International locations include Turkey, Jordan, Israel, Cameroon, Kenya, Nigeria, Rwanda, South Africa, and Mexico.

16. See ITU, "Connect a School, Connect a Community Toolkit: Uruguay Case Study," http://www.connectaschool.org/en/schools/connectivity/devices/section_5.7/case_studies/Uruguay.

17. ITU, "Connect a School, Connect a Community Toolkit, Module 2: Disseminating Low-Cost Computing Devices in Schools," http://www.itu.int/ITU-D/sis/Connect_a_school/Modules/Mod2.pdf.

18. ITU, World Telecommunications/ICT Indicators database, http://www.itu.int/ITU-D/ict/statistics/at_glance/KeyTelecom.html.

19. For example, in 2008, Radio Shack in the United States offered a laptop for US$99, along with an AT&T data card and a two-year service contract for US$60 a month (Kraemer 2008).

20. The commercial information presented was current as of the preparation of this report. Commercial service offerings in this sector are subject to frequent change.

21. "The Switch to LTE: When's the Tipping Point?," *Motorola eZine*, http://www.motorola.com/web/Business/Solutions/Industry%20Solutions/Service%20Providers/Network%20Operators/_Documents/_static%20files/LTE%20Tipping%20Point.pdf?localeId=33.

22. "Broadband Laptop Bundles: The Best Deals," *My Broadband*, July 2010, http://mybroadband.co.za/news/broadband/13694-Broadband-laptop-bundles-The-best-deals.html.

23. Ultimately, the subsidies were not paid exactly this way; the initial connections were subsidized by government at the rate of €800, and service for the first several years was free.

24. ECLAC, "Observatory for the Information Society in Latin America and the Caribbean (OSILAC)," http://www.cepal.org/tic/flash/, as cited in ECLAC (2010, 32).

25. India, Department of Information Technology, "National e-Governance Plan," http://mit.gov.in/content/national-e-governance-plan.

26. Kerala State IT Mtission, "Friends," http://itmission.kerala.gov.in/ksitm-e-governance-projects/82-friends.html.

27. For an example of some of the regulatory obligations relating to Internet Protocol telephony in Australia, see Australian Communications and Media Authority, "VoIP for Service Providers," http://www.acma.gov.au/WEB/STANDARD/pc=PC_310067.

28. See BT, "BT Vision," http://www.productsandservices.bt.com/consumerProducts/displayCategory.do;JSESSIONID_ecommerce=WWG1Nf2LyzDZSCd20Z7Kwc8MqZQMj4ZFbcrChsfrTGbw2MRYPvv2!-229543251?categoryId=CON-TV-I.

29. American Library Association, "Civic Participation and e-Government," http://www.ala.org/ala/issuesadvocacy/egovtoolkit/civicparticipation/index.cfm.

30. Colombia, Ministerio de Tecnologías de la Información y las Comunicaciones, "Vive Digital Colombia," http://vivedigital.gov.co/.

31. Maintaining the security and privacy of patient health information and records is critical. More information on safeguards that have been developed can be found at the following: for the United States, http://www.hhs.gov/ocr/privacy/; for Canada, http://www.ipc.on.ca/english/Home-Page/; for the United Nations, http://www.hon.ch/home1.html.

32. telecomAfrica, "mHealth: Pushing Frontiers of Health Care in Developing Countries," February 16, 2011, http://telecomafrica.org/?p=780.

33. Ericsson, "Ericsson and Apollo Hospitals to Bring Healthcare Access to Rural India," Press Release, June 5, 2008, http://www.ericsson.com/ericsson/press/releases/20080605-1225191.shtml.

34. For more information, visit http://g3ict.org.

35. Facebook, "Statistics," https://www.facebook.com/press/info.php?statistics.

36. ITU, "ITU ICT EYE: Dynamic Reports for Mobile Broadband and Fixed Broadband Subscriptions per 100 Inhabitants," http://www.itu.int/ITU-D/ICTEYE/Reporting/DynamicReportWizard.aspx.

37. "Infographic: Facebook's Indonesia Users Overtake the UK," *The Guardian*, April 6, 2011, http://www.guardian.co.uk/media/pda/2011/apr/06/facebook-statistics.

38. For example, a study among mobile users in 10 Southeast Asian nations found that Facebook was the top site in five of them, the second ranked site in three, the third ranked in one, and not among the top 10 in only one of the countries. Opera Software, "State of the Mobile Web," January 2011, http://www.opera.com/smw/2011/01/#snapshot.

39. comScore, "Indonesia, Brazil, and Venezuela Lead Global Surge in Twitter Usage," Press Release, August 11, 2010, http://www.comscore.com/Press_Events/

Press_Releases/2010/8/Indonesia_Brazil_and_Venezuela_Lead_Global_Surge_in_Twitter_Usage.

40. Wikipedia, "Wikipedia: About," http://en.wikipedia.org/wiki/Wikipedia:About.

41. WikiMedia, "List of Wikipedias," as of March 23, 2011, http://meta.wikimedia.org/wiki/List_of_Wikipedias.

42. Pimienta, Prado, and Blanco (2009, 35) compare the presence on the Internet of English with European languages. For every 100 pages in English on the Internet in 2007, there were 8 in Spanish, 10 in French, 6 in Italian, 3 in Portuguese, and 13 in German.

43. For example, there is no standardized keyboard layout for Pashto, an Indo-Iranian language spoken by about 25 million people in Afghanistan, India, the Islamic Republic of Iran, Pakistan, Tajikistan, the United Arab Emirates, and the United Kingdom. There is a standard for Pashto text encoding, so some progress has been made. However, there is no standard interface terminology translation in Pashto, which makes achieving digital literacy more challenging. See Hussain, Durrani, and Gul (2005).

44. Sri Lanka's ICT Agency has a Local Languages Initiative to enable ICT in languages such as Sinhala or Tamil (http://www.icta.lk/en/programmes/pli-development/68-projects/557-local-languages-initiative-lli.html).

45. Wikipedia, "YouTube," http://en.wikipedia.org/wiki/YouTube.

References

Atkinson, Robert D., Daniel K. Correa, and Julie A. Hedlund. 2008. "Explaining International Broadband Leadership." Information Technology and Innovation Foundation, Washington, DC, May.

Australia, MCEECDYA (Ministerial Council for Education, Early Childhood Development, and Youth Affairs). 2008a. *Melbourne Declaration on Educational Goals for Young Australians.* Carlton, Australia: Curriculum Corporation.

———. 2008b. *National Assessment Program: ICT Literacy Years 6 and 10 Report 2008.* Carlton: Curriculum Corporation. http://www.mceecdya.edu.au/verve/_resources/NAP-ICTL_report_2008.pdf.

Battisti, Daniela. n.d. "Broadband Policies: Focus on the Italian Government Action Plan." OECD, Paris. http://www.oecd.org/dataoecd/35/45/1936957.ppt.

Byrne, Joseph. 2009. "Cellular Modems Gain Favor." Linley Group, February 1. http://www.linleygroup.com/news_detail.php?num=42.

CIS (Center for Information and Society). 2009. "Turkey: Public Access Study Summary." CIS, University of Washington, Seattle. http://faculty.washington.edu/rgomez/projects/landscape/country-reports/Turkey/1Page_Turkey.pdf.

Colombia, Fedesarrollo (Fundación para la Educación Superior y el Desarrollo). 2010. "Evaluación de impacto de los programas de formación de técnicos y tecnólogos y formación especializada del recurso humano vinculado a las empresas del servicio nacional de aprendizaje: SENA." Fedesarrollo, Bogotá, July.

Cooper, Brendan. 2010. "Giving Africa's Neediest Communities IT Skills . . . and Hope." *Cisco News*, August 23. http://newsroom.cisco.com/dlls/2010/ts_082310b .html.

Council of the European Union. 2008. "Vienna 'E-Inclusion' Ministerial Conference Conclusions." Presidency of the European Union, Brussels, December 2. http://ec.europa.eu/information_society/events/e-inclusion/2008/doc/ conclusions.pdf.

CVTelecom. 2010. "As telecomunicações ao serviço da Saúde." *Antena* 55 (September): 9. http://www.grupocvt.com.cv/boletins.

Dessoff, Alan. 2010. "Cutting Technology Costs with Refurbished Computers." *District Administration*, January 1. http://www.districtadministration.com/ viewarticle.aspx?articleid=2264.

DuBravac, Shawn G. 2007. "Broadband in America: Access, Use, and Outlook." Consumer Electronics Association, Arlington, VA, July. http://www.ce.org/pdf/ cea_broadband_america.pdf.

ECLAC (Economic Commission for Latin America and the Caribbean). 2010. "Monitoring of the Plan of Action eLAC2010: Advances and Challenges of the Information Society in Latin America and the Caribbean." United Nations, ECLAC, Santiago.

Educational Testing Service. 2002. "Digital Transformation: A Framework for ICT Literacy." ETS, Princeton, NJ. http://www.ets.org/research/policy_research_ reports/ict-report.

ESA (Economics and Statistics Administration) and NTIA (National Telecommunications and Information Administration). 2010. "Exploring the Digital Nation: Home Broadband Internet Adoption in the United States." U.S. Department of Commerce, Washington, DC, November. http://www.esa.doc.gov/sites/default/ files/reports/documents/report.pdf.

European Commission. 2008. *E-Inclusion Ministerial Conference, Conference Report: November 30–December 2.* Brussels: European Commission. http:// ec.europa.eu/information_society/events/e-inclusion/2008/doc/final_report. pdf.

Favaro, Edgardo, Samia Melhem, and Brian Winter. 2008. "Small States, Smart Solutions: Improving Connectivity and Increasing the Effectiveness of Public Services." In *E-Government in Cape Verde,* ch. 6. Washington, DC: World Bank. http://siteresources.worldbank.org/INTDEBTDEPT/Resources/468980-120697 4166266/4833916-1206989877225/SmallStatesComplete.pdf.

Gobry, Pascal-Emmanuel. 2011. "HUGE: Mobile Broadband Will Hit 1 Billion Users in 2011." *Business Insider*, January 12. http://www.businessinsider.com/ mobile-broadband-will-hit-1-billion-users-in-2011-2011-1.

Gupta, P. N., Ajay Kr. Singh, Vaneeta Malhotra, and Lavanya Rastogi. 2003. "Role of IT Education in India: Challenges and Quality Perspectives." *Delhi Business Review* 4 (2, July–December): 77–93. http://www.delhibusinessreview.org/ v_4n2/v4n2h.pdf.

Hauge, Janice Alane, and James E. Prieger. 2009. "Demand-Side Programs to Stimulate Adoption of Broadband: What Works?" October 14. http://ssrn.com/ abstract=1492342.

Hernandez, Janet, Daniel Leza, and Kari Ballot-Lena. 2010. "ICT Regulation in the Digital Economy." 2010 GSR Discussion Paper, Telecommunications Management Group, Inc., ITU, Geneva. http://www.itu.int/ITU-D/treg/Events/Seminars/GSR/GSR10/documents/GSR10-paper3.pdf.

Hussain, Samad, Nadir Durrani, and Sana Gul. 2005. "Pan-Localization, Survey of Language Computing in Asia." Center for Research in Urdu Language Processing, National University of Computer and Emerging Sciences, Lahore. http://www.panl10n.net/english/outputs/Survey/Pashto.pdf.

Indosat. 2010. *Unleashing Our Potential: 2010 Annual Report*. Jakarta: Indosat.

Kim, Yongsoo, Tim Kelly, and Siddhartha Raja. 2010. "Building Broadband: Strategies and Policies for the Developing World." Global Information and Communication Technologies Department, World Bank, Washington, DC, June. http://www.infodev.org/en/Publication.1045.html.

Kraemer, Brian. 2008. "Radio Shack's $99 Acer Aspire Notebook Too Good to Be True." *CRN*, December 12. http://www.crn.com/blogs-op-ed/the-channel-wire/212500037/radio-shacks-99-acer-aspire-notebook-too-good-to-be-true.htm.

Kramer, Kenneth, Jason Dedrick, and Prakul Sharma. 2009. "One Laptop per Child: Vision vs. Reality." *Communications of the ACM* 52 (6, June): 66–73. http://dl.acm.org/citation.cfm?id=1516063.

Kuchinskas, Susan. 2007. "A Beginner's Guide to Web 2.0 Tools for Business." *BNET*, May. http://www.bnet.com/article/a-beginners-guide-to-web-20-tools-for-business/66096.

Mansfield, Ian. 2010. "Social Networking Dominates UK Mobile Broadband Traffic." *Cellular-News*, October 28. http://www.cellular-news.com/story/46136.php.

Merrill, Duane. 2009. "Mashups: The New Breed of Web App." *IBM developerWorks*, July 24. http://public.dhe.ibm.com/software/dw/xml/x-mashups-pdf.pdf.

Muller, Rudolph. 2011. "The Best Entry-Level Broadband Deal in SA." *My Broadband,* April. http://mybroadband.co.za/news/broadband/19633-The-best-entry-level-broadband-deal.html.

Obura, Fredrick. 2010. "ICT Board Sh320m Grant to Promote Local Content." *The Standard,* June 1. http://www.standardmedia.co.ke/InsidePage.php?id=2000010692&cid=14&story=ICT%20Board%20Sh320m%20grant%20to%20promote%20local%20content.

OECD (Organisation for Economic Co-operation and Development). 2008. "Broadband Growth and Policies in OECD Countries." Paper prepared for the OECD "Ministerial Meeting on the Growth of the Internet," Paris. http://www.oecd.org/dataoecd/32/57/40629067.pdf.

Oestmann, Sonja. 2003. "Mobile Operators: Their Contribution to Universal Service and Public Access." Intelecon Research and Consultancy, Vancouver, January.

OFCOM. 2011. Statistique officielle des télécommunications 2009. March 15, p. 58. http://www.bakom.admin.ch/dokumentation/zahlen/00744/00746/index.html?lang=fr&download=NHzLpZeg7t,lnp6I0NTU042l2Z6ln1ae2IZn4Z2qZpnO2Yuq2Z6gpJCDeoF6fmym162epYbg2c_JjKbNoKSn6A--.

O'Reilly, Tim. 2005. "What Is Web 2.0: Design Patterns and Business Models for the Next Generation of Software." *O'Reilly*, September 30. http://oreilly.com/web2/archive/what-is-web-20.hPal, Joyojeet. 2007. "Examining e-literacy Using Telecenters as Public Spending: The Case of Akshaya." Proceedings of the second IEEE/ACM "International Conference on Information and Communication Technologies and Development," Bangalore, December 15–16.

Pal, Joyojeet. 2007. "Examining e-literacy using Telecenters as Public Spending: The Case of Akshaya." Proceedings of the 2nd IEEE/ACM International Conference on Information and Communication Technologies and Development, December 15-16, Bangalore, India.

Pew Internet and American Life Project. 2010. *Home Broadband 2010: Trends in Broadband Adoption*. Washington, DC: Pew Internet and American Life Project, August. http://www.pewinternet.org/Reports/2010/Home-Broadband-2010/Part-1/Most-non-internet-users-have-limited-exposure-to-online-life.aspx.

Pimienta, Daniel, Daniel Prado, and Álvaro Blanco. 2009. *Twelve Years of Measuring Linguistic Diversity in the Internet: Balance and Perspectives*. Paris: United Nations Educational, Scientific, and Cultural Organization.

Redecker, Christine, Kirsti Ala-Mutka, Margherita Bacigalupo, Anusca Ferrari, and Yves Punie. 2009. "Learning 2.0: The Impact of Web 2.0 Innovations on Education and Training in Europe." Joint Research Centre, European Commission, Brussels.

Robertson, R. John. 2011. "Technical Standards in Education, Part 7: Web 2.0, Sharing, and the Open Agenda." *IBM developerWorks*, April 5. http://public.dhe.ibm.com/software/dw/industry/ind-edustand7/ind-edustand7-pdf.pdf.

Rodrigues, A. J. 2009. "ICT Research Capacity: Challenges and Strategies in Kenya." School of Computing and Informatics, University of Nairobi, March. http://euroafrica-ict.org.sigma-orionis.com/downloads/rwanda/Rodriques.pdf.

San Román, Edwin. 2009. "Bringing Broadband Access to Rural Areas: A Step by Step Approach for Regulators, Policy Makers, and Universal Access Program Administrators; the Experience of the Dominican Republic." Paper prepared for the ninth "Global Symposium for Regulators," Beirut, November 10–12.

Scott, Jeremy. 2011. "25 Jawdropping YouTube Video Facts, Figures, and Statistics." *Reelseo*, April. http://www.reelseo.com/youtube-statistics/.

Sri Lanka, Department of Census and Statistics. 2009. "Computer Literacy Survey." Department of Census and Statistics, Colombo. http://www.statistics.gov.lk/CLS/BuletinComputerLiteracy_2009.pdf.

Switzerland, OFCOM (Office Fédéral de la Communication). 2009. "Statistique officielle des télécommunications 2009." OFCOM, Bienne, Switzerland.

Thia, Tyler. 2011. "Social Media Most Evolved in Singapore." *ZDNet Asia*, February 10. http://www.zdnetasia.com/social-media-most-evolved-in-s-pore-62206580.htm.

Trucano, Michael. 2010. "Learning from a Randomized Evaluation of OLPC in Peru." "World Bank Blog on ICT Use in Education," December 10. http://blogs.worldbank.org/edutech/OLPC-peru.

UNESCO (United Nations Educational, Scientific, and Cultural Organization). n.d. "Information and Communication Technology in Education Statistics (ICT4E

Stats).” Institute for Statistics, Paris. http://siteresources.worldbank.org/
EDUCATION/Resources/278200-1121703274255/1439264-1247694138107/
6305512-1321460039533/05_NuneVoskanyan_InformationandCommunication
TechnologyinEducationStatistics.pdf.

United States, FCC (Federal Communications Commission). 2010. “National
Broadband Plan.” FCC, Washington, DC.

Vital Wave Consulting. 2009. “mHealth for Development: The Opportunity of
Mobile Technology for Healthcare in the Developing World.” United Nations
Foundation, Washington, DC; Vodafone Foundation, Newbury, U.K. http://www
.vitalwaveconsulting.com/pdf/mHealth.pdf.

Waterfield, Charles. 2004. “Virtual Conference on Electronic Banking for the Poor:
Final Report.” MicroSave, Nairobi. http://www.gdrc.org/icm/040412%20
Ebanking%20Conference.doc.

Wilhelm, Alex. 2010. “Chinese: The New Dominant Language of the Internet.” *The
Next Web,* December 21. http://thenextweb.com/asia/2010/12/21/chinese-the-
new-dominant-language-of-the-internet-infographic/.

World Bank. 2010. “Facilitating Broadband Development: Funding Options.” Global
Information and Communication Technologies, World Bank, Washington, DC,
August. http://www.itu.int/ITU-D/asp/CMS/Events/2010/ABBMN/S1A_Ms_
Tenzin_Norbhu.pdf.

CHAPTER 7

Global Footprints: Stories from and for the Developing World

Developing nations face many barriers to broadband development, on both the demand and supply sides.[1] These include a shortage of wireline infrastructure, constrained inter- and intramodal competition, low income, and limited awareness. Developing countries also often face challenges such as weak regulatory and legal frameworks and significant differences between rural and urban areas. As a result, they typically lag behind developed economies in broadband penetration (figure 7.1), although there are exceptions such as some nations in the Caribbean or the Arab Gulf states.

Many studies show that broadband can enable economic growth (see chapter 1). Broadband is also an agent for economic, social, and political development as well as a platform for innovation, an enabler of small and medium enterprise (SME) growth, and a facilitator of new firm foundation. This is particularly relevant for countries facing the challenge of development and looking to raise the standard of living of their citizens and foster social, human, and political progress. In that regard, it is useful to look at international objectives for promoting development and examine how broadband can be part of the strategy to achieve these goals.

Broadband has taken on increased relevance within the development community because of its potential to reduce poverty and better enable countries to participate in the global information society. International

Figure 7.1 Global Broadband Subscriptions per 100 People, Wireline and Wireless (Active), by Region and Income Level, 2010

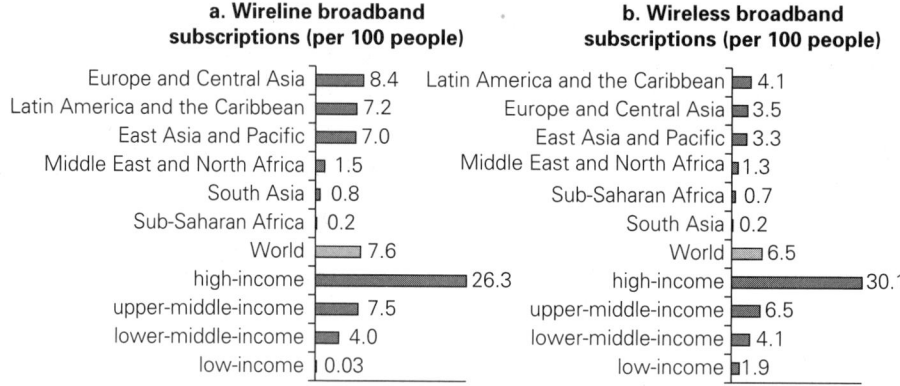

a. Wireline broadband subscriptions (per 100 people)

Europe and Central Asia	8.4
Latin America and the Caribbean	7.2
East Asia and Pacific	7.0
Middle East and North Africa	1.5
South Asia	0.8
Sub-Saharan Africa	0.2
World	7.6
high-income	26.3
upper-middle-income	7.5
lower-middle-income	4.0
low-income	0.03

b. Wireless broadband subscriptions (per 100 people)

Latin America and the Caribbean	4.1
Europe and Central Asia	3.5
East Asia and Pacific	3.3
Middle East and North Africa	1.3
Sub-Saharan Africa	0.7
South Asia	0.2
World	6.5
high-income	30.1
upper-middle-income	6.5
lower-middle-income	4.1
low-income	1.9

Source: World Bank analysis based on data from the International Telecommunication Union and ictDATA.org.

Note: In the case of wireless broadband, refers to subscriptions providing at least 256 kilobits per second (kbit/s) download speed. In the case of wireless broadband, refers to active subscriptions (using wireless broadband networks to access the Internet). Regions refer to developing-country members only (that is, non-high-income economies).

agreements on development and information and communication technologies (ICTs) provide a context for the significance of broadband in developing countries.

This chapter looks at the Millennium Development Goals (MDGs) and World Summit on the Information Society (WSIS) targets as a global roadmap for developing-country policy makers. It also reviews broadband bottlenecks and opportunities in developing nations, summarizes the broadband status of developing regions, identifies regional and national policies for boosting broadband penetration, and identifies groups of countries that face specific income, geographic, or other limiting conditions. The last section of the chapter provides summaries of broadband experiences in selected countries.

Broadband and Global Goals for Developing Countries

In September 2000, governments adopted the Millennium Declaration, committing their nations to reducing poverty monitored through measurable targets (box 7.1). The targets have a 2015 deadline and are known as the MDGs.[2] Several reports have illustrated how ICTs can help to achieve the

Box 7.1: The Eight Millennium Development Goals

- Eradicate extreme poverty and hunger.
- Achieve universal primary education.
- Promote gender equality and empower women.
- Reduce child mortality.
- Improve maternal health.

- Combat HIV/AIDS (human immunodeficiency virus/acquired immunodeficiency syndrome), malaria, and other diseases.
- Ensure environmental sustainability.
- Develop a global partnership for development.

Source: United Nations, MDG Monitor.

MDGs (for example, see ITU 2003; Broadband Commission 2010). Broadband is no different, and its impact on the MDGs may be greater than that of any other ICT. For example, one of the barriers to achieving Goal 2 on universal primary education is the lack of primary school teachers. Broadband, in particular, can facilitate fast-track teacher training through distance education and e-learning. In addition, three of the MDGs are related to health, and high-speed networks can have an impact through applications such as telemedicine. The importance of ICTs for achieving the MDGs is also highlighted by Goal 8 on developing a global partnership for development, specifically Target 8.F: "In cooperation with the private sector, make available the benefits of new technologies, especially information and communications."[3] As an ICT itself, as well as a "pipe" capable of delivering ICTs, broadband may be considered an integral part of Target 8.F.

The WSIS, which was held in two phases, in 2003 in Geneva and in 2005 in Tunis, set an internationally agreed agenda for the adoption of ICTs worldwide and illustrated the level of global political commitment to deploying broadband networks across different sectors.[4] The Declaration of Principles identifies ICTs as an "essential foundation for the information society," noting, "A well-developed information and communication network infrastructure and applications, adapted to regional, national, and local conditions, easily accessible and affordable, and making greater use of broadband and other innovative technologies where possible, can accelerate the social and economic progress of countries, and the well-being of all individuals, communities, and peoples." WSIS adopted 10 targets addressing connectivity across different sectors (box 7.2). The International Telecommunication Union (ITU) has reviewed progress toward the WSIS

Box 7.2: The 10 WSIS Targets

- Connect villages with ICTs and establish community access points.
- Connect universities, colleges, secondary schools, and primary schools with ICTs.
- Connect scientific and research centers with ICTs.
- Connect public libraries, cultural centers, museums, post offices, and archives with ICTs.
- Connect health centers and hospitals with ICTs.
- Connect all local and central government departments and establish websites and e-mail addresses.

- Adapt all primary and secondary school curricula to meet the challenges of the information society.
- Ensure that all of the world's population has access to television and radio services.
- Encourage the development of content and put in place technical conditions in order to facilitate the presence and use of all world languages on the Internet.
- Ensure that more than half of the world's inhabitants have access to ICTs within their reach.

Source: WSIS 2003.

targets and emphasized that most should be considered as having a broadband component:

> It is widely recognized that ICTs are increasingly important for economic and social development. Indeed, today the Internet is considered as a general-purpose technology and access to broadband is regarded as a basic infrastructure, in the same way as electricity or roads. . . . Such developments need to be taken into consideration when reviewing the WSIS targets and their achievement, and appropriate adjustments to the targets need to be made, especially to include broadband Internet (ITU 2010b, 2).

Taken together, the MDGs and WSIS targets provide a global roadmap for developing-country policy makers. Broadband can help to achieve the MDGs and thus place high-speed networks within the context of overall national development goals, while the WSIS targets can aid the monitoring of broadband deployment across different sectors. In other words, broadband is not an end unto itself but a means to an end (for example, broadband can be a means to achieving universal primary education).

Broadband Bottlenecks and Opportunities in Developing Regions

Improving access to broadband networks requires addressing supply- and demand-side bottlenecks. On the supply side, there are two broadband routes with different characteristics and market developments: wireline and wireless. The three main wireline broadband technologies in use are digital subscriber line (DSL), cable modem, and fiber to the premises (FTTP); figure 7.2. DSL is the predominant technology, accounting for almost two-thirds of wireline broadband subscriptions in 2010. Broadband access over cable television (CATV) networks is used by one in five subscriptions around the world. FTTP accounts for just 16 percent of global wireline broadband, but its share has grown since 2005, while the shares of DSL and cable modem have dropped.

Wireline broadband requires an underlying wired infrastructure. In the case of DSL, this consists of the copper lines used to connect subscribers to the telephone network. In the case of cable modem, it is the coaxial cable used to provide television (TV) access to subscribers (see chapter 5). Fiber optic broadband uses flexible glass enclosed by cables running directly to the home or building. Significant investments are required in order to

Figure 7.2   Global Distribution of Wireline Broadband Subscriptions, 2005 and 2010

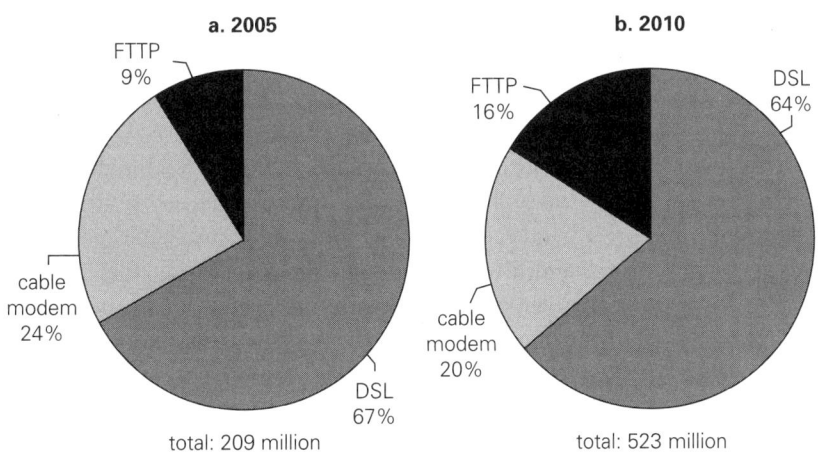

Source: Point-Topic.

deploy any wireline infrastructure. For developed countries that have been building out telephone and CATV infrastructure for decades, investment costs in these technologies have often already been recouped. Taking advantage of networks with greater capacity, however, requires additional investments in fiber optic networks. Many developing countries lack extensive wireline infrastructure, and investments in telephone, cable, and fiber optic networks often require new up-front costs.

Given these constraints, the wireless broadband route appears more promising for many developing nations and is especially attractive for serving nonurban populations. Although the deployment costs of mobile broadband are less than those of wireline, they are still significant. Converting mobile networks to broadband readiness requires investment in spectrum and equipment by operators and the purchase of new devices by users. This results in high costs, at least initially, making mobile broadband more expensive for end users than current wireless services. Other wireless options include technologies such as fixed wireless and satellite. Like mobile broadband, investments in spectrum and equipment are needed for terrestrial fixed wireless technologies, and it may not be feasible to leverage the existing mobile infrastructure in terms of towers and backbone networks. Satellite broadband is an option, particularly for remote locations, but it is more costly than other solutions for mass deployment and has usage limitations for some applications.

Conditions vary across the developing world, and each country is endowed with differing levels of communication networks. Some, such as Costa Rica or Croatia, have a relatively well-developed wireline telephone network that could support broadband deployment, while others, such as China and Romania, have widespread CATV networks that are able to provide a measure of facilities-based competition to telephone service operators. The challenge in such cases is to create incentives so that existing networks can be used to offer broadband services in competition with one another. In other countries, the challenge is to roll out broadband-capable networks from scratch. Lithuania, for instance, has focused on greenfield deployment of fiber to the premises; by 2010 around a quarter of Lithuanian homes had fiber broadband access, ranking the country sixth in the world.[5]

Diversity in broadband infrastructure creates a higher degree of intermodal competition. Therefore, countries should consider how they could leverage existing infrastructure to create greater competition in the broadband market. In 2009, the world was only using a little over one-fifth of telephone lines for DSL and around a third of CATV connections for cable broadband; just over 10 percent of mobile subscriptions were broadband (figure 7.3).

Figure 7.3 Broadband Connections Relative to Underlying Infrastructure in 2008 or 2009, by Region

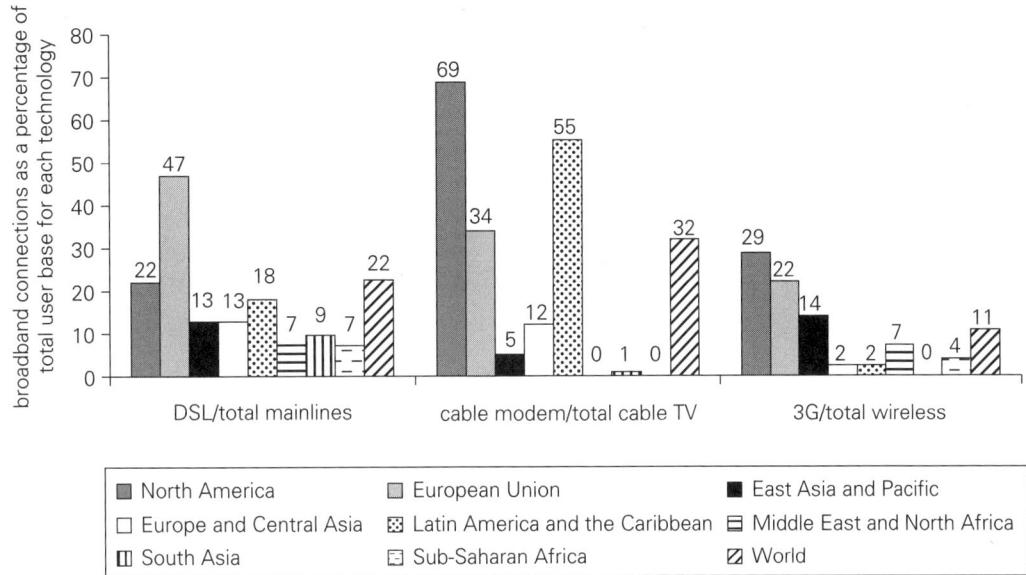

Sources: Data for DSL from TeleGeography's GlobalComms database (for 2008), data for third-generation (3G) wireless from Wireless Intelligence (for 2008), and data for cable modems from ictDATA.org (for 2009).

Broadband is also dependent on demand-side constraints such as accessibility to and affordability of broadband services as well as awareness of its benefits. Services, applications, and content are key drivers: they need to be interesting, in the local language, and locally relevant. If these demand-side issues are not tackled, a country risks creating a mismatch between supply and demand and will not be able to fulfill its broadband potential. As shown in chapter 1, a country's level of income affects the ability to pay for broadband services, while education levels affect awareness. Figure 7.4 illustrates the significant relationship between broadband take-up and the United Nations Development Programme's Human Development Index.

Although developing nations face supply- and demand-side bottlenecks in their broadband markets, they represent some of the fastest-growing markets and offer great potential as ICT uptake and broadband deployments grow. According to Point-Topic, a broadband market analyst company, the countries ranked as the top 10 fastest-growing broadband markets are all emerging economies, and all saw more than 20 percent growth in the number of broadband subscriptions in 2010 (figure 7.5).

Figure 7.4   Broadband and Human Development, 2010

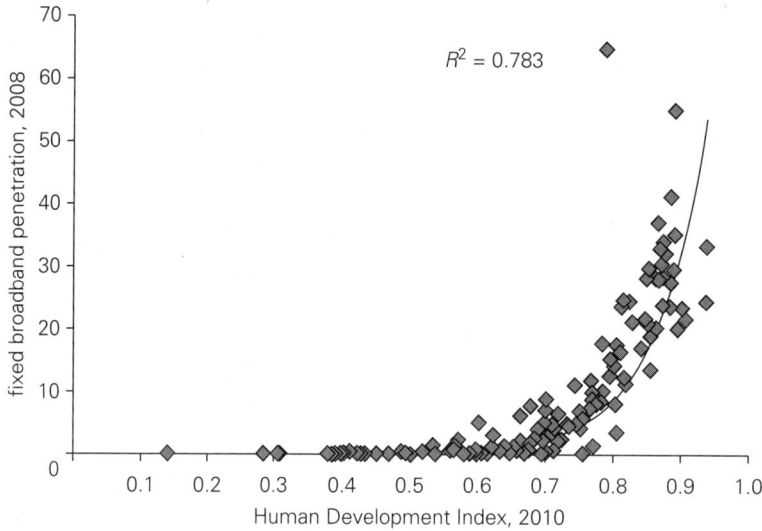

Source: United Nations Development Programme, http://hdr.undp.org/en/statistics/.

Figure 7.5   Growth in Wireline Broadband Subscriptions in the Countries with the Fastest-Growing Broadband Markets, 2010

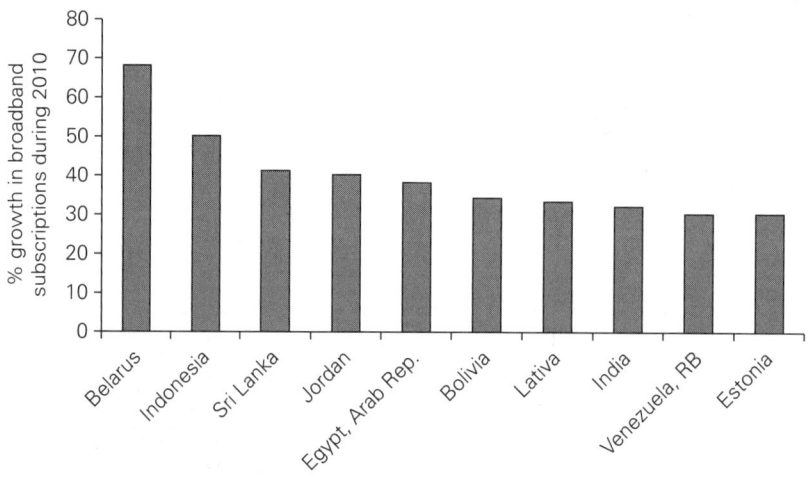

Source: Broadband Forum, citing Point-Topic, http://www.broadband-forum.org/news/download/press releeases/2011/India2011.pdf.

Regional Developments

This section highlights broadband status in developing regions around the world, including East Asia and the Pacific, Europe and Central Asia, Latin America and the Caribbean, Middle East and North Africa, South Asia, and Sub-Saharan Africa.

East Asia and the Pacific

East Asia and the Pacific region is home to world broadband leaders such as the Republic of Korea; Hong Kong SAR, China; and Japan, where super-high-speed access is increasingly becoming the norm and competitive markets have been stimulated through disruptive new entrants (box 7.3). However, a wide broadband divide distinguishes "the mostly high-income countries that are broadband leaders from the mostly middle- and low-income countries that are broadband challenged" (ESCAP 2010).

Box 7.3: The Third Man: Encouraging Disruption in Broadband Markets

Growth in some of the more successful developed-economy broadband markets has been triggered by the entry of brand new disruptive operators. These new service providers tend to be the third player entering the market, shaking up duopolies of DSL and cable broadband operators or a dominant incumbent and a major wireless operator. This is the case in the developed East Asian economies of Korea, Japan, and Hong Kong SAR, China, where new operators entered the broadband market with innovative business plans and models, unsettling the market and triggering a beneficial stimulus to broadband growth.

- Hanaro entered the market in 1999 as a facilities-based telephone operator in competition with the incumbent Korea Telecom. Soon after entry, Hanaro began offering broadband DSL services resulting in intensive competition, a major factor in Korea's rise as a top-ranked broadband country. Hanaro had captured a fifth of the broadband market by 2010.

- Softbank entered the Japanese broadband market in 2001 by leasing unbundled local loop lines from the incumbent telephone operators, and in 2004, it obtained a facilities license and began deploying its own infrastructure. Softbank acquired Japan's third largest mobile operator in 2006, allowing it to enter the mobile broadband market. Marketing its service as Yahoo!BB, Softbank had an 11 percent share of the broadband market in 2010, and over a third of its subscribers were getting speeds of 50 megabits per second (Mbit/s). According to the company,

(continued)

Box 7.3 *continued*

"It is not an exaggeration to say that the wireline broadband service in Japan was created by the Softbank Group."

- Hong Kong Broadband Network (HKBN) entered the market in 2000 after it was awarded a fixed wireless license. The city's compact high-rise building environment shaped HKBN's technological strategy of installing in-building wiring; communications between buildings and HKBN's routers and switches were carried out using wireless transmission through rooftop antennas. HKBN was able to penetrate the market quickly and shook up the quasi-duopoly between the incumbent wireline operator and CATV company for broadband provision. HKBN later acquired a wireline license and once again is shaking up the market by deploying fiber optic to the home. It had a 25 percent share of the wireline broadband market by 2010.

The process of disruption has also occurred in some European markets where alternative operators initially entered using the infrastructure of incumbent operators and then, having established a foothold, began investing in their own infrastructure. This is the case in France and Italy:

- Free started as a dial-up operator in France in 1999 and began providing broadband services in 2002 using asymmetric DSL (ADSL) over France Telecom's unbundled local loop (ULL). In 2006, it began rolling out its own fiber to the home (FTTH) network and intends to cover 4 million homes by 2012, representing an investment of about €1 billion. Free has been providing triple-play services since December 2003. Its Internet Protocol television (IPTV) service offers over 300 channels, and Free's broadband speeds range between 22 and 28 Mbit/s. In 2009 it was awarded the country's fourth third-generation (3G) license. Free had 22 percent of the French wireline broadband market in 2010.

- In Italy, FASTWEB started by deploying a fiber optic network in Milan. In 2001, it began providing triple-play services using DSL over Telecom Italia's infrastructure. The company has partnered with other operators in a Fiber for Italy project where they will pool resources to provide FTTH in Italy's 15 largest cities, an investment expected to cost €2.5 billion. Meanwhile, FASTWEB has also been building its own FTTH network that passes nearly 2 million homes, offering speeds of up to 100 Mbit/s. FASTWEB had 13 percent of the wireline broadband market in 2010.

Market-disruptive operators are spreading to emerging and developing economies:

- Starnet entered the Moldovan market in 2003, providing ADSL over the incumbent's telephone network. In 2006, Starnet began providing voice over broadband (VoB) and also started the construction of its fiber optic network. In 2009, IPTV was added to its portfolio, and by the end of 2010, Starnet had captured one-quarter of the wireline broadband market.

- In Morocco, Wana was awarded wireless broadband spectrum in 2006. A company owned by national investors, it launched services in 2007 using high-speed Evolution Data Optimized (EV-DO) technology.

(continued)

This resulted in intense competition with the existing mobile operators and led to the rapid adoption of 3G services, which soon passed wireline broadband subscriptions. By the end of 2010, there were 1.4 million 3G subscribers in Morocco, almost three times the number of wireline broadband connections. Wana had 41 percent of the mobile broadband market.

The lesson for developing countries is that, while it is critical to open broadband markets to competition, it is just as important to introduce brand new operators. Setting aside spectrum for a new operator and lowering other market entry barriers, particularly those relating to the ability to provide convergent services, can encourage this.

Source: Adapted from ITU 2008 and operating reports of the companies discussed and regulatory authorities for broadband market shares; Softbank 2007; FASTWEB, "2010 NGN and Executive Customers Highlights of the Year," http://company.fastweb.it/index.php?sid=6.

Several developing countries in the region have deployed telephone and CATV network infrastructures, but often they are not adequately upgraded for wireline broadband access. For example, the region's developed economies have been successful in developing broadband access through CATV network infrastructure. This is not the case in the region's developing nations. Despite large CATV markets in some countries such as China, the Philippines, and Thailand, broadband competition from CATV providers is generally low. One reason is that networks have not been upgraded to support broadband access via cable modem. For example, despite having the world's largest CATV market, with almost 175 million subscribers in 2009, China has relatively few cable modem subscriptions, and only about a quarter of its subscriptions are digital. This is likely to change with China's new Triple Network project announced in 2010.[6] The project aims to enhance convergence among telecommunications, Internet, and broadcast networks by reducing barriers so that each market segment can provide any broadband service.

Most East Asian nations have licensed mobile broadband spectrum, and in several of the region's developing nations, mobile broadband subscriptions exceed wireline subscriptions. In Indonesia, Telkom had 3.8 million mobile broadband subscriptions using data cards, compared with 1.6 million wireline broadband subscriptions in December 2010 (Telkom Indonesia 2010). Mobile broadband coverage, however, still needs to be extended throughout the region, mainly from urban to rural areas.

Malaysia's 2006 Information, Communications, and Multimedia Services 886 Strategy set several goals for broadband services, including an increase in broadband penetration to 25 percent of households by the end of 2006 and 75 percent by the end of 2010 (Kim, Kelly, and Raja 2010). Despite growth, the ambitious target for 2010 has not yet been met, and the government is now focusing on fixed wireless, 3G mobile, and fiber to the home platforms to boost broadband adoption. To that end, it is funding a fiber optic network that will connect about 2.2 million urban households by 2012. The network will be rolled out by Telekom Malaysia under a public-private partnership (PPP). The government will invest RM 2.4 billion (US$700 million) in the project over 10 years, with Telekom Malaysia covering the remaining costs. The total cost of the project is estimated to be RM 11.3 billion (US$3.3 billion).

Connecting the Pacific region with broadband is a major challenge due to its unique geographic challenges. It is critical for Pacific economies to gain access to adequate bandwidth for supporting broadband development. Many of the island nations in the region are widely dispersed, and backbone networks are limited. Most countries rely on high-cost, limited-capacity satellites, and only a few economies have access to fiber optic submarine cables. The subregion has also been slow to develop mobile broadband, a consequence of previously limited competition in mobile markets. However, several countries now have competitive mobile markets, which should spur deployment of high-speed wireless networks (Howes and Morris 2008). Vietnam has made impressive strides in boosting international high-speed connectivity and broadband use. The case of Vietnam is highlighted in the final section of this chapter.

Europe and Central Asia

The region is well positioned to promote broadband adoption, with populations enjoying relatively high levels of education and significant existing wireline and cable television network build-out. However, the region is highly diverse with regard to geography, integration, and income, making it difficult to reach a common vision for broadband strategy. It ranges from countries with large sea coasts to landlocked nations and from the Baltic States to the Balkans and Eastern Europe to Central Asia, from members of the Commonwealth of Independent States to the European Union, and from low-income to high-income economies.

Several countries in the region adopted broadband strategies within the framework of national ICT plans. Most of the plans were launched in the early to mid-2000s and coincided with significant increases in broadband penetration. For example, in 2005, Moldova adopted its Information Society Strategy, which incorporated tracking indicators to monitor the impact of policies and programs on broadband access (Moldova, Government of 2005). Broadband penetration in Moldovan households rose from less than 1 percent in 2003 to 17 percent by 2009 (Moldova, ANRCETI 2010). International bandwidth availability rose significantly in the landlocked country following completion of an optical fiber connection to Romania (box 7.4).

Some of the countries in the region rank among the top countries in the world in broadband deployment and average download speeds. However, many landlocked countries in Central Asia face the challenge of ensuring that regional broadband backbones keep up with the region's growing ICT needs. Within that context, the Economic and Social Commission for Asia and the Pacific (ESCAP) undertook a feasibility study in four countries: Kazakhstan, the Kyrgyz Republic, Tajikistan, and Uzbekistan (ESCAP

Box 7.4: Impact of Improved Access to International Connectivity: The Case of Moldova

Until April 2010, Moldova's international connectivity market was entirely controlled by state-owned incumbent Moldtelecom. Due to this and because Moldova is a landlocked country, private firms did not have direct access to the Internet. At that time, the government reformed its policy and procedures to open the market to competition. By July 2010, three companies—mobile telephony provider Orange and Internet service providers (ISPs) Starnet and Norma—successfully applied to construct and operate cross-border fiber optic cables and gain direct access to carriers via Romania.

The impact of liberalization on availability, prices, and quality was immediate. International Internet bandwidth available in Moldova went from 13 gigabits per second (Gbit/s) in December 2009 to over 50 Gbit/s

in July 2010. In response, Moldtelecom dropped the prices for wholesale connectivity by a third over that same time, with some of this drop coming in anticipation of the liberalization in late 2009. Retail subscribers in some parts of the country have already seen their available bandwidth double, while subscription rates have remained the same.

As Moldova looks to establish its position as an ICT hub in Eurasia, this move marks the first step toward connecting Moldova's fledgling information technology (IT)–based services to global markets. Improved connectivity will allow SMEs to connect with new markets at lower prices and enhance their competitiveness. However, the country needs to inject greater competition by removing all entry barriers.

Source: World Bank analysis.

2009). Results of the study were issued in the report "Broadband for Central Asia and the Road Ahead," which included the findings that these countries have low Internet access speeds coupled with high costs to consumers for broadband services, which has affected adoption (ESCAP 2009). As such, ESCAP recommended that expansion of a regional broadband network is "a fundamental element" to satisfying broadband demand in these countries.

Turkey's government recognizes the importance of a vibrant telecommunications market and is keen to promote the spread of broadband. For instance, many educational institutions now have broadband access. The Information Society Strategy for 2006–10 aimed to develop regulation for effective competition and to expand broadband access. Targets included extending broadband coverage to 95 percent of the population by 2010 and reducing tariffs to 2 percent of per capita income. The regulator has also looked at issuing licenses for the operation of broadband fixed wireless access networks in the 2.4 gigahertz (GHz) and 3.5 GHz bands. The case of Turkey is highlighted in a case study later in this chapter.

Latin America and the Caribbean

The Latin America and the Caribbean region has a relatively high number of wireline telephone lines and CATV subscribers compared to other developing regions. Cable broadband has been particularly successful, with over half of the subscribers having a broadband subscription. In contrast, the number of telephone lines being used for broadband (via DSL) is relatively low.

Mobile broadband development initially lagged compared to other regions. One factor related to delays in the award of new spectrum bands used specifically for 3G services. However, this was mitigated somewhat by policies throughout the region that allow operators to use their existing 850/900 megahertz (MHz) spectrum, originally allocated for voice, for high-speed mobile data services. Compared to the typical frequencies awarded in many countries for mobile broadband, these frequencies support wider coverage with fewer base stations so that investment costs are lower (Roetter 2009).

On the demand side, Latin America and the Caribbean fares favorably compared to other developing regions. Education levels are relatively high, and the existence of common languages throughout many countries—Spanish in Latin America and English in much of the Caribbean—results in access to considerable content, spurring demand. Despite relatively high per capita income for a developing region, incomes are highly skewed, and affordability remains an issue. For example, over half of Mexican households reported that they did not have Internet access in 2009 because they could not afford it (Mexico, INEGI 2009).

In November 2010, ministers at the Third Ministerial Conference on the Information Society of Latin America and the Caribbean adopted eLAC2015, a regional roadmap for the information society highlighting six goals for universal broadband access in the region (box 7.5). eLAC2015 considers broadband pivotal, noting,

---

Box 7.5: eLAC2015 Universal Broadband Access Goals

- *Goal 1.* Increase direct investment in broadband connectivity to make it available in all public establishments.

- *Goal 2.* Advance toward universal availability of affordably priced broadband connectivity in homes, enterprises, and public access centers to ensure that, by 2015, at least 50 percent of the Latin American and Caribbean population has access to multiple convergent interactive and interoperable services.

- *Goal 3.* Coordinate efforts to bring down the costs of international links by means of a larger and more efficient regional and subregional broadband infrastructure, the inclusion of (at least) the necessary ducts for fiber optic cables in regional infrastructure projects; the creation of Internet exchange points (IXPs); the promotion of innovation and local content production; and the attraction of content suppliers and distributors.

- *Goal 4.* Collaborate and coordinate with all regional stakeholders including academia and business, the technical community, and organizations working in the field, such as the Latin American and Caribbean Internet Addresses Registry and the Internet Society, to ensure that Internet Protocol version 6 (IPv6) is broadly deployed in the region by 2015; and implement, as soon as possible, national plans to make government public services portals in Latin America and the Caribbean accessible over IPv6 and to make public sector networks native IPv6 capable.

- *Goal 5.* Harmonize indicators that provide an overview of the situation of broadband in the region, in terms of both penetration and uses of applications, in accordance with international standards.

- *Goal 6.* Promote ICT access and use by persons with disabilities, with emphasis on the development of applications that take into account standards and criteria on inclusion and accessibility; in this connection, promote compliance by all government web portals with the web accessibility standards established by the World Wide Web Consortium.

Source: eLAC2015.

For the countries of Latin America and the Caribbean, the universalization of broadband access in the twenty-first century is as important for growth and equality as were electric power and road infrastructures in the twentieth century. Broadband is an essential service for the economic and social development of the countries of the region, and it is indispensable for progress, equality, and democracy. That is why the strategic goal is for broadband Internet access to be available to all of the citizens of Latin America and the Caribbean (ECLAC 2010).

Chile was the first Latin American country to announce a national broadband strategy. The strategy identified ICT as a priority for economic development. Chile has also planned and implemented ICT policies from both the supply and demand sides. The demand-side strategy has included programs for e-literacy, e-government, and ICT diffusion. For example, almost all taxes are filed electronically, and government e-procurement more than doubled the volume of transactions processed between 2005 and 2008. The government has also promoted broadband use by municipalities. By 2008, almost all municipalities had Internet access, and 80 percent had websites. In order to reach the objectives of a digital Chile, the government's broadband goal is to double broadband connections and complete nationwide coverage by 2012.

Brazil is one of the few countries in the region with a specific broadband plan, while St. Kitts and Nevis has the highest broadband penetration in the region. The cases of these countries are highlighted at the end of this chapter.

Middle East and North Africa

The Middle East and North Africa region is relatively well equipped with wireline telephony for a developing region, and most wireline broadband is primarily via ADSL. Nonetheless, prospects for wireline broadband are constrained. Few alternative wireline operators have deployed copper line infrastructure, and local loop unbundling (LLU), for the most part, is not available across the region. Further, the development of intermodal competition through CATV is inhibited by the popularity of satellite television, widely available at no charge through the informal market. Most new entrants to the traditional telephony market have been wireless based.

A report analyzing the main factors affecting broadband demand in many of the countries in the region identified challenges hindering broadband deployment and suggested recommendations to overcome them (ESCWA 2007). Challenges include high retail prices, poor regional and international connectivity, limited wireline access infrastructure, lack of and restrictions on content, high cost of personal computers, and limited competition. The report's overarching conclusion was the need for convergence—through bundled offers and transition to IP-based networks—which would trigger mass broadband adoption.

Many, but not all, of the countries have awarded spectrum for mobile broadband services. Morocco, for example, was one of the first countries to award 3G frequencies in the region. It did so through a beauty contest, which lowered spectrum costs for operators. Some of the spectrum was awarded to a new operator, shaking up the existing duopoly and triggering intense competition in the mobile broadband market (box 7.3). As a result, mobile broadband subscriptions in Morocco have surpassed wireline connections. The country has adopted the Maroc Numérique 2013 Strategy with targets for providing broadband to all schools and one-third of households by 2013 (Morocco, Ministry of Industry, Commerce, and New Technologies n.d.). Morocco is highlighted in a case study at the end of this chapter.

Most countries in the region share a common language, which facilitates collaboration on developing digital Arab content to improve demand for broadband.[7] The Jordanian minister of information and communications technology has outlined the importance of the content industry as a main driver of Internet penetration, especially as it relates to local and Arabic content. The digital content industry in Jordan received a boost in 2009, when chipmaker Intel announced plans to invest in two digital content companies: Jeeran and ShooFeeTV.[8] The funding will be used to help both companies to pursue regional growth as well as extend their product offerings. Jeeran is the largest user-generated content site in the Arab world, reaching 1 million members and 7 million unique visitors per month.[9] ShooFeeTV provides online information for more than 120 Arab satellite channels, including listings, programming information, celebrity news, pictures, and video clips.[10] Global social networking sites such as Facebook and Twitter have also grown in popularity, as reflected in their extensive use during the so-called Arab Spring in 2011. The number of Facebook users in the Arab region grew 78 percent in 2010, while in Tunisia the proportion of Facebook users increased 8 percent in the first two weeks of January 2011 following the beginning of demonstrations.[11]

South Asia

South Asia faces severe supply- and demand-side constraints in promoting broadband access. In absolute terms, there is a significant base of wireline telephone lines and cable television subscribers. India has the third largest wireline telephone network (measured by subscriptions) in the developing world, and Pakistan has the fourteenth largest. In terms of CATV subscriptions, India ranks second and Pakistan ranks third among developing nations. Nevertheless, wireline infrastructure is relatively limited compared to other regions, and the number of telephone lines and CATV connections for broadband services is relatively low. Some countries have been late to award mobile broadband spectrum that would trigger intermodal broadband competition. On the demand side, the region is the second poorest developing region after Sub-Saharan Africa, and levels of education are relatively low.

India was the first country in the region to adopt a broadband policy in 2004 (India, Ministry of Communications and Information Technology 2004). However, it has not achieved the goals set. The country published a consultative document on a new broadband policy, and in December 2010 the Telecommunications Regulatory Authority of India (TRAI) issued broadband recommendations.[12] A key strategy is to develop an open-access national fiber optic backbone network connecting all localities with more than 500 inhabitants by 2013.

Pakistan published a broadband policy in 2004 (see Pakistan, Ministry of Information Technology 2004). But broadband deployment has not lived up to expectations—the number of broadband subscribers in 2007 was only half of the level targeted for that year and well short of the half million targeted for 2010. In an effort to accelerate broadband take-up, a universal service fund (USF) is being used to subsidize the deployment of broadband throughout the country.[13]

Other South Asian nations have also adopted or are developing broadband plans (ITU 2010a). However, programs that would address demand-side affordability issues are limited (table 7.1).

Sri Lanka, which was one of the first countries in South Asia to award 3G spectrum, has the second highest penetration, the lowest tariffs, and the fastest mobile broadband speeds in the region. Sri Lanka's broadband experience is highlighted in a case study at the end of this chapter.

Sub-Saharan Africa

The Sub-Saharan Africa region faces tremendous barriers in broadening access to broadband. It starts from a very low base, with limited wireline telephone networks and practically no CATV networks on the supply side, coupled with demand-side bottlenecks including the lowest per capita income and fewest years of schooling of all developing regions.

Table 7.1 Broadband Plans and Policies in Selected South Asian Nations

Country	Is there a national broadband plan?	Does universal service include broadband?	Are there other financing mechanisms for broadband?	Are there social tariffs for broadband subscribers?
Afghanistan	Under development	No	Telecommunications Development Fund	No
Bangladesh	Yes	No, but foreseen in the National Broadband Policy	No	No
Bhutan	Yes	No	No	No
Maldives	No	No	No	Yes, for education
Nepal	Under development	Yes, in rural areas. Universal service obligation was imposed on the incumbent wireline operator and financed through the Universal Service Fund and interconnection charges	Tax exemption for telecom equipment imported for rural services	No

Source: ITU 2010a.

Over the past decade, a large amount of private investment, driven by sector liberalization and competition and major advances in cellular technology, has brought mobile services within reach of the majority of Africa's population. The region's focus, thus far, on mobile networks to address an immediate service need has heightened the need for development of backbone networks capable of supporting broadband. This has created a major bottleneck in the rollout of high-bandwidth services and in the upgrading of cellular networks to provide value added services (Williams 2009). Overcoming this infrastructure hurdle is an important element in shaping the structure and policy framework of the telecommunications services sector. Without it, broadband will remain expensive and limited to businesses and high-income customers. Backbone constraints will also limit access speeds, affecting quality.

The backbone deficit has been acutely felt in international bandwidth. Due to limited local content, most Internet traffic is directed at countries outside Sub-Saharan Africa. Unfortunately, a lack of international high-speed fiber optic capacity has meant that even where countries have been able to deploy local access broadband infrastructure, performance is affected by slow international connectivity. Where connectivity exists, cable theft continues to be a major problem for reliability. In addition, there are relatively few national IXPs in the region, forcing even intraregion traffic to be hauled outside the region for switching and then sent back.

Until 2009, South Atlantic 3/South Africa Far East (SAT3/SAFE) was the only major regional submarine optic cable serving the continent, and it was limited to a few countries on the west coast. Other countries had to use more costly and slower satellite links. This has changed dramatically since the arrival of several new undersea cable systems—The East African Marine System (TEAMS), Southern and East African Cable System (SEACOM), and Eastern Africa Submarine Cable System (EASSy)—including the first system to the region's east coast (that is, TEAMS). Total capacity rose by a factor of 8.5 in 2009, and additional planned cables are expected to increase undersea capacity to over 20 terabits per second (Tbit/s) by 2012.[14]

International connectivity is just part of the supply chain. Sub-Saharan African countries also need to ensure that bandwidth gets disbursed throughout the country, and, in the case of the region's landlocked countries, national backbones must be in place to connect to neighboring countries. PPPs may be helpful to generate the necessary investment and to ensure an effective and open-access operating arrangement. The Kenyan government, for example, has supported open access to backbone infrastructure in various ways. It encouraged operators to participate in the TEAMS undersea cable and has also pursued public-private partnerships for national backbone construction. It is now contemplating the same for the construction of broadband wireless networks using Long-Term Evolution (LTE) technology. See the Kenya case study at the end of this chapter.

At the local access level, mobile broadband holds great promise. However, outside of a few countries, the region has yet to exploit this on a significant scale. Around two dozen Sub-Saharan African countries had commercially deployed 3G networks at the end of 2010, with around 9 million subscriptions.

Few African countries have elaborated a specific broadband policy. If mentioned at all, broadband is touched upon in overall sector strategies. One exception is South Africa, where the Broadband Policy for South Africa was published in July 2010 (South Africa, Department of Communications 2010). Defining broadband as speeds of at least 256 kbit/s, the government has identified two targets for 2019: all inhabitants to be within 2 kilometers of a public broadband access point and a household broadband penetration rate of 15 percent.

Countries in Special Circumstances

In addition to regional groups, countries are also classified by particular economic, geographic, and political situations. This section identifies several groupings relevant to the international development community and how the specific characteristics of that group can affect broadband development.

Least Developed Countries

The United Nations created the least developed countries (LDCs) category in 1971 to recognize the existence of a group of countries with severe poverty and weak economic, institutional, and human resources.[15] This group consists of 49 countries with a combined population of 815 million in 2008. Of these, 15 are located in Africa, 12 in Asia, two in Latin America, and two in Central and Eastern Europe. Around half are either small islands states or landlocked.

LDCs face tremendous supply- and demand-side challenges in deploying broadband networks. The existing level of wireline infrastructure is low, as are demand-side indicators such as incomes and education levels. The capacity for developing effective broadband strategies and policies is also limited due to institutional weaknesses and insufficient human resources in ministries and regulators.

New technologies such as broadband can help LDCs to overcome development challenges and move away from their dependence on primary commodities and low-skill manufacturing (UNCTAD 2010). There has been some urgency to deploy broadband networks in order to mitigate LDCs falling further behind technologically and becoming even more marginalized in the world economy (UNCTAD 2007). The development of international and national backbones is a main priority that will likely require innovative PPPs. Wireless broadband holds great promise given the significant increase in mobile networks in the LDCs and the lower costs of deploying wireless broadband compared to wireline infrastructures. LDCs will need to introduce greater competition and allocate spectrum for wireless broadband services in order to encourage the deployment of these technologies.

Landlocked Developing Countries

Landlocked developing countries (LLDCs),[16] predominantly located in Sub-Saharan Africa and Asia, "face severe challenges to growth and development due to a wide range of factors, including a poor physical infrastructure, weak institutional and productive capacities, small domestic markets, remoteness from world markets, and a high vulnerability to external shocks."[17] There are 31 LLDCs, with a total population of 370 million in 2008.

Given their status as LLDCs, the main obstacle for these countries is distance from key ports, causing high transaction costs and reducing international competitiveness. These geographic conditions pose a supply-side challenge for LLDCs in terms of global connectivity through high-speed fiber networks.

"Virtual coastlines" can be created for LLDCs through the connection of national backbones to countries directly linked to undersea cables. This connectivity can then be brought to "virtual landing stations" in the LLDC where all ISPs gain cost-based access to international bandwidth. Rwanda has created a virtual landing station, where optic fiber cables from undersea landing stations in Kenya and Tanzania (Rwanda's "virtual coastline") are terminated (Kanamugire n.d.).

Access to high-speed international bandwidth will require regional cooperation and PPPs to spur investment in national backbones and ensure onward connectivity to neighboring countries with undersea fiber optic cable. According to an ESCAP study on Central Asia, countries must cooperate to expedite and ensure effective regional connectivity (ESCAP 2009). Broadband backbone infrastructure that transcends borders requires interconnection. Along with management and maintenance, the need for interconnection affects all the countries benefiting from the network.

Small Island Developing States

The United Nations has recognized the particular problems of small island developing states (SIDSs) since 1994.[18] According to the United Nations Conference on Trade and Development (UNCTAD), SIDSs face "a greater risk of marginalization from the global economy than many other developing countries" due to their small size, remoteness, and vulnerability to external shocks.[19] They are also susceptible to natural disasters such as tsunamis and damaging environmental changes such as sea-level rise. There are 38 United Nations members classified as SIDS, with a population of 55 million in 2008. Over one-quarter of SIDSs are also LDCs.

Broadband connectivity can help to overcome these challenges in several ways, such as economic diversification through establishment of IT-enabled industries, creating a virtual closeness to the rest of the world, and real-time weather modeling and monitoring. Additionally, tourism has a big economic impact in many SIDSs, and broadband plays a vital role for various travel applications such as reservation systems and marketing.

The SIDSs are geographically diverse, with different broadband supply and demand challenges. On the demand side, many SIDSs have relatively small populations, which may deter investment. However, the small geographic areas of SIDSs often make it easier and cheaper to deploy networks quickly with a high degree of coverage, and a growing number of SIDSs are achieving universal mobile service.[20] On the supply side, most of the Caribbean SIDSs are located in a condensed area, crisscrossed by a number of undersea fiber optic cable networks. Pacific SIDSs tend to be more spread out. Since there are far fewer options for access to undersea fiber optic cables, most Pacific SIDSs are dependent on more expensive satellite solutions. Some Pacific SIDSs, such as Fiji, are served by undersea cables and therefore are in a position to become a potential fiber hub to neighbors (World Bank 2009).

Most of the Caribbean SIDSs introduced competition in telecommunications networks several years ago, whereas the Pacific countries have done so only more recently. Mobile broadband has yet to have a significant impact in most SIDSs due to a lack of spectrum allocation and uncertain demand.

The Eastern Caribbean Telecommunications Authority (ECTEL) was established as a regional regulator for countries in that subregion. ECTEL overcomes human resource limitations of each country staffing its own full-fledged regulatory institution and harmonizes subregional policies. ECTEL recently moved to make high-speed Internet more accessible by designating the 700 MHz band for broadband wireless services (EC-TEL 2009). St. Kitts and Nevis, a Caribbean SIDS, is profiled in a broadband country case study at the end of this chapter.

Postconflict Countries

Postconflict countries refer to nations where war and civil strife have led to the destruction of institutions and economic facilities. There is no official definition of a postconflict economy, but such economies are often locations where civil conflicts have necessitated the intervention of peacekeeping missions.[21] ICTs can play a beneficial role in helping to reconstruct these countries by attracting foreign investment, generating employment, enhancing education prospects, and creating linkages to the global economy.[22] Given the often poor or destroyed telecommunications infrastructure, postconflict countries can leapfrog to state-of-the-art, next-generation networks. However, this will require a liberalized telecommunications regime that encourages convergence and investment in IP networks.

In Afghanistan, for example, years of civil strife destroyed much of the economy, shutting down most government institutions, including schools. A project sponsored by the North Atlantic Treaty Organization has installed broadband access in universities using satellite technology.[23] This has overcome shortages of learning materials and teachers since professors and students can download teaching information and use online learning tools. In East Timor, the Australian government has been assisting with the development of the new country's media sector by providing journalists with the ability to upload and research news through the establishment of broadband centers.[24]

One notable development in some postconflict countries is the stunning result of private sector investment in ICTs. Private investors have been willing to take risks in highly unstable environments such as Afghanistan and Iraq. Starting from a very low base, these countries now have growing levels of mobile access and are expanding into wireless broadband solutions. The case of Sri Lanka, a country emerging from a decades-long civil conflict, is highlighted in a broadband study at the end of this chapter.

Broadband Experiences in Selected Countries

This section summarizes the results of various countries' broadband experiences, as commissioned for the Broadband Strategies Toolkit.[25] The countries studied cover a range of regions and development status, as shown in figure 7.6. Additionally, examples of efforts to address supply-side and demand-side issues in each of the studied countries are outlined in table 7.2.

Brazil

Brazil is the world's sixth most populous nation, so unsurprisingly, it is among the top countries ranked by total number of broadband subscriptions (Jensen 2011). At the end of 2010, Brazil was in ninth position, with 15 million fixed broadband subscribers as well as 20 million mobile broadband subscribers. Despite the size of the Brazilian broadband network, penetration is relatively low given its large population. Two key constraints include a shortage of fixed broadband infrastructure and wide income disparities in the country. The level of wireline infrastructure is relatively low for fixed broadband services. Competition in the fixed telephone line sector is low, and penetration has been falling due to mobile substitution. Fixed broadband is also available through cable modem, but growth has been limited due

Figure 7.6 Broadband Country Summaries

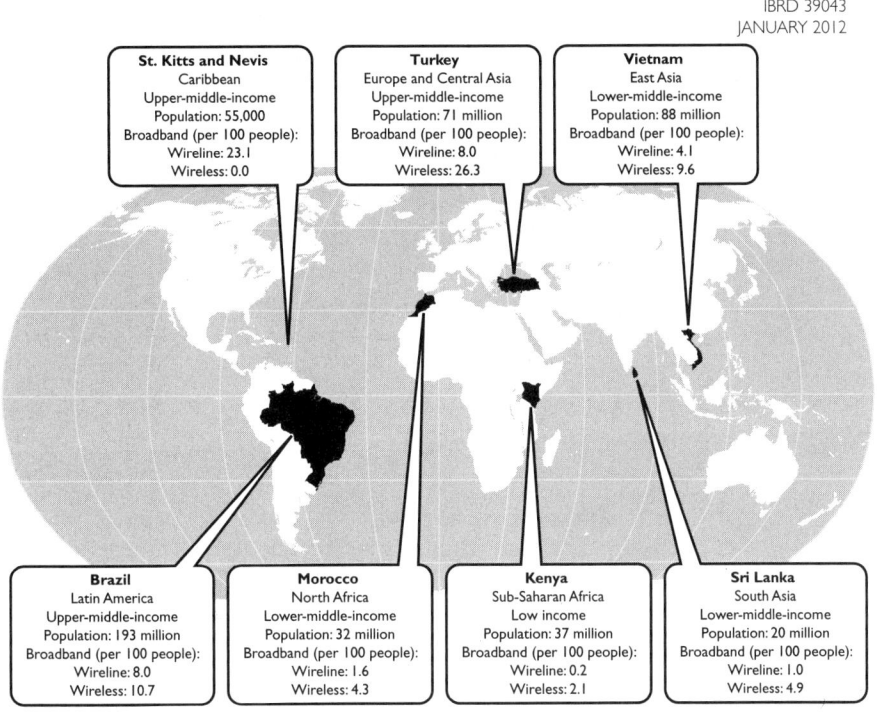

IBRD 39043
JANUARY 2012

St. Kitts and Nevis
Caribbean
Upper-middle-income
Population: 55,000
Broadband (per 100 people):
Wireline: 23.1
Wireless: 0.0

Turkey
Europe and Central Asia
Upper-middle-income
Population: 71 million
Broadband (per 100 people):
Wireline: 8.0
Wireless: 26.3

Vietnam
East Asia
Lower-middle-income
Population: 88 million
Broadband (per 100 people):
Wireline: 4.1
Wireless: 9.6

Brazil
Latin America
Upper-middle-income
Population: 193 million
Broadband (per 100 people):
Wireline: 8.0
Wireless: 10.7

Morocco
North Africa
Lower-middle-income
Population: 32 million
Broadband (per 100 people):
Wireline: 1.6
Wireless: 4.3

Kenya
Sub-Saharan Africa
Low income
Population: 37 million
Broadband (per 100 people):
Wireline: 0.2
Wireless: 2.1

Sri Lanka
South Asia
Lower-middle-income
Population: 20 million
Broadband (per 100 people):
Wireline: 1.0
Wireless: 4.9

This map was produced by the Map Design Unit of The World Bank. The boundaries, colors, denominations and any other information shown on this map do not imply, on the part of The World Bank Group, any judgment on the legal status of any territory, or any endorsement or acceptance of such boundaries.

Source: World Bank adapted from national regulatory authorities and statistical offices.

Note: All figures are for 2010. Wireless broadband penetration refers to subscriptions and not active Internet use.

to lack of scale and growing preference for satellite-delivered multichannel television. With the relatively high penetration of mobile networks—subscription penetration exceeded 100 percent by December 2010—3G services are expanding rapidly to fill the demand for broadband. As a result, wireless access is likely to be the main growth area for broadband in Brazil now that constraints on the availability of radio spectrum have been addressed and the 450 MHz band has been released for rural communications.

The other major constraint is the big variation in income levels across the country. Broadband access is uneven—at one end of the spectrum, the level of access is high in industrialized urban areas, mostly in the southeast of the country. Here, the nation has recorded some of the world's highest levels of Internet use, and, in particular, Brazilians have been early users of social

Table 7.2 Examples of Policies and Programs for Broadband Development, by Country, Region, and Economic Level

Economic level, region, and country	Supply: infrastructure	Services	Demand Applications	Users
Low-income economy				
Sub-Saharan Africa: Kenya	Kenyan government encouraged local operators to participate in undersea TEAMS cable through PPP.	VoIP has been legal since 2006 with liberal licensing for ISPs.	Judiciary Telepresence project connects judges and courts.	The Kenya ICT Board establishes and funds Pasha Digital Villages with broadband access for communities.
Lower-middle-income economy				
East Asia and the Pacific: Vietnam	Several fixed and mobile broadband operators have been licensed.	A broadband network connecting over 1,000 educational institutions has been installed in Ho Chi Minh City.	Several plans and programs exist for promoting software and digital content industries.	Procurement of digital information devices for households with financial difficulties is supported through the USF and spectrum auction proceeds.
Middle East and North Africa: Morocco	License was granted to new operator Wana, which is now the second largest broadband operator.	The Genie Program is installing broadband multimedia computer labs in all schools, affecting 6 million students.	The Idarati (E-Government) Program led to 97% of administrative units having a website with some 200 services online.	Laptops are subsidized for engineering students and teachers.

South Asia: Sri Lanka	Sri Lanka was among the first in the region to award mobile broadband spectrum and has the lowest prices and highest speeds.	The regulator compiles broadband quality of service statistics showing difference between advertised and actual speeds.	The E–Sri Lanka Program has resulted in 112 online services and some 4 million people conducting transactions with government online.	The Easy Seva project used PPPs to install more than 50 public Internet facilities in rural areas connected with mobile broadband.

Upper-middle-income economy

Europe and Central Asia: Turkey	Incumbent is required to provide wholesale broadband access to its fixed telephone network.	Broadcast firms are allowed to provide broadband, and incumbent is allowed to provide IPTV.	Share of government services provided online to total public services reached 66% in 2010.	Government has provided 1,850 public Internet access points to provide ICT access and ICT competency to citizens.
Latin America and the Caribbean: St. Kitts and Nevis	Licensed cable TV and fixed wireless operators are to provide broadband services in competition with incumbent.	VoIP services such as Vonage, MagicJack, and Skype are used extensively by residential consumers.	Government encourages local portals developed by entrepreneurs; SKNVibes gets 2 million hits a month.	Students in the final grade of high school are provided with laptops; operators bid on providing Internet access to these students on a pay-as-you-go basis.

Source: Adapted from World Bank, "Broadband Strategies Toolkit, Module 3: Country Case Studies of Broadband in the Developing World," http://www.broadbandtoolkit.org.

networking services. At the other end of the spectrum, there are vast hinterlands of unconnected rural areas, most particularly in the less wealthy northern and western parts of the country. This pattern of uneven access also repeats itself at the local level. Most cities have wealthy areas with high levels of household broadband access, while close by in the *favelas* (informal townships), there is almost no fixed broadband and people must depend on cybercafés or relatively slow and expensive 3G connections.

The federal government has had little success in addressing the digital divide using the Universal Service Fund, although state and municipal initiatives have improved public access. The private sector has invested heavily in telecommunications, but Brazil's vast size and low population density in the rural areas make it difficult to achieve pervasive nationwide broadband.

In a renewed effort to address the continued disparities in broadband access, the government began a major new infrastructure development initiative in mid-2010, setting ambitious targets to triple broadband uptake by 2014. Called the National Broadband Program, it aims to provide broadband access for low-income households and in areas where private operators have little commercial interest. The US$6.1 billion project aims to cover 4,000 cities and towns—40 million homes—with broadband at a speed equal to or greater than 512 kbit/s for about US$20 per month. The initial focus has been on addressing the deficiencies in the national backbone and ensuring that sufficient fiber infrastructure is in place. The old state-owned operator, Telebras, has been revitalized as manager to integrate existing resources, including utilizing the fiber networks of oil and electricity utilities.

With the national broadband plan and steadily rising economic prosperity for the less wealthy, as well as infrastructure projects associated with the soccer World Cup in 2014 and the Olympics in 2016, the prospects for wider adoption of broadband in Brazil have improved.

Kenya

Kenya has a natural geographic advantage, being strategically located on the east coast of Africa and well positioned vis-à-vis the Arab Gulf states (Msimang 2011). Its government-led "build it and they will come" approach to broadband development has leveraged the country's geographic location and played a major role in dramatically increasing fiber optic backbone capacity. Many of Kenya's milestones have been realized in less than five years. Connections were made to three fiber optic submarine cables by the end of 2010, changing the face of the broadband market. The country has gone from relying on satellite for international capacity at the beginning of 2009 to having access to capacity of almost 4 Tbit/s over fiber toward the end of 2010.

Although the landing of the cables is merely a first step, it has already resulted in an 80 percent decrease in wholesale bandwidth costs (although reliability is sometimes a problem). Lower prices and greater availability are expected to increase access to the Internet as well as to promote the continued spread of sophisticated mobile applications and services and consequently to improve opportunities for the creation of and access to information and knowledge. Affordable broadband is expected to increase Kenya's competitiveness, particularly in the business process outsourcing industry, and to encourage entrepreneurship and innovation.

Kenya is also emerging as a mobile broadband hub. This builds on its success with the M-PESA mobile money platform. Mobile broadband was launched in 2008 and far outnumbers wireline subscriptions. LTE is being tested, and construction of a wholesale backbone network is also being considered. A regional mobile application laboratory is being established in Nairobi with the assistance of the World Bank, with the aim of fostering the development of mobile applications and locally relevant content.

With an estimated wireline and mobile broadband penetration rate of two subscriptions per 100 people in 2010, Kenya still has significant progress to make with respect to broadband uptake. Stimulating demand and usage by citizens and the public and private sectors remains a challenge. Kenya, largely through the government, has taken an innovative and proactive approach to putting the user at the center and addressing the other elements of the broadband ecosystem, such as education, literacy, applications, and content. This has been done through progressive regulation, the promotion of polices relating to ICT in education, the subsidization of relevant content and application projects, and the facilitation of creative public-private partnerships.

Much of Kenya's success comes from four important factors: (a) a clear national approach of how broadband fits into its Vision 2030 development goals; (b) strong leadership and direction; (c) a credible regulatory, policy, and institutional framework; and (d) approaches that leverage the strength of the public and private sectors through PPPs. Elements of these traits permeate all aspects of the broadband ecosystem. Although there have been a few setbacks in the pace of implementation and overlaps in the policy and institutional framework, the Kenyan broadband experience is inspiring, particularly its potential to transform economic and social activity.

Morocco

Morocco is a lower-middle-income economy in the northwest of Africa (Constant 2011). Its 2009 economic growth rate of 4.2 percent surpassed the Middle East and North Africa average of 2.3 percent, yet the country remains vulnerable to economic shocks, high illiteracy, and high unemployment as well as increasing pressure on natural resources. Despite these challenges, the country continues to make gradual progress in human development and economic indicators through investment in diversification and sound macroeconomic policies.

For example, Morocco invests more on ICT than any other country in the region: in 2008, 12.5 percent of its gross domestic product was spent on ICTs compared to the regional average of 5.8 percent. As a result, the telecommunications market has advanced rapidly, with the spread of mobile phones emerging as a bright spot in the country's ICT sector. Penetration rose 20 percentage points in 2010 to reach over one mobile subscription per person.

Morocco was one of the first countries in the region to award 3G frequencies, which took place in 2006. Unlike most other countries that awarded 3G frequencies through an auction, Morocco chose a beauty contest, resulting in lower costs for operators. Some of the spectrum was awarded to a new operator, shaking up the existing duopoly and triggering intense competition in the mobile broadband market. As a result, mobile broadband, which launched in 2007, surpassed fixed broadband connections by 2009 and made up almost three-quarters of all broadband connections in 2010.

Exchanging videos and music, social networking, and Internet telephony are the main uses driving people to broadband, with a combined increase of 25 percentage points in 2010. Considering that the majority of Internet access is over mobile broadband, growing usage is beginning to impose constraints on networks, affecting quality. Further, only about one-quarter of households in Morocco have a broadband connection. Consequently, a significant number of Internet surfers use cybercafés where pay-as-you-go pricing is cheaper than a home subscription.

Broadband availability in large enterprises is widespread, and most companies have a website. However, the use of online transactions is limited, with only around one-fifth of the population buying or selling goods and services over the Internet. Broadband use by micro, small, and medium enterprises with fewer than 10 employees is much more limited.

The country has adopted the Maroc Numérique 2013 (Digital Morocco) Strategy to enhance e-government services and overcome current limitations. Targets include creating employment opportunities within the sector and providing broadband to all schools and one-third of households by 2013. In addition, some 400 community access centers are to be created. The development of local content is a key strategy, including increasing the availability of e-government to some 90 online applications. In an effort to get more small enterprises to adopt broadband, free training will be provided, including sensitizing businesses about the benefits of high-speed Internet for increasing productivity and competitiveness.

St. Kitts and Nevis

The island nation's approach to developing broadband is encouraging. At the end of 2010, wireline broadband subscription rates stood at almost 30 percent, the highest rate among all countries in the Latin American and Caribbean region. This achievement in broadband can be attributed in part to the small physical size of St. Kitts and Nevis, which has enabled faster rollout of the physical infrastructure, facilitated marketing, and promoted maximum impact for government-led ICT policy initiatives (Anius 2011). Among the Caribbean islands, however, "smallness" is certainly not unique, and several other factors have contributed to this achievement.

The phrase "strength in depth" is borrowed from the world of soccer, the most popular sport on the island. The term is used to underscore the point that the strength of the island's achievement in the broadband sector lies in its commitment to nurturing the foundational components of the broadband ecosystem. Promoting basic education and digital literacy, building technology awareness, facilitating access to basic technologies, and encouraging a competitive telecommunications environment are but a few examples of where the country has developed its core strengths.

The St. Kitts and Nevis broadband ecosystem includes the following key strategies:

- Competitive environment through efficient legislation and regulation

- Regional coordination, particularly for design of policy frameworks

- Government as facilitator by providing strong leadership in the ICT sector

- Government as leader by promoting service demand through content provision

- Universal service for broadening access to technologies

- Public-private partnerships to catalyze and strengthen broadband initiatives.

The deployment of a second submarine fiber network in 2006 has introduced competition in international backbone capacity that should further enhance the broadband sector. However, as in any ecosystem, sustainability and growth can be threatened by internal weaknesses. Some of these weaknesses have served as lessons learned and were adjusted at the national level; others continue to pose a challenge to the islands. Costly services, an unstable power supply, quality of service issues, lack of high-speed mobile networks, and deficiencies in the availability of local content and applications that create network value for citizens are some of the challenges for future growth of the broadband sector in St. Kitts and Nevis.

In general, the country has been successful in promoting uptake of broadband Internet by taking measured approaches, which may be of relevance to discussions on broadband strategies pertaining to other small island developing states.

Sri Lanka

Sri Lanka, an island nation located in the Indian Ocean just south of India, has recently experienced rapid growth in the availability and use of mobile broadband services (Galpaya 2011). A key factor is mobile broadband spectrum availability. 3G frequencies were made available as far back as 2003 for testing, and commercial 3G services were launched in 2005. Early access to spectrum enabled operators to gain experience and constantly innovate to stay competitive. As a result, Sri Lanka has the fastest mobile broadband technologies in South Asia.

The government's e-development agenda has also triggered broadband uptake. E–Sri Lanka is a cross-sector ICT development program financed in part by the World Bank. A series of comprehensive supply- and demand-side projects has helped to create awareness about broadband in the country. For example, one project set up a network of nearly 500 rural telecenters, while a least-cost subsidy scheme has been planned to build and operate a fiber backbone in rural areas as well as to establish a comprehensive e-government program. Additionally, operators have been motivated to invest in network infrastructure in light of projected demand.

Beginning in the early 1980s, Sri Lanka was plagued by a violent ethnic conflict, which forced a large portion of the minority Tamil population to leave Sri Lanka and seek refuge in other countries. This large migrant population generated high demand for Internet services in order to communicate with relatives remaining in Sri Lanka. Demand for Internet telephony was unusually high in conflict zones, with Internet cafés catering to the demand.

Innovative business models have contributed to widening access to services. The development of wireless broadband rides on the wave of extremely high mobile voice growth. Intense competition forced operators to innovate in such a way as to be able to serve even the poorest consumers profitably. Network costs were reduced drastically by sharing passive and active infrastructure and outsourcing key parts of the operation. Distribution costs were minimized through e-reloads, eliminating the need to print and distribute top-up cards for prepaid users. Small top-up values attracted consumers with low and variable incomes to the market. This "budget telecom model" enabled operators to earn positive margins, even though average revenue per user was low. This same model is now being applied to mobile broadband in Sri Lanka. By enabling prepaid, very low value recharge, and promotional discounts for students, the youth segment is being brought into the mobile broadband market. These early adopters are, in turn, spreading interest about the benefits of high-speed wireless networks.

The downside of the budget telecom model is that quality is sometimes sacrificed for price. Compared with the developed world, Sri Lankan consumers get less broadband value for the money they spend. Part of the reason is that advertised broadband speeds are theoretically possible, but rarely delivered in reality. Another bottleneck is international connectivity. A significant portion of Internet traffic is routed outside of the country, and wholesale international connectivity prices are relatively high, making Internet capacity a sought-after resource.

Turkey

Throughout history, Turkey has been a prominent center of commerce because of its land connections to the continents of Europe, Asia, and Africa and the sea surrounding it on three sides (Cagatay 2011). An Organisation for Economic Co-operation (OECD) member, it has long awaited European Union (EU) membership. As an upper-middle-income economy, Turkey suffers from comparison with these mainly high-income groupings. Its fixed broadband penetration stood at 9.4 subscriptions per 100 inhabitants in June 2010 compared to the OECD average of 24.2, and 34 percent of Turkish homes had a broadband connection compared to the EU average of 61 percent in 2010.

These statistics disguise the fact that, compared to other countries in its income group, Turkey is doing relatively well. It has a higher broadband penetration than recent EU members and the fourth largest fixed broadband network among upper-middle-income economies.

E-government initiatives have been a major driving force for development of the broadband ecosystem. This has triggered demand by enterprises in the ICT sector and motivated citizens to increase Internet usage. Ensuring a shared vision among political leaders and technocrats has also been an important factor in pushing e-government programs. Political leaders saw e-government as a central instrument that would support public reforms and larger changes in the political system. A central organizational structure was formulated to develop strategies and put public money into the pipeline for a set of strategically important projects with high value and high transaction costs.

The high-tempo growth of Turkish economy in the last decade is another supportive factor. Various market-oriented reforms have been complemented with a proactive foreign policy resulting in large sums of overseas capital flowing into the country. Communications, software, and hardware segments of ICT industries have expanded rapidly. This has included significant investment in upgrading mobile networks to broadband. Broadband mobile networks were only launched in 2009, yet by the end of 2010 around a quarter of the population was already capable of accessing high-speed wireless services.

The Turkish population has largely embraced social networking. The country is the fourth largest Facebook market in the world. In addition, local content is growing, and Turkish websites are getting more popular and increasingly diversified.

Nevertheless, the country continues to face economic and social barriers to its ability to absorb broadband technologies effectively on a large scale and better utilize them for leveraging competitiveness. Fixed broadband competition is limited and dominated by DSL technology. ICT skill gaps among SMEs and the less educated need to be addressed with the participation of private initiatives. The lack of a suitable national accounting framework for more detailed analysis hinders international benchmarking in ICTs and innovation.

If Turkey can overcome these barriers, the results could be considerable. According to its Digital Vision roadmap, broadband could boost economic growth by 0.8 to 1.7 percentage points per year. The economic momentum enabled by an enhanced broadband ecosystem would create between 180,000 and 380,000 new jobs each year.

Vietnam

With some 86 million inhabitants, Vietnam is the thirteenth most populated country in the world. Its land area is larger than Italy and almost the size of

Germany (Tuan 2011). Wireline broadband has grown over 1,000 percent since 2005, and, with 3.6 million subscriptions in 2010, it had the ninth largest network among developing counties. Its wireline broadband penetration is the sixth highest among lower-middle-income economies, with 4.4 subscriptions per 100 people.

Solid economic growth has coincided with increased broadband usage. This has been accompanied by opening of the economy, which has attracted investment from foreign capital. Liberalization of the telecommunications sector has led to growing competition, with 11 enterprises providing infrastructure. Service providers have developed modern IP-based networks with extensive fiber optic backbones. Incomes have risen so that more people can afford broadband. This, in turn, has created a virtuous circle, with explosive demand creating a larger market, resulting in economies of scale and lower prices. Another factor driving fixed broadband growth is that Vietnam was a latecomer to mobile broadband. Major mobile operators did not launch their networks until 2009, with around 15 percent of mobile subscribers having 3G capability toward the end of 2010.

Despite these successes, Vietnam faces challenges in broadening broadband access, particularly in rural areas, where some 70 percent of the population resides. Young people in urban areas "live" with high-speed Internet access; however, less than 1 percent of rural households had any type of Internet access in 2008.

Most businesses are focused on using the Internet for basic needs such as sending and receiving e-mail and finding information, while more advanced applications such as e-commerce are not used as widely. Despite rising Internet access in households, many users have yet to exploit broadband applications fully. Survey data indicate that a computer's Internet connection in Vietnam is used to search for personal information and serve children's learning. The lack of relevant content and fragmented information are problems; a public information network with a unified portal, equipped with an automatic translation engine and rich multimedia content covering health, education, culture, and agriculture, is lacking.

The cost of fiber optic access is only economical in new urban areas and for large enterprises, so DSL remains the fixed broadband choice of households. But copper lines provide less quality than fiber, and it is difficult to upgrade the transmission capacity. At the same time, telecom enterprises have recently been focusing on developing mobile broadband to the detriment of the fixed network.

The large number of operators has led to overlap in investment in the access network. Interconnection is difficult because operators use a variety of technologies, affecting standardization of the national telecommunications infrastructure. Intense competition has resulted in price wars that threaten long-term sustainability. Service providers are looking to reduce duplication by cooperating on shared infrastructure, but so far no specific measures have been implemented.

While Vietnam has had tremendous achievements in broadband, there are challenges arising from its rapid growth: (a) development of width (for example, the number of subscribers) needs to be coupled with development of depth (for example, service quality); (b) differences in the level of broadband between regions can contribute to widening gaps; and (c) the rapid development of broadband can cause policy problems affecting social life, security, and politics.

Notes

1. This chapter classifies developing economies into geographic and economic groupings according to World Bank regional and income classifications. See World Bank, "How We Classify Countries," http://data.worldbank.org/about/country-classifications.

2. United Nations, "Millennium Development Goals," http://www.un.org/millenniumgoals/.

3. MDG Monitor, "Tracking the MDGs," http://www.mdgmonitor.org/browse_goal.cfm.

4. ITU, "WSIS," http://www.itu.int/wsis/index.html.

5. FTTH Council, "Global FTTH Councils' Latest Country Ranking Shows Further Momentum on All-Fiber Deployments," Press Release, February 10, 2011, http://www.ftthcouncil.org/en/newsroom/2011/02/10/global-ftth-councils-latest-country-ranking-shows-further-momentum-on-all-fiber-.

6. "Triple Network Project Launched," *People's Daily Online*, July 2, 2010, http://english.peopledaily.com.cn/90001/90778/90860/7050112.html.

7. See ESCWA, "Final Meeting of the Project on 'Promotion of the Digital Arabic Content Industry through Incubation,'" http://www.escwa.un.org/divisions/projects/dac/index.asp.

8. Intel, "Intel Capital to Invest in Two Digital Content Companies in Jordan," Press Release, May 17, 2009, http://www.intel.com/capital/news/releases/090519.htm.

9. http://www.jeeran.com/.

10. http://www.shoofeetv.com.

11. Dubai School of Government, "Arab Social Media Report," January 2011, http://www.dsg.ae/NEWSANDEVENTS/UpcomingEvents/ASMRHome.aspx.

12. TRAI (2010); TRAI, "TRAI Issues Recommendations on 'National Broadband Plan,'" Press Release, December 8, 2010, http://www.trai.gov.in.

13. See Pakistan Universal Service Fund, "Broadband Programme," http://www
.usf.org.pk/Broadband-Programme.aspx.14. See "African Undersea Cables,"
http://manypossibilities.net/african-undersea-cables/.

14. See African Undersea Cables, http://manypossibilities.net/african-undersea-
cables/.

15. LDCs are identified through three criteria: income per capita, human capital,
and economic vulnerability. For the methodology, see UN-OHRLLS, "Criteria
for Identification of LDCs," http://www.unohrlls.org/en/ldc/related/59/. For
the list of LDCs, see UN-OHRLLS, "Country Profiles," http://www.unohrlls
.org/en/ldc/related/62/.

16. For a list of LLDCs, see UN-OHRLLS, "Country Profiles," http://www.unohrlls
.org/en/lldc/39/.

17. See UNCTAD, "UN Recognition of the Problems of Land-Locked Developing
Countries," http://www.unctad.org/Templates/Page.
asp?intItemID=3619&lang=1.

18. For a list of SIDSs, see UN-OHRLLS, "Country Profiles," http://www.unohrlls
.org/en/sids/44/.

19. See UNCTAD, "UN Recognition of the Problems of Small Island Developing
States," http://www.unctad.org/Templates/Page.asp?intItemID=3620&lang=1.

20. See ictDATA, "Samoa Mobilized," September 24, 2010, http://www.ictdata
.org/2010/09/samoa-mobilized.html.

21. There is no official definition of a postconflict economy. They are often
locations where civil conflicts have necessitated the intervention of peacekeeping troops. For a list of locations where United Nations peacekeeping troops are
stationed, see http://www.un.org/en/peacekeeping/.

22. The World Bank–financed E–Sri Lanka Development Project argucs, "ICT can
promote peace efforts by providing connectivity and electronic delivery of
much needed information and public services, bridging space, time, and
promoting understanding between the North and East and the rest of the
country." See World Bank, "Sri Lanka: E-Lanka Development Project,"
December 1, 2003, http://go.worldbank.org/567ZZUWMD0.

23. See NATO Chronicles, "Broadband for Afghanistan," Episode 2, March 2010,
http://www.natochronicles.org/#/en/episode2.

24. See AusAID, "Governance Activities: East Timor," http://www.ausaid.gov.au/
country/east-timor/governance.cfm.

25. http://www.broadbandtoolkit.org.

References

Anius, Diana. 2011. "Broadband in St. Kitts and Nevis: Strength in Depth."
World Bank, IFC, and *Info*Dev, Washington, DC. http://www.infodev.org/en/
Publication.1109.html.

Broadband Commission. 2010. "Broadband and the Interlinked and Interdepen-
dent MDG Agenda." In *A 2010 Leadership Imperative: The Future Built on
Broadband*, ch. 4. Geneva: ITU; Paris: UNESCO.

Cagatay, Telli. 2011. "Broadband in Turkey: Compared to What?" World Bank, IFC, and *Info*Dev, Washington, DC. http://www.infodev.org/en/Publication.1132.html.

Constant, Samantha. 2011. "Broadband in Morocco: Political Will Meets Socio-Economic Reality." World Bank, IFC, and *Info*Dev, Washington, DC. http://www.infodev.org/en/Publication.1125.html.

ECLAC (Economic Commission for Latin America and the Caribbean). 2010. "Plan of Action for the Information and Knowledge Society in Latin America and the Caribbean (eLAC2015)." United Nations, Santiago, November. http://www.eclac.cl/cgi-bin/getProd.asp?xml=/socinfo/noticias/documentosdetrabajo/5/41775/P41775.xml&xsl=/socinfo/tpl-i/p38f.xsl&base=/socinfo/tpl/top-bottom.xsl.

EC-TEL (European Conference on Technology Enhanced Learning). 2009. "Policy on the Allocation and Assignment of Frequencies in the 700 MHz Band." ECTEL, Heerlen, the Netherlands. http://www.ectel.int/pdf/consultations/2010/700%20Mhz%20Band%20Plan%20and%20Policy.pdf.

ESCAP (Economic and Social Commission for Asia and the Pacific). 2009. "Broadband for Central Asia and the Road Ahead." ID/TP-09-05, United Nations, Information and Communications Technology and Disaster Risk Reduction Division, Bangkok. http://www.unescap.org/idd/working%20papers/IDD_TP_09_05_of_WP_7_2_909.pdf.

———. 2010. "Broadband Development in Asia and the Pacific." United Nations, Bangkok, August 13. http://www.unescap.org/idd/events/cict-2010/CICT2_INF5.pdf.

ESCWA (Economic and Social Commission for Western Asia). 2007. "Broadband for Development in the ESCWA Region." United Nations, Beirut. http://www.alcatel-lucent.com/wps/portal/!ut/p/kcxml/04_Sj9SPykssy0xPLMnMz0vM0Y_QjzKLd4w3MfQFSYGYRq6m-pEoYgbxjgiRIH1vfV-P_NxU_QD9gtzQiHJHR0UAAD_zXg!!/delta/base64xml/L0lJayEvUUd3QndJQSEvNElVRkNBISEvNl9BXzRDUi9lbl93dw!!?LMSG_CABINET=Docs_and_Resource_Ctr&LMSG_CONTENT_FILE=News_Releases_2007/News_Article_000148.

Galpaya, Helani. 2011. "Broadband in Sri Lanka: Glass Half Full or Half Empty?" World Bank, IFC, and *Info*Dev, Washington, DC. http://www.infodev.org/en/Publication.1113.html.

Howes, Stephen, and Matt Morris. 2008. *Pacific Economic Survey 2008: Connecting the Region*. Canberra: Australian Agency for International Development. http://www.ausaid.gov.au/publications/pdf/pacific_economic_survey08.pdf.

India, Ministry of Communications and Information Technology. 2004. "Broadband Policy 2004." Ministry of Communications and Information Technology, New Delhi, December 13. http://www.dot.gov.in/ntp/broadbandpolicy2004.htm.

ITU (International Telecommunication Union). 2003. "ICTs and the Millennium Development Goals." In *World Telecommunication Development Report 2003: Access Indicators for the Information Society*, ch. 4. Geneva: ITU.

———. 2008. "Asia-Pacific Telecommunication Indicators; Broadband in Asia-Pacific: Too Much, Too Little?" ITU, Geneva.

———. 2010a. *Stimulating Universal Access to Broadband in Afghanistan, Bangladesh, Bhutan, Maldives, and Nepal*. Geneva: ITU. http://www.itu.int/pub/ D-HDB-UNIVERSA;-2010.

———. 2010b. *World Telecommunications/ICT Development Report: Monitoring the WSIS Targets*. Geneva: ITU. http://www.itu.int/pub/D-IND-WTDR-2010.

Jensen, Mike. 2011. "Broadband in Brazil: A Multi-Pronged Public Sector Approach to Digital Inclusion." World Bank, IFC, and *Info*Dev, Washington, DC. http://www.infodev.org/en/Article.774.html.

Kanamugire, David. n.d. "The Role of Governmental Institutions in Fostering ICT Research Capacity." Ministry of Science and Technology, Rwanda. http://euroafrica-ict.org.sigma-orionis.com/downloads/rwanda/Kanamugire.pdf.

Kim, Yongsoo, Tim Kelly, and Siddhartha Raja. 2010. "Building Broadband: Strategies and Policies for the Developing World." Global Information and Communication Technologies Department, World Bank, Washington, DC, January. http://www.infodev.org/en/Publication.756.html.

Mexico, INEGI (Instituto Nacional de Estadística y Geografía). 2009. "Estadística sobre disponibilidad y uso de tecnología de información y comunicaciones en los hogares, 2009." INEGI, Mexico, DF. http://www.inegi.org.mx/prod_serv/contenidos/espanol/bvinegi/productos/encuestas/especiales/endutih/ENDUTIH_2009.pdf.

Moldova, ANRCETI (National Regulatory Agency for Electronic Communications and Information Technologies). 2010. *Report on Activity of the National Regulatory Agency for Electronic Communications and Information Technology and Evolution of Electronic Communications Markets in 2009*. Chisinau: ANRCETI. http://en.anrceti.md/files/filefield/2009_RAPORT_(ENG).pdf.

Moldova, Government of. 2005. "National Strategy on Building Information Society: e-Moldova." http://en.e-moldova.md/Sites/emoldova_en/Uploads/strat_ENG.36370AF841D74D74B3969D0FA3FBE6D2.pdf.

Morocco, Ministry of Industry, Commerce, and New Technologies. n.d. "Maroc numerique 2013: Stratégie nationale pour la société de l'information et l'économie numérique 2009–2013." Ministry of Industry, Commerce, and New Technologies, Rabat. http://www.egov.ma/Documents/Maroc%20Numeric%202013.pdf.

Msimang, Mandla. 2011. "Broadband in Kenya: Build It and They Will Come." World Bank, IFC, and *Info*Dev, Washington, DC. http://www.infodev.org/en/Publication.1113.html.

Pakistan, Ministry of Information Technology. 2004. "Broadband Policy." Ministry of Information Technology, Islamabad, December 22. http://www.pta.gov.pk/media/bbp.pdf.

Roetter, Martyn. 2009. "Mobile Broadband, Competition, and Spectrum Caps." Paper prepared for the GSMA, Arthur D. Little, Boston, January. http://www.gsmworld.com/documents/Spectrum_Caps_Report_Jan09.pdf.

Softbank. 2007. *Annual Report 2007: Redrawing the Map*. Tokyo: Softbank Corporation. http://www.softbank.co.jp/en/irinfo/library/annual_reports/.

South Africa, Department of Communications. 2010. "Broadband Policy for South Africa." *Government Gazette* 33377 (July 13): 3–21. http://www.info.gov.za/view/DownloadFileAction?id=127922.

Telkom Indonesia. 2010. "Info Memo: The Year End 2010 Results (Audited)." Telkom Indonesia, Bandung. http://www.telkom.co.id/investor-relation/reports/info-memo/.

TRAI (Telecom Regulatory Authority of India). 2010. "Consultation Paper on National Broadband Plan." TRAI, New Delhi. http://www.trai.gov.in.

Tuan, Tran Minh. 2011. "Broadband in Vietnam: Forging Its Own Path." World Bank, IFC, and *Info*Dev, Washington, DC. http://www.infodev.org/en/Publication.1127.html.

UNCTAD (United Nations Conference on Trade and Development). 2007. *The Least Developed Countries Report 2007.* Geneva: United Nations. http://www.unctad.org/en/docs/ldc2007_en.pdf.

———. 2010. *The Least Developed Countries Report 2010.* Geneva: United Nations. http://www.unctad.org/templates/webflyer.asp?docid=14129&intItemID=5737&lang=1&mode=downloads.

Williams, Mark. 2009. *Broadband for Africa Policy for Promoting the Development of Backbone Networks.* Washington, DC: World Bank, August. http://www.infodev.org/en/Publication.526.html.

World Bank. 2009. "Regional Telecoms Backbone Network Assessment and Implementation Options Study." World Bank, Washington, DC, January. http://www.itu.int/ITU-D/asp/CMS/Events/2009/PacMinForum/doc/POLY_WB_GeneralReport_v3%5B1%5D.0.pdf.

WSIS (World Summit on the Information Society). 2003. "Geneva Plan of Action." Document WSIS-03 /GENEVA/DOC/5-E, ITU, Geneva, December 12. http://www.itu.int/wsis/docs/geneva/official/poa.html.

APPENDIX A

Weblinks to National Broadband Plans

Country	Title of the plan	Weblink
Brazil	Programa Nacional de Banda Larga (National Broadband Program or National Broadband Plan)	http://www4.planalto.gov.br/brasilco nectado/forum-brasil-conectado/ documentos/3o-fbc/documento-base-do-programa-nacional-de-banda-larga
Colombia	Plan Vive Digital Colombia (Colombia Digital Living Plan)	http://www.mintic.gov.co/vivedigital/ pdfs/material.pdf
Finland	Making Broadband Available to Everyone: The National Plan of Action to Improve the Infrastructure of the Information Society	http://www.lvm.fi/c/document_library/ get_file?folderId=57092&name= DLFE-4311.pdf
France	France Numérique 2012: Plan de Dével-oppement de l'Économie Numérique (Digital France 2012)	http://lesrapports.ladocumentationfran caise.fr/BRP/084000664/0000.pdf
Japan	E-Japan Strategy in 2001 (updated at intervals)	http://www.kantei.go.jp/foreign/policy/ it/index_e.html
Oman	Digital Oman Strategy	http://www.ita.gov.om/ITAPortal/ITA/ strategy.aspx?NID=646&PID= 2285&LID=113
Singapore	Intelligent Nation 2015 (iN2015)	http://www.ida.gov.sg/images/content/ About%20us/About_Us_level1/_iN2015/ pdf/realisingthevisionin2015.pdf

(continued next page)

Country	Title of the plan	Weblink
South Africa	Broadband Policy for South Africa	http://thornton.co.za/resources/ gg33377_nn617_pg3-24.pdf
United States	National Broadband Plan: Connecting America	http://www.broadband.gov/ download-plan/

Source: World Bank.

Policies and Programs for Promoting Broadband in Developing Countries

Table B.1 Infrastructure Policies and Programs

Sector and program	Supply and demand impacts	Description	Example
Promote investment and market entry	Supply: all levels Demand: access, affordability	The first step of broadband policy implementation is to foster competition with minimal market entry barriers. Lowering or removing entry barriers in broadband markets drives competition. A key consideration is technological neutrality. The rapid development and diffusion of broadband is largely due to competition between technologies such as digital subscriber line (DSL), cable modem, fiber optics, and wireless. To enjoy the full benefits of such competition, governments should not influence the technological choices of providers without good reason.	The Thai government considers that international connectivity could be a bottleneck and for that reason issues automatic licenses for international gateway services (Thailand, National Telecommunications Commission 2010).
Promote international coordination	Supply: all levels	Coordination among countries can affect all levels of the broadband supply chain by lowering costs through common technical standards and facilitating the development of international, regional, and national backbones. A high level of global and regional cooperation already exists in areas such as equipment standards and frequency coordination. Regional harmonization in broadband regulatory approaches can help to reduce uncertainty and attract investment.	The Eastern Caribbean Telecommunications Authority (ECTEL) is a regulatory body for its five member states. It coordinates policy in several areas, including aspects related to broadband such as frequencies for broadband wireless access, wholesale access to networks, and quality of service.[a]
Reduce administrative burdens and provide incentives for research and development, pilots, and network rollout	Supply: all levels Demand: access, affordability	High license fees, taxes, and burdensome administrative processes can discourage investment in the broadband sector, especially when the market is nascent and the returns are uncertain. Measures such as providing investors with tax benefits and low-interest, long-term loans can promote investment in network development. Likewise, allowing operators to use broadband spectrum for pilots prior to formal allocation provides an opportunity to test the feasibility of different frequencies and gain valuable experience.	In order to encourage broadband connectivity, India removed licensing requirements for use of Wi-Fi and WiMAX in the 2.4–2.4835 GHz band.[b]

| Allocate and assign spectrum | Supply: domestic backbone, local access Demand: access | Allocating the appropriate spectrum can significantly alter the business case for wireless broadband. Governments should manage their radio spectrum appropriately to reduce entry barriers, promote competition, and enable the introduction of innovative technologies. Given the rapid development of wireless broadband technologies, governments should allow providers to obtain new frequencies by expanding available frequency bands. They should implement management policies that encourage efficient use and shift spectrum from low-value uses to broadband. Spectrum managers should also keep in mind the effect of their spectrum allocations on business economics: higher bands make mobile communication more difficult and more expensive. Spectrum should be assigned on a technology- and service-neutral basis. This approach is critical to enabling all the different types of applications of broadband services: voice, video, and data can all be provided by wireless broadband technologies. Finally, operators should be allowed to use their existing spectrum for mobile broadband services. | Widespread policies throughout Latin America allow the existing 850/900 megahertz spectrum, originally allocated for voice, to be used for high-speed mobile data services. These frequencies also support wider coverage with fewer base stations so that investment costs are lower and more people can gain access (Roetter 2009). |
| Facilitate infrastructure sharing | Supply: international connectivity, domestic backbone, local access Demand: access, affordability | Civil works (for example, trenches, ducts, and cables) are the biggest sunk cost in broadband network construction in both the access and the backbone segments. The costs of backbone network construction can be cut by establishing legal grounds for open access to the passive infrastructure (conduits, ducts, and poles) of other services (roads, railways, and power supply facilities). Similarly, when contractors construct | In Thailand, operators signed a Memorandum of Understanding on infrastructure and network sharing in November 2010 in support of the country's National Broadband Policy (Aphiphunya 2010). According to the government, the agreement will lead to more efficient use of networks. |

(continued next page)

Table B.1 *(continued)*

Sector and program	Supply and demand impacts	Description	Example
		other types of new infrastructure, the government can require them to build passive infrastructure that communications service providers can access on a nondiscriminatory basis. Another option is to require the installation of basic infrastructure, such as ducts, when homes and offices are constructed or renovated. Finally, governments can permit or facilitate joint construction of backbone and subscriber networks among providers.	
Establish Internet exchanges	Supply: international connectivity Demand: affordability	There are many advantages to local routing of Internet traffic via a common exchange point, including substantial cost savings by eliminating the need to put all traffic through more expensive long-distance links to the rest of the world. In addition, local links are faster because of the reduced latency in traffic, which makes fewer hops to get to its destination.	The Rwanda Internet Exchange has been operational since mid-2004. In October 2003, SIDA (Swedish International Development Agency) began an initiative to assist Rwanda in establishing a national Internet exchange point (IXP). Rwanda fulfilled the prerequisites needed for SIDA assistance, including the presence of a neutral body to host the peering point, the existence of at least two independent Internet service providers (ISPs) in the country, and a team of technicians from the various Internet providers trained in the techniques of setting up and maintaining a peering point. Each network operator provides a circuit from its backbone and a router that connects to the IXP switch (Jenson 2009).
Public-private partnerships (PPPs) for deployment of open-access broadband networks	Supply: international connectivity, domestic backbone	Network construction is the highest entry barrier in the communication industry, requiring significant financial resources. Construction of domestic and international backbone networks is essential to ensure that high-quality, low-cost connectivity is available. Businesses might initially avoid investing in backbone networks because	The Kenyan government has been aggressively promoting the development of broadband backbones through PPPs. It took an active role in The East African Marine System (TEAMS), an undersea fiber optic cable linking Mombasa in Kenya and Fujairah in the United Arab Emirates. The government encouraged operators in Kenya

		they are unsure of the returns on their invest-ments. Governments can partner with the private sector to provide up-front support in order to reduce risks or act as an anchor tenant to induce investment.	to join it in taking an 85% stake in the cable, which was launched in 2009. More than 10 operators have an ownership interest in TEAMS, guaranteeing them access at wholesale rates. Kenya also encouraged PPPs for building the national fiber backbone and is considering the same for Long-Term Evolution (LTE) networks (World Bank 2011).
Coordinate access to rights-of-way	Supply: domestic backbone, local access Demand: affordability	Obtaining the rights-of-way necessary to deploy broadband infrastructure can be a complex process, adding to costs and delaying deploy-ment.	Canada's Telecommunications Act includes provisions to facilitate operators' access to public property.[c]
Facilitate open access to critical infrastructure	Supply: international connectivity, domestic backbone, local access Demand: affordability	Critical infrastructure consists of essential network elements or services that are typically owned by a single or small number of suppliers. These include facilities such as international and national fiber optic backbones and fixed local access networks that cannot be easily replicated. Facilitating open access to these facilities through options such as an obligation for providers to provide wholesale access or structural separation of wholesale and retail activities can stimulate competition and lower retail broadband prices.	The European Commission requires incumbent operators to offer unbundled access to their fixed telephone networks (European Parliament and Council of the European Union 2000).

Source: World Bank.

a. See the ECTEL website, http://www.ectel.int.

b. See Indonesia, Department of Telecommunications, "Indian Telecom Sector," http://www.dot.gov.in/osp/Brochure/Brochure.htm.

c. See "ICT Regulation Toolkit: Practice Note, Sharing Rights-of-Way," http://www.ictregulationtoolkit.org/en/PracticeNote.aspx?id=3245.

Table B.2   Services Policies and Programs

Sector and program	Supply and demand impacts	Description	Example
Connect schools to broadband networks	Supply: domestic backbone, local access Demand: access, affordability, awareness	School connectivity provides many benefits, including access to an ever-growing volume of educational information, opportunities for collaboration, and the use of online applications. It provides students and teachers hands-on experience for developing ICT skills. Schools can also be leveraged to provide connectivity in off-hours to the rest of the community.	In Chile, the Center for Education and Technology within the Ministry of Education administers Enlaces, the country's initiative to improve education in subsidized state schools using ICTs.[a] Enlaces provides access to the Internet to approximately 75% of students in schools that are enrolled in the project, 67% of which have a broadband connection.
Use government as an anchor tenant	Supply: domestic backbone, local access Demand: access	One of biggest expenses in providing broadband connectivity in rural areas is the "middle mile," the portion connecting a town to the Internet backbone. Once the backbone connection to government institutions is established, it can be leveraged to provide retail broadband services to local residences and businesses. Broadband-connected government institutions thus become "anchor points" from which broadband connectivity can be shared with the surrounding community.	The United States has recommended that broadband connectivity in federal offices located around the country should be used to extend broadband access to unserved and underserved communities.[b]
Monitor service quality	Demand: attraction	Broadband service providers often advertise broadband speeds that are higher than the bandwidths actually experienced by the user. Differences between advertised and actual speed can affect users' confidence in the quality of broadband services. This lack of confidence can be overcome through regular reporting of service quality levels.	The Telecommunications Regulatory Authority of Bahrain publishes quarterly results of its broadband quality of service monitoring (Bahrain, Telecommunications Regulatory Authority 2011). It carries out a predefined set of tests around the clock. The results are stored in a centralized database. Actual versus advertised speeds for different ISPs are tested based on access to local and international websites. The measurements supplement information already available to consumers with respect to prices and advertised speeds.

Create an enabling environment for intermodal competition	Supply: local access Demand: attraction	Convergence allows for the provision of voice, data, and broadcast services over telephone, broadcast, mobile, and Internet networks. Governments should a low any type of network to offer any type of broadband service in order to intensify competition. This includes the legalization of voice over broadband and television over Internet Protocol services.	Chile allows telecom and television operators to provide voice, data, and video services. Cable television operators account for almost half of broadband lines and around one-fifth of voice subscriptions.[c]
Ensure nondiscriminatory access for service, application, and content providers	Demand: attraction	It is critical to ensure that all broadband providers of services, applications, and content have fair access to broadband networks. "Network neutrality" helps to achieve this by preventing broadband operators from blocking or degrading access to specific content except when requested by the user.	Chile's Internet and Network Neutrality Law prohibits operators from blocking applications or content unless requested by the user. Intensive users are required to subscribe to a broadband plan that reflects the cost of their usage.[d]
Consider expanding universal service obligation to include broadband	Supply: local access Demand: access, affordability	In some countries, the type and quality of telecommunications services that must be made available to subscribers are defined in laws. The inclusion of broadband in such definitions would require operators to make broadband available on demand.	In July 2010, the Communications Market Act in Finland was revised to include a reasonably priced Internet connection in the definition of universal service. According to the Ministry of Transport and Communications, "Telecom operators defined as universal service providers must be able to provide every permanent residence and business office with access to a reasonably priced and high-quality connection with a downstream rate of at least 1 Mbit/s."[e] The connection can be either fixed or wireless.

Source: World Bank.

a. See "Connect a School, Connect a Community," http://connectaschool.org/en/schools/connectivity/regulation/Section_6.1_Chile_case_study.

b. See "Government Performance," http://www.broadband.gov/plan/14-government-performance/#r14-1.

c. See Chile, SUBTEL, "Información estadística," http://www.subtel.cl/prontus_subtel/site/artic/20070212/pags/20070212182348.html.

d. Chile, SUBTEL, "Reglamento de neutralidad recoge todos los beneficios y derechos de los usuarios consagrados en al Ley de Internet," Press Release, January 20, 2011, http://www.subtel.cl/prontus_subtel/site/artic/20110117/pags/20110117093211.html.

e. Finland, Ministry of Transport and Communications, "1 Mbit Internet Access a Universal Service in Finland from the Beginning of July," Press Release, June 29, 2010, http://www.mintc.fi/web/en/pressreleases/view/1169259.

Table B.3 Applications and Content Policies and Procedures

Sector and program	Supply and demand impacts	Description	Example
Undertake government-led demand aggregation, with government agencies as early adopters and innovators	Supply: domestic backbone, local access Demand: access	In many countries, pockets of broadband demand exist that are too small to obtain adequate broadband service at favorable prices. By pooling that demand together, a larger market can be created, providing incentives for broadband operators to supply the market.	In Italy, an agency of the Ministry of Treasury has aggregated government demand for broadband, leading to a sharp reduction in the prices paid (Battisti 2002).
Provide e-government applications	Demand: attraction	Computerizing public information and providing e-government services through broadband networks are essential. E-government encourages citizens to subscribe to broadband services.	In Colombia, all municipalities have a website. Colombia is the first Latin American country to accomplish this. The Colombian e-government portal is linked to some 3,000 websites, with information about 3,000 administrative processes of which 541 could be accomplished completely online as of December 2009. Citizen use of e-government services doubled in 2009 to over half a million visits per month.[a]
Promote adoption by industry	Supply: international connectivity, local access Demand: affordability, awareness	Support for broadband-related industries increases demand for supply-side components, enhancing infrastructure investment and helping to create long-term sustainable demand for broadband services. Providing training and incentives for small and medium enterprises can help them to get broadband connected to improve their productivity and widen their market opportunities.	In Vietnam, the government supports software parks by developing basic infrastructure and incubation and securing domestic and foreign investment for tenants.[b]
Promote creation of digital content	Demand: attraction	Support for content creation relevant to local needs and in national languages can help to attract people to use broadband.	The Jordanian government has facilitated foreign investment in the digital creation industry. In 2009, chipmaker Intel announced an investment in two Jordanian digital content companies: Jeeran and

Support secure e-transactions	Demand: security	Online transactions are an important part of the broadband environment. Transactions must be secure and legal to encourage the development of two-way interactive e-commerce, e-government, and telemedicine applications. This means that legal systems need to recognize electronic signatures and transactions. Information security such as encryption technologies and antihacking software, are also critical for a stable and safe broadband atmosphere.	ShooFeeTV.[c] The funding will be used to help both companies to pursue regional growth as well as extend their product offerings. The Association of South East Asian Nations (ASEAN) published a reference framework for e-commerce back in 2001 and has since guided the creation and harmonization of e-commerce laws in the region. By April 2008, eight of its 10 members had enacted e-commerce legislation enabling the legal recognition of online transactions to support applications such as online retailing and Internet banking.[d] ASEAN is the first developing region in the world to implement a harmonized e-commerce legal framework throughout member countries.
Implement reasonable intellectual property protections	Demand: security	One enabler of content and media development is the creation of an intellectual property rights regime that protects creators' interests while enabling others to use and improve those creations. Such rights need to balance the interests of creators with the larger goals of enabling knowledge sharing, fair use, and adaptation. This is particularly relevant for the development of e-learning and distance education applications.	Creative Commons licenses allow creators to specify which rights they wish to reserve, thereby allowing a range of possibilities between full copyright and the public domain.[e]

Source: World Bank.

a. Colombia, Ministry of ICT, "Así marcha el programa," Press Release, May 10, 2010, http://programa.gobiernoenlinea.gov.co/noticias.shtml?apc=e1c1-8x=2480.

b. See "'Leaders of High-Tech Parks from the Asian Science Park Association (ASPA) Gathered in Ha Noi," http://www.hhtp.gov.vn/69d40b41_c573_4726_b03c_4f86b90969e1_cms_204.hhtp.

c. Intel, "Intel Capital to Invest in Two Digital Content Companies in Jordan," Press Release, May 17, 2009, http://www.intel.com/capital/news/releases/090519.htm.

d. Galexia, "Harmonisation of E-Commerce Legal Infrastructure in ASEAN," April 2008, http://www.galexia.com/public/research/articles/research_articles-art53.html.

e. See the Creative Commons website, http://creativecommons.org/.

Table B.4 Policies and Procedures for Users

Sector and program	Supply and demand impacts	Description	Example
Provide low-cost user devices in education	Supply: access device Demand: affordability, awareness	The spread of low-cost computers in schools typically includes an ecosystem for operating and maintaining the devices, which often involves providing broadband access in schools in order to download software and support the Wi-Fi capability of the devices. The provision of low-cost educational computers also develops ICT skills at an early age, helping to create demand for broadband.	Uruguay has supplied Wi-Fi-enabled laptops to all primary school children (Brechner 2009). One of the goals of the Uruguayan plan was to boost overall household computer ownership by leveraging the students taking the laptops home after school. This has resulted in 220,000 new homes with computers, including 110,000 in the lowest-income families.
Develop digital literacy programs for citizens	Demand: awareness	To raise public awareness of the benefits of broadband services and promote their use, governments should provide training on how to use computers and the Internet. This training can contribute to the rapid and widespread penetration of broadband. In the short run, such training generates demand. It can also be a step toward universal service when the program targets underserved groups. ICT training for children and students can change their learning behavior and interests and, by extension, alter their parents' views of ICT and broadband.	In Colombia, the Compartel Program within the Ministry of ICT devoted around Col$153 billion (US$84 million) in 2009 for teaching free computer literacy courses at some 1,670 Internet centers around the country. The courses taught around 200,000 people about basic computer tools, Internet navigation, e-mail, search engines, chat, and ICT applications. In addition, teachers use virtual training and video conferencing at the centers to offer courses in other subjects. The centers are often located in educational institutions with access provided to the local community for training during nonschool hours.[a]
Address content and security concerns	Demand: security	Many users are leery of broadband Internet access because of objectionable content and security concerns. These concerns can be alleviated through programs that educate users about perceived risks, child online protection, and how to use the Internet safely.	The regulator in Qatar has created a site for children, teenagers, teachers, and parents providing tips for safe online surfing.[b]

Expand access to underserved communities with Universal Service Fund (USF) support	Supply: national backbone, local access Demand: access	USFs—typically financed by contributions from telecom operators—were initially created to facilitate the development of telephone infrastructure in rural and other underserved localities. Given that broadband connectivity can provide many beneficial services in addition to voice telephony, countries should consider broadening the scope of USFs to cover broadband deployment in underserved areas.	Pakistan's USF is funded by a 1.5% levy on telecom operator revenues. Broadband projects are eligible for funding and include the connection of schools through broadband computer labs and extending domestic fiber optic backbones to rural areas.[c]
Community access centers	Supply: local access Demand: affordability, awareness, attraction	Citizens in underserved communities do not use broadband because they have no access, cannot afford it, or are not aware of its benefits. Creating facilities for public broadband use can alleviate these barriers by establishing a place of access, offering free or low-cost tariffs, and offering training. This can include adapting existing public facilities such as libraries or using schools after-hours for community access.	In Malaysia, the government established community broadband centers to provide collective high-speed Internet to underserved areas identified under the Universal Service Provision Program.[d] Each center is outfitted with computers connected to broadband and provides training.
Facilitate affordability of broadband devices	Supply: devices Demand: affordability	Computers, mobile phones, and data cards for broadband use are expensive for many citizens of developing countries. Countries could consider developing policies and programs that make user devices more affordable for people who want to buy them but lack the means to do so. This includes reducing or eliminating taxes on broadband-enabled devices and subsidizing or offering low- or zero-interest loans for their purchase.	In March 2009, China announced that it had selected 14 vendors to offer low-priced personal computers (PCs) in rural areas. All the PCs in the winning bid are priced from US$290 to US$510. This approach is part of the National Home Appliance Subsidy Program for rural areas. About 57% of the rural population—about 200 million households—will be eligible for a 13% subsidy if they purchase one of those PCs (He and Ye 2009).

Source: World Bank.

a. See Compartel, "Más de 200 mil alfabetizados digitalmente en los 'Nuevos Telecentros Compartel,'" http://archivo.mintic.gov.co/mincom/faces/index.jsp?id=19037.

b. See ictQATAR's, "Stay Safe Online," http://www.safespace.qa/csk/en/home.aspx.

c. See Pakistan Universal Service Fund, "Company Profile," http://www.usf.org.pk/Company.aspx.

d. See Malaysian Communications and Multimedia Commission, "Community Broadband Centres," http://www.skmm.gov.my/index.php?c=public&v=art_view&art_id=34.

References

Aphiphunya, Prasert. 2010. "Broadband Development in Thailand." Presentation at the Committee on Information and Communications Technology meeting, United Nations, ESCAP, Bangkok, November 24. http://www.unescap.org/idd/events/cict-2010/Mr-Prasert-NTC.pdf.

Bahrain, Telecommunications Regulatory Authority. 2011. "Broadband Quality of Service Report." Telecommunications Regulatory Authority, Manama, January. http://www.tra.org.bh/en/marketQuality.asp.

Battisti, Daniela. 2002. "Demand Aggregation to Encourage Infrastructure Rollout to Underserved Regions." Paper presented at the WPIE/OECD workshop, "Public Sector Broadband Procurement," Paris, December 4. http://www.oecd.org/dataoecd/41/60/2491219.pdf.

Brechner, Miguel. 2009. "Plan Ceibal: One Laptop per Child and per Teacher." Paper presented at the seminar, "Reinventing the Classroom," Inter-American Development Bank, Washington, DC, September 15. http://events.iadb.org/calendar/eventDetail.aspx?lang=en&id=1444&.

European Parliament and Council of the European Union. 2000. "Regulation (EC) No. 2887/2000 of the European Parliament and of the Council of 18 December 2000 on Unbundled Access to the Local Loop." *Official Journal of the European Communities,* December 30. http://eur-lex.europa.eu/LexUriServ/LexUriServ.do?uri=OJ:L:2000:336:0004:0004:EN:PDF.

He, Eileen, and Simon Ye. 2009. "Rural China PC Program Will Increase PC Shipments in 2009." Gartner, Stamford, CT. http://www.gartner.com/DisplayDocument?id=909330.

Jensen, Mike. 2009. "Promoting the Use of Internet Exchange Points: A Guide to Policy, Management, and Technical Issues." Internet Society, Reston, VA. http://www.isoc.org/internet/issues/docs/promote-ixp-guide.pdf.

Roetter, Martyn. 2009. "Mobile Broadband, Competition, and Spectrum Caps." Paper prepared for the GSMA, Arthur D. Little, Boston, January. http://www.gsmworld.com/documents/Spectrum_Caps_Report_Jan09.pdf.

Thailand, National Telecommunications Commission. 2010. "Enabling Open Networks." Presentation at the 2010 "Global Symposium for Regulators," International Telecommunication Union, Dakar, November 10–12. http://www.itu.int/ITU-D/treg/Events/Seminars/GSR/GSR10/consultation/contributions/Thailand.pdf.

World Bank. 2011. "Kenya Broadband Case Study." World Bank, Washington, DC.

INDEX

freedom of opinion and expression, 139–40, 146–47n40

Frequency Division Duplexing (FDD), 233, 234, 236

Frequency Division Multiple Access (FDMA), 233t5.5

FRIENDS. *See* Fast Reliable Instant Efficient Network Disbursement of Services (FRIENDS), India

FTTB. *See* fiber to the building or business (FTTB)

FTTC. *See* fiber to the curb (FTTC)

FTTH. *See* fiber to the home (FTTH)

FTTN. *See* fiber to the node (FTTN)

FTTP. *See* fiber to the premises (FTTP)

functional separation, 58–59, 122–23, 145n23

funding
    for access facilities, 268–69
    for broadband strategies, 257
    direct government funding, 70–71
    for school connectivity, 254–55
    See also financing broadband development; fiscal support; investments

Fundo de Universalização dos Serviços de Telecommunicações (FUST), Brazil, 183

G

gambling online, 140

GDP. *See* gross domestic product (GDP)

general authorizations, 94–95

General Packet Radio Service (GPRS), 198, 230, 243nn30–31

general-purpose technology (GPT), 4, 50
    and cloud computing, 13–14, 31n6
    e-government applications, 17–19
    and health care sector, 16–17, 18t1.2
    and human capital, 15–16
    overview, 11–12
    and research and development, 12–13
    and retail and services sectors, 14–15
    and supply chain management, 15

geography, as challenge to service, 306, 317, 322–23

geostationary communication satellites, 202

Gigabit passive optical network (GPON), 228t5.3

Global Partnership on Output-Based Aid (GPOBA), 181

Global Privacy Enforcement Network, 138, 146n36

Global System for Mobile Communications (GSM), 97, 102, 230, 233

goals
    for broadband in development countries, 296–98
    for broadband in United Kingdom, 69–70
    cross-sector nature of, 62–63
    See also policies and policy making

Google, 134, 145–46n30, 210

Google Docs, 282

governance, 27–28, 177

governments, 131, 134, 140, 258
    access and connectivity issues
        funding for digital connectivity, 162–63
        and infrastructure development, 322
        instruments to achieve access, 155–56
        interventions for metropolitan connectivity, 219, 221
        interventions to drive access, 163–67
        policies to improve universal access, 167–69, 190n15
        subsidies to support access, 171–78, 190nn17–19
    broadband development, 164, 260
        direct intervention in, 70–71
        elements to consider when creating policies for, 48b2.3
        Kenya, 323
        promotion of, 2, 30, 38–44, 59–62
        providing national level focal point for, 49–51
        role in facilitation of, 51–52, 63–64
        services to promote participation in, 272–754
        sources of funding for, 178–81, 190n21, 191nn23–25
    citizen participation in, 17–19
    and data collection issues, 134–35
    and economic justification for fiscal support, 66–69, 85n9
    and fiber optic backbone networks, 217–18
    measures to monitor services, 270–71
    and private sector, 42–43, 52, 54, 169–71
    programs to increase purchase of devices, 261, 263, 264b6.5
    and public input into policies, 45–46

GPON. *See* Gigabit passive optical network (GPON)

GPRS. *See* General Packet Radio Service (GPRS)

promoting of, 284–86, 289*nn*42–44
and social networking, 280–81
Sub-Saharan Africa, 313
local governments
and access to broadband, 165
and broadband development, 73, 74*b*2.5
and broadband infrastructure, 167
local loop unbundling (LLU), 59, 116–17,
238–39
Long-Term Evolution (LTE)-Advanced, 236
Long-Term Evolution (LTE) technology,
116, 198, 236, 314
low-and-middle income countries, 156–58
LTE. *See* Long-Term Evolution (LTE)

M

macroeconomics
and broadband absorptive capacity,
27–28
See also economies
Malaysia, 248, 306
broadband strategies in, 47*b*2.2
employment growth link to
broadband, 10
use of PPPs for broadband development,
72–73
Wi-Fi, 237
Malaysian Communications and
Multimedia Commission (MCMC), 10,
248
Malaysian Information, Communications,
and Multimedia Services (MyICMS),
47*b*2.2, 72–73
Maldives, 210, 211*b*5.1
markets, 2, 159, 164–66, 171, 176
advantage of LLU in, 117
backbone markets, 76, 86*n*16
and broadband development, 19–24,
155–56, 247
for CATV networks, 305
competition to promote growth of,
54–55, 181, 182
disruptions of by new entrants, 303–5
and international connectivity, 112–13
mechanisms for spectrum, 118–19
private-led, 156–57, 169, 189*n*1
and regulatory framework, 167
secondary markets for spectrum
trading, 100
and subsidies, 172–74, 177
Mauritius, 14
m-banking services, 277–78
McKinsey and Company, 5

MCMC. *See* Malaysian Communications and
Multimedia Commission (MCMC)
MDGs. *See* Millennium Development Goals
(MDGs)
measurement
of adoption of Internet, 76, 77–79
of availability, 76–77
of digital literacy in Australia, 256*b*6.2
and evaluation, 82–84
overview of broadband indicators, 76,
86*nn*16–17
of pricing policies, 76, 77*f*2.4, 81–82,
83*t*2.6
of quality, 76, 77*f*2.4, 79–81, 87*n*20
See also monitoring
medical outcomes, 16–17, 18*t*1.2
medical records, 275–77, 288*n*31
Metis Community, 178, 190*n*21
metropolitan connectivity, 115–16, 196, 197,
218–21
Mexico, 136, 230, 232*b*5.2
m-health, 16, 17*b*1.3, 276–77
microwave systems, 203–4, 214, 217
Middle East and North Africa region
broadband development in, 310–11
See also specific country
middle-income countries, broadband
strategies in, 47*b*2.2
"middle-mile" infrastructure, 115–16, 218
military campuses, and Internet
connectivity, 257
Millennium Development Goals (MDGs),
296–98
mobile application laboratory, 323
mobile broadband services
Brazil, 319
contrasted with wireline services, 128–29
East Asia, 305
Kenya, 323
Latin America and Caribbean, 308
Morocco, 311, 324
and service neutrality, 93
Sub-Saharan Africa, 314
subscriptions for, 19–20, 82, 83*t*2.6,
305, 311
Turkey, 328
and use of Wi-Fi, 238
Vietnam, 329
mobile devices, 5–6, 157, 250, 263–67, 280,
287*n*20
mobile education labs, 16
mobile health (m-health), 16, 17*b*1.3, 276–77
mobile money services, 277

mobile health services in, 17b1.3
network-sharing agreements, 57–58
Nigerian Communications Commission
(NCC), 112–13
NITEL, 112–13
Norma, Moldova, 307b7.4
North Atlantic Treaty Organization, 318
Norway, 218, 236
NPV. *See* net present value (NPV)

O

O_2, 120
OBA. *See* output-based aid (OBA)
OECD. *See* Organisation for Economic
Co-operation and Development
(OECD)
OFDM. *See* Orthogonal Frequency Division
Multiplexing (OFDM)
OLCP. *See* One Laptop per Child (OLCP)
Initiative
Oman, weblink to national broadband
plans, 335
One Laptop per Child (OLCP) Initiative,
263
online content. *See* content
online service providers (OSPs), 141
online transactions, 147n40, 324, 329
Openreach, 123
open-source technologies, 282–83, 284
operating rights, 176
optical network units, 228
Orange, 120, 307b7.4
Organisation for Economic Co-operation
and Development (OECD)
and costs of backbone networks, 216–17
and e-government services, 273
initiatives to promote broadband
investment, 65
network coverage data, 77
and privacy issues, 138, 146n35
as source of statistics, 84
subscription definition, 78
Orthogonal Frequency Division
Multiplexing (OFDM), 228–29, 236
OSPs. *See* online service providers (OSPs)
output-based aid (OBA), 181, 191n25

P

Pacific region. *See* East Asia and Pacific
region
Packet One Networks, 237
packet-switched transmission, 230, 232t5.4
Pakistan, 217, 311–12

Partnership on Measuring ICTs for
Development, 76, 86n17
Pashto language, 289n43
passive optical network (PON), 228
passive sharing of infrastructure, 119–21
patents, protection of, 143
pay-as-you-go pricing, 324
pay or play mechanisms, 190n15
pay phones, 175
PDAs. *See* personal digital assistants
(PDAs)
peering and transit arrangements, 108–10,
144n15, 197, 207–8, 210
peer-to-peer technologies, 142, 270
penetration rates, 5–8
Chile, 310
growth of, 19
link to employment rate, 10
mobile devices, 263–64
See also subscriptions
personal computers (PCs), 141
affordability of, 261–63, 265b6.5
for promoting digital literacy, 264b6.5
shipments in, 21
personal digital assistants (PDAs), 248
personally identifiable information
(PII), 136
Peru, 104, 165, 166–67b4.3, 184
Pew Internet American Life Project,
259–60
Pew Research Center, 286n2
Philippines, 164
PII. *See* personally identifiable
information (PII)
pilot projects, 49
funding for in Canada, 178–79
for universal service in Mongolia, 173b4.4
ping time, Bahrain, 81f2.7, 240
See also latency times
PLMNs. *See* public land mobile networks
(PLMNs)
points of presence (POPs), 183, 186b4.7,
209–10
Point-Topic, 301
point-to-point connections, 203, 204, 219,
227–28
policies and policy making, 110
for applications and content, 344–45tB.3
and backbone networks, 217
and converged networks and services,
92–93
and economic justification of fiscal
support, 66–69, 85nn9

transparency
  and accounting methods, 58–59
  in data collection, 136–37
  as factor in competition for subsidies, 177
  of ISPs, 125
  in spectrum allocation, 100
  of traffic management policies, 126–27, 128
  for UASFs, 183, 185, 187
TRA. *See* Telecommunications Regulatory Authority (TRA), Bahrain
Tribal Digital Village, 286n3
Triple Network project, China, 305
triple-play offers, 200
tromboning, 104, 106
Tunisia, 311
Turkey, 82
  access to broadband, 308
  broadband strategies in, 47b2.2, 319f7.6, 321f7.2, 327–28
  broadband subscriptions, 78–79, 80f2.6
  public Internet access centers, 255, 257
Turkish Information Communications and Technology Authority, 78–79, 80f2.6
Turk Telecom, 255, 257
TV. *See* television (TV)
Twitter, 139, 280, 311

U

UAF. *See* Universal Access Fund (UAF) Company, Jamaica
UASFs. *See* Universal Access and Service Funds (UASFs)
UAS. *See* universal access and service (UAS)
Uganda, 186b4.7, 210, 217
ULL. *See* unbundled local loop (ULL)
UMTS. *See* Universal Mobile Telecommunications System (UMTS) networks
unbundled local loop (ULL), 303–4b7.3
UNCTAD. *See* United Nations Conference on Trade and Development (UNCTAD)
underserved areas, 39, 169, 184, 344, 346
  and broadband promotion, 40, 44, 54, 62t2.4, 166–67b4.3
  Canada, 158–59
  European Union, 70b2.4
  government interventions for, 71
  incentives to extend fiber optic cable to, 217
  local connectivity for, 116
  See also rural areas

unified authorizations, 94–95
United Kingdom, 258
  advertised versus actual speeds in, 239f5.13
  broadband development issues, 49, 69–70, 165, 167
  and functional separation, 123
  and infrastructure sharing, 120
  pilot projects in, 49
  television service in, 272
United Nations, Human Rights Council, 139
United Nations Conference on Trade and Development (UNCTAD), 316
United Nations Development Programme, Human Development Index, 301–2
United States, 10, 81, 104, 105
  banking online, 277
  Commerce Department, 259–60
  data privacy protection, 135
  DMCA, 141
  e-government services, 19, 274, 275
  Federal Communications Commission, 87n20, 110, 128, 186b4.6
  Federal Trade Commission, 135, 136
  and LTE deployment, 236
  National Broadband Plan, 99, 110, 160, 161, 259, 336
  online gambling, 140
  reasons for not adopting Internet in, 41–42, 85n2
  reform of USF in, 186b4.6
  and satellite technology, 243n44
  subsidies for mobile devices, 265
  use of BAK, 108
  use of USFs, 73–74
  weblink to national broadband plan, 336
Universal Access and Service Funds (UASFs), 163, 179–85, 186, 187, 188, 191n23
universal access and service (UAS), 160–63, 167–68
  mechanisms to drive support for build-out, 169–71
  Mongolia, 173b4.4
  review of funds to programs, 187–88
  subsidies for fiscal support for, 171–78, 190nn17–19
Universal Access Fund (UAF) Company, Jamaica, 160
universal broadband access, 155
  in Latin America and Caribbean, 309–10
  levels of, 158–60
  mechanisms to drive, 163–69, 190n15
  overview, 156–63, 189n1, 189n9
  strategies for, 160–65, 167–68, 189n9

wireless access technologies, 229–38, 243nn30–31

wireless broadband, 117–19, 300
  Australia, 71
  Brazil, 319
  increase in users of, 19, 98
  in LDCs, 315
  Morocco, 304–5*b*7.3, 311
  See also mobile broadband services

Wireless Fidelity (WiFi) technologies, 17*b*1.3, 72, 237–38, 243*n*42
  privacy issues, 134, 145–46*n*30
  and spectrum, 240
  and use of unlicensed spectrum, 102–3

WirelessMAN-Advanced, 236

wireline broadband services, 116–17, 299
  Brazil, 318
  contrasted with mobile services, 128–29
  infrastructure for, 299–302
  and local connectivity, 221–29
  and right-of-way, 58
  and shared infrastructure, 121
  South Asia, 311–12
  Vietnam, 329

wireline broadband subscriptions, 19–20
  Australia, 71
  in emerging economies, 301–2

  Indonesia, 305
  measurement of, 78–79
  prices for in select countries, 82, 83*t*2.6

wireline telephones, 157, 300, 310

World Bank, 31*n*14, 323, 326
  Private-Public Infrastructure Advisory Facility (PPIAF), 181
  study linking GDP and broadband penetration, 5–6
  study on broadband in Korea, 163
  UAS study, 168

World Health Organization, 190*n*18

World Summit on the Information Society (WSIS), 296, 297–98

World Trade Organization, 140

Worldwide Interoperability for Microwave Access (WiMAX), 17*b*1.3, 47*b*2.2, 204, 234–35

WSIS. *See* World Summit on the Information Society (WSIS)

Y

Yahoo! Inc., 147*n*40

YouTube, 279, 280, 282, 285–86

Z

Zimbabwe, 211